RESEARCH METHODS
IN
MARINE BIOLOGY

edited by

Carl Schlieper

Professor in Marine Zoology
University of Kiel

with contributions from numerous scientists

Translated from the German by
Elizabeth Drucker

UNIVERSITY OF WASHINGTON PRESS
SEATTLE

Library of Congress Cataloging in Publication Data.

Main entry under title:

Research methods in marine biology.

(Biology series)
Translation of 'Methoden der meeresbiologischen Forschung'.
Includes bibliographies.
1. Marine biology—Technique. I. Schlieper, Carl, ed. [DNLM: 1. Marine
Biology. QH 90.57 S344m 1972] QH91.57.A1M413 1972   574.92'07'2
72-6089
ISBN 0-295-95234-2

List of Contributors

Dr. Gerhard DREBES

Biologische Anstalt Helgoland, Meeresstation
2192 Helgoland

Dr. Jürgen FLÜCHTER

Biologische Anstalt Helgoland, Meeresstation
2192 Helgoland

Prof. Dr. Sebastian GERLACH

Institut für Meeresforschung
285 Bremerhaven,
Am Handelshafen 12

Prof. Dr. Fritz GESSNER

Institut für Meereskunde der Universität
23 Kiel, Hohenbergstraße 2

Dr. Klaus GRASSHOFF

Institut für Meereskunde der Universität
23 Kiel, Hohenbergstraße 2

Prof. Dr. Karl Gottlieb GRELL

Zoologisches Institut der Universität
74 Tübingen, Hölderlinstraße 12

Dr. Wilfried GUNKEL

Biologische Anstalt Helgoland, Meeresstation
2192 Helgoland

Prof. Dr. Carl HAUENSCHILD

Zoologisches Institut der TH
33 Braunschweig, Pockelsstraße 10a

Dr. Willy HÖHNK

Institut für Meeresforschung
285 Bremerhaven,
Am Handelshafen 12

Prof. Dr. Rudolf KÄNDLER

Institut für Meereskunde der Universität
23 Kiel, Hohenbergstraße 2

Dr. Peter KORNMANN

Biologische Anstalt Helgoland, Meeresstation
2192 Helgoland

Dr. Jürgen LENZ

Institut für Meereskunde der Universität
23 Kiel, Hohenbergstraße 2

Prof. Dr. Gerhard RHEINHEIMER — Institut für Meereskunde der Universität
23 Kiel, Hohenbergstraße 2

Prof. Dr. Carl SCHLIEPER — Institut für Meereskunde der Universität
23 Kiel, Hohenbergstraße 2

Dr. Heinz SCHWENKE — Institut für Meereskunde der Universität
23 Kiel, Hohenbergstraße 2

Dr. Gerold SIEDLER — Institut für Meereskunde der Universität
23 Kiel, Niemannsweg 11

Dr. Gotram UHLIG — Biologische Anstalt Helgoland, Meeresstation
2192 Helgoland

Dr. Erich ZIEGELMEIER — Biologische Anstalt Helgoland, Litoralstation
2282 List (Sylt)

Preface

Research in marine biology is not necessarily identical with the biological study of marine organisms. Rather, marine biology is the science of life in the ocean and the laws governing this life. Apart from the systematic survey of the fauna and flora, which is indispensable as its base, marine biology is concerned in particular with the dependence of marine organisms on abiotic and biotic environmental factors. Accordingly, this science consists in the first place of ecological research, which is devoted to the descriptive and experimental analysis of life processes in the ocean. An area of marine research which – apart from marine fishery biology – has been internationally promoted in recent years is biological oceanography. Together with physical and chemical oceanography, it establishes – among other things – values for the description and investigation of oceanic water masses and their dynamics. One of the principal aims of modern marine biology is indeed the complete quantitative understanding of the food and energy cycle in the ocean in all its stages.

Marine biological research has experienced an unanticipated development since the end of the Second World War. The recognition of the fact that wide areas of the deep sea, the largest life region of our planet, were still *terra incognita* for us stimulated new interest in oceanographic research – and consequently in marine biological research as well – in all civilized states. The desire to utilize the nutritional reserves of the sea – to combat hunger in particular and especially in the presence of protein deficiency – worked in the same direction. Numerous new textbooks, monographs, and periodicals concerned with this newly-disclosed scientific field have appeared in rapid sequence in recent years. I should like to mention in particular the bulky *Treatise in Marine Ecology* (Washington 1957) of J. W. Hedgepeth, the textbooks of H. B. Moore *Marine Ecology* (New York 1958), J. M. Pérès' *Biological Oceanography and Marine Biology* (Paris 1959, 1963), and *Marine Biology* (Berlin 1965 and London, 1970) by H. Friedrich.

These books, apart from many others not mentioned, are an excellent introduction. However, the young scientist who wishes to work in this field also requires a practical guide to the principal research methods. This is particularly true since modern marine biology has made use or developed many research procedures which are unknown or little used in other areas of biological research. This book has been written to remedy this lack.

In planning and preparing this work intended for students and young scientists it has not been our ambition in any way to present a complete collection of methods. Rather, we attempted to describe individual basic methods and their possible applications. The principal chemical and physical methods of sea water analysis which are of importance for the marine biologist were not left out. The bibliography at the end of every section should enable the reader to work further. At the end of the book we added a comparative list of various types of apparatus and substances in order to widen the field of application of the methods.

Not all the Sections are of equal length, but this does not reflect a value judgement in any way. The length of each section depends on the complexity of the research methods which require either a brief or a more detailed description, and is also determined by the fact that research has gone further in some areas than in others. For example, the Section dealing with methods in fisheries biology is unusually long because marine fisheries biology is one of the oldest and most developed areas of the science of marine biology. We decided not to include a special separate section describing the mathematical procedures required for the interpretation of the research results in marine biology, since these are in any case among the tools of every natural scientist.

In the name of all the co-authors, the editor would like to thank the publishers and his co-workers for their co-operation and their sympathetic collaboration in the publication of this work.

Kiel, end 1967   C. SCHLIEPER

Contents

RESEARCH METHODS
IN
MARINE BIOLOGY

Section I

HYDROGRAPHIC METHODS

A. Chemical Methods

K. GRASSHOFF

The *salinity* of a seawater sample is defined by Forch, Knudsen, and Sörensen as:

'The salinity of a seawater sample is the weight of solid materials in grams (*in vacuo*) contained in 1 kg water when all the carbonate has been oxidized, all the bromide and iodide have been replaced by chloride, all organic matter has been oxidized and the residue has been dried at 480°C to constant weight.'

The salinity is given in ‰. The relationship between the salinity and the chloride content of a given seawater sample is expressed by the following equation:

$$S‰ = 1 \cdot 80655 \ Cl‰*$$

The 'chlorinity' of a water sample is understood to mean the weight in grams (*in vacuo*) of chloride contained in 1 kg water when all the bromide and iodide have been replaced by an equivalent amount of chloride. A more precise definition of chlorinity is given by the equation:

$$Cl‰ = 0 \cdot 3285234 \cdot (Ag) \cdot in \ g,$$

(Ag) being the atomic weight of silver (*in vacuo*) which, as silver ion, is used to precipitate all the chloride, bromide and iodide from 1 kg water (see Table 1).

Another value used to describe the chlorinity is 'chlorisity', which is the chloride content (or bromide and iodide content) per litre. 'Chlorisity' is usually given at 20°C. This value is used to form relationships with other chemical values. The chlorosity $(Cl/l_{(20)})$ is converted into S‰ according to the following equation:

$$S‰ = 0 \cdot 03 + (1 \cdot 8050 \cdot Cl/l_{(20)} \cdot {}^1/Q_{(20)})$$

* This relationship is not always valid in coastal areas.

TABLE 1

Concentration of the principal ions appearing in seawater with various degrees of salinity
(from J. P. Riley and G. Skirrow, *Chemical Oceanography*, Academic Press, 1965)

Concentration in g/Kg seawater

Salinity ($^o/_{oo}$)	Na^+	Mg^{++}	Ca^{++}	K^+	Sr^{++}	B	Cl^-	SO_4^{--}	Br^-	F^-	HCO_3^-
5	1·537	0·135	0·059	0·056	0·001	0·001	2·765	0·388	0·009	0·0002	0·020
10	3·074	0·370	0·118	0·111	0·002	0·001	5·530	0·775	0·019	0·0004	0·041
15	4·611	0·555	0·177	0·167	0·003	0·002	8·294	1·162	0·029	0·0006	0·061
20	6·148	0·739	0·236	0·221	0·004	0·002	11·059	1·550	0·038	0·0008	0·081
25	7·685	0·924	0·295	0·277	0·006	0·003	13·824	1·937	0·048	0·0009	0·101
30	9·222	1·109	0·354	0·332	0·007	0·003	16·589	2·324	0·057	0·0011	0·122
35	10·759	1·294	0·413	0·387	0·008	0·004	19·354	2·712	0·067	0·0013	0·142
40	12·296	1·479	0·472	0·442	0·009	0·005	22·118	3·099	0·077	0·0015	0·162

Concentration in g/l seawater at 20°C

Salinity ($^o/_{oo}$)	Na^+	Mg^{++}	Ca^{++}	K^+	Sr^{++}	B	Cl^-	SO_4^{--}	Br^-	F^-	HCO_3^-
5	1·540	0·135	0·059	0·056	0·001	0·001	2·771	0·389	0·009	0·0002	0·020
10	3·092	0·372	0·119	0·112	0·002	0·001	5·562	0·780	0·019	0·0004	0·041
15	4·656	0·560	0·179	0·169	0·003	0·002	8·374	1·173	0·029	0·0006	0·062
20	6·231	0·749	0·239	0·224	0·004	0·002	11·207	1·571	0·039	0·0008	0·082
25	7·817	0·940	0·300	0·282	0·006	0·003	14·062	1·970	0·049	0·0009	0·103
30	9·416	1·132	0·361	0·339	0·007	0·003	16·937	2·373	0·058	0·0011	0·124
35	11·029	1·326	0·423	0·379	0·008	0·004	19·839	2·780	0·069	0·0013	0·146
40	12·647	1·521	0·485	0·455	0·009	0·005	22·750	3·188	0·079	0·0015	0·167

In the 'Hydrographic Tables' based on the measurements of Knudsen and Forch, the values for S are calculated for all existing chlorinity levels. (The 'Knudsen Tables' will be replaced by new tables in the near future.)

1. Determination of the Salinity by the Mohr-Knudsen Method

Range:
0 to $41 \cdot 5\%_0$ S.

Principle of the Measurement

The halogen ions (except for fluoride) are precipitated from a standard volume (15 ml) of seawater with silver ions. The solution products of AgI, AgBr and AgCl ($pK_L = 16$; $12 \cdot 4$; $9 \cdot 96$) are sufficiently small to allow highly accurate direct precipitation titration. The end point of the titration is indicated by the precipitation of red silver chromate. The solution product of silver chromate ($pK_L = 11 \cdot 7$)* is much larger than those of silver halogenide, but smaller than that of silver carbonate. Titration can therefore be carried out in a neutral solution. The silver nitrate solution is standardized with 15 ml standard seawater which has a precisely known chloride content. The standard seawater is produced in Charlottenlund, Copenhagen. The use of this water as titrimetric standard is obligatory in oceanography. (In order to economize on the expensive normal water it is possible to use a sub-standard water; its salinity is compared with the normal water by repeated titration and is continuously checked.)

Attainable Accuracy: $\pm 0 \cdot 01\%_0$ Cl; $\pm 0 \cdot 02\%_0$ S.

The above accuracy can only be obtained when the specifications are rigorously observed, the glassware is kept perfectly clean and the burette stopcocks are absolutely tight. The room temperature should not fluctuate by more than $\pm 1°C$ during the determination.

The following rules must be observed during the titration: all samples and the standard water must be treated in exactly the same manner. Precisely the same quantity of water must always be titrated. The solutions, sample and glassware must be kept at the same uniform temperature. The same amount of time must be allowed for each titration. The colour change of the indicator must be absolutely the same for each titration; this can only be attained if good lighting is used and stirring is uniform and vigorous. The

* Here apparently smaller than with AgCl, but note $L = (Ag)^2 . (CrO4 -)$.

burette must always be read immediately after the colour change of the indicator. A titration should never take longer than 4 minutes.

The titre of the silver nitrate solution is selected in such a manner that the consumption in double cubic centimetres agrees with the chlorinity or only departs from it slightly. Additive corrections for slight deviations are given in the Knudsen Hydrographic Tables.

In order to obtain the greatest possible accuracy, Knudsen introduced special pipettes and burettes (Fig. 1).

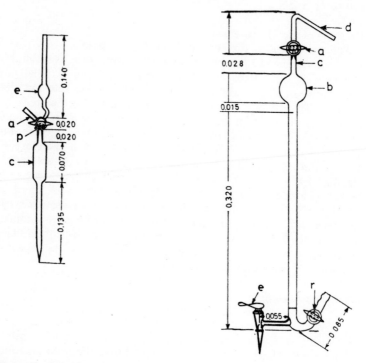

Figure 1. Knudsen pipette and burette for the chloride determination in seawater (according to M. Oxner)

Reagents

Silver nitrate solutions: 37·11 g silver nitrate is dissolved in 1 l water. This solution has approximately the correct titre when the Knudsen pipette volume is exactly 15 ml. If the pipette volume deviates from the theoretical value, which can be determined by weighing out, the silver nitrate will be dissolved in $(1 + a)$ l water.

Pipette volume	a
14·80	+ 0·014
14·90	+ 0·007
15·00	− 0·000
15·10	− 0·007
15·20	− 0·014

The water used must be entirely free of halide.

Indicator solution: 8 g potassium chromate is dissolved in 100 ml water.

Determination Procedure

1. All solutions and samples are matured for several hours (preferably overnight) in the laboratory.

2. The silver nitrate solution in the stock bottle is agitated and all the condensation water drops are combined with the solution. The silver nitrate solution standing in the burette is drained off and the burette is filled once with fresh solution and rinsed out. The solution must drain evenly from the burette and must not 'cling'. The only remedy for this is cleaning. (The most careful titration becomes worthless if the glassware is not clean!) The glassware is best cleaned with 1:1 dilute nitric acid, rinsed with water, rinsed with methanolic caustic soda solution and washed out with distilled water. The stopcocks must subsequently be re-greased. Silicone grease is to be avoided at all costs! Too much grease immediately contaminates the burette!

3. After the water sample has been drained into the beaker the tip of the pipette is held against the wall of the beaker for a further 15 seconds (do not knock). The supernatant water in the nozzle of the pipette is *poured out through the nozzle*. The pipette is then filled with the next sample and suspended in the rack. The indicator solution is added to the sample.

4. The titration container is placed below the outlet of the burette and the magnetic stirrer is turned on. The contents of the container are stirred vigorously until approximately 2 dccm before the end point. The burette is closed and vigorously stirred for a little while, without splashing. Titration is then continued until approximately 1 dccm before the end point. The agitator is turned off again, the burette stirred vigorously, and titration is continued until approximately 0·2 dccm before the end point. After closing the stopcock almost completely the silver nitrate is added dropwise until the initial indicator colour change becomes visible in the 'stirring funnel'. At the very end the solution is added in single drops

only, and titration is continued until the so-called 'first red colour change' of the indicator appears.

5. The last part of the silver nitrate solution is added in divided drops. For this purpose remove small quantities of the solution from the tip of the stopcock with the glass rod and put them into the container by immersing the glass rod.

6. The burette reading is recorded immediately after the final colour change. One must be careful always to take the recording from the lower edge of the meniscus, free from parallax.

7. If the salinity is unknown and varies greatly during the series, it will be necessary to carry out a low-precision titration for each sample before making the high-precision titration.

Note

The indicator colour change is best recognized in the stirring funnel when lateral lighting is used. The titration will reach the level of accuracy specified above when the routine is carried out almost automatically. Only with practice can the indicator colour change be recognized perfectly.

One standard or sub-standard sample should be inserted after every ten titrations.

2. MICROCHLORIDE DETERMINATION

Range

0 to 25‰ Cl.

Accuracy of the Determination

$\pm 0 \cdot 04$‰, if all the steps are carried out with the greatest care. Since only small volumes are titrated, 'clinging' burettes and pipettes are particularly undesirable.

Principle of the Determination

The end point of the precipitation titration $Cl' + Ag^+ \rightarrow AgCl$ is shown by an absorption indicator.

Reagents

Silver nitrate solution: $17 \cdot 818$ g $AgNO_3$ is dissolved in 500 ml water.

Indicator solution: $7 \cdot 5$ mg sodium fluorescein is dissolved in 5 l distilled water.

Standard chloride solution: $32 \cdot 725$ g Na Cl is dissolved in 1 l water. The solution has a chlorinity of $19 \cdot 380$‰.

Determination Procedure

25 ml of a solution of uranin in distilled water is transferred by pipette into a 50 ml wide-neck Erlenmeyer flask or into a 100 ml beaker.

1 ml of the water sample is measured with a wash pipette and placed in the container. (The final drop is temporarily left in the pipette.) Using a 5 ml microburette, titrate in a thin stream – dropwise at the end – to light pink, then wash the rest into the weakly over-titrated solution and titrate precisely to the point of colourlessness.

TABLE 2A
Conversion table g Cl/Kg into g Cl/l and vice versa
(according to Kalle, 1951)

g Cl/l	− k =	g Cl/kg	g Cl/l	− k =	g Cl/kg
0·00	0·00	0·00	16·64	0·35	16·29
2·41		2·40	16·88		16·52
6·23	0·05	6·18	17·80	0·40	17·40
6·84		6·78	18·02		17·61
8·87	0·10	8·77	18·89	0·45	18·44
9·30		9·19	19·10		18·64
10·87	0·15	10·72	19·92	0·50	19·42
11·23		11·07	2ʊ·12		19·61
12·56	0·20	12·36	20·91	0·55	20·36
12·88		12·67	21·10		20·54
14·05	0·25	13·80	21·85	0·60	21·25
14·33		14·07	22·03		21·42
15·40	0·30	15·10	22·75	0·65	22·10
15·66		15·30	22·93		22·27

TABLE 2B
Preparation of artificial seawater 35⁰/₀₀ S
(from G. Dietrich and K. Kalle, *General Oceanography*, Berlin 1957)

Two separate solutions are made up by dissolving the following salts:

Solution A		Solution B	
NaCl	239·0 g	$Na_2SO_4 . 10 H_2O$ ...	90·6 g
$MgCl_2 . 6 H_2O$	108·3 g	$NaHCO_3$	0·2 g
$CaCl_2$ anhy.	11·5 g	NaF	0·003 g
$SrCl_2 . 6 H_2O$	0·04 g	H_3BO_3	0·027 g
KCl	6·82 g		
KBr	0·99 g		
Aqua dist.	8560 ml	Aqua dist.	1000 ml

Solution B is added in a thin stream to solution A while stirring vigorously and filtration is carried out through a fine-pore filter (SS Blauband) after approximately 24 hours.

The burette reading is taken immediately after the end of the titration, by magnifying lens.

Calibration of the Silver Nitrate Solution

Either 'normal water' or the standard chloride solution can be used for the calibration. Instead of the water sample 1 ml of the standard solution is taken; otherwise the procedure is as above.

Factor Determination of the Silver Nitrate Solution

A 'normal water' sample is titrated as described above; the Cl content of the normal water is then converted from g Cl/kg into g Cl/l by adding a correction, shown in Table 2.

Example: for the titration of $1 \cdot 00$ ml secondary normal water from $19 \cdot 363$ g Cl/kg we require $2 \cdot 064$ ml AgNO₃ solution.
Thus: $2 \cdot 064$ ml corresponds to   19·363

$$+ \ \frac{0 \cdot 49}{19 \cdot 853} \ \text{g Cl/l}$$

or 1,000 ml AgNO₃ solution corresponds to $\dfrac{19 \cdot 853}{2 \cdot 064} = 9 \cdot 617$ g Cl/l.

This AgNO₃ solution is used for the titration of the test solution. The Cl content/litre calculated from the consumption of AgNO₃ is converted into g Cl/kg by subtraction of the correction.

Notes

The titration should never be carried out in direct sunlight since the precipitating AgCl rapidly liberates Ag, which disturbs the colour change of the titration because of its dark tint.

---

3. Oxygen Determination by the Winkler Procedure

Principle of the Determination

Concentrated solutions of divalent manganese and alkaline potassium iodide are added to the water sample. Both solutions are thoroughly mixed with the water sample under vacuum. White manganous hydroxide is first formed, and then oxidized to manganic hydroxide by the molecularly dissolved oxygen. The brown, very voluminous manganic hydroxide settles on the bottom of the specimen bottle. The settling rate depends on the salinity.

The solid phase is now covered with a layer of sulphuric acid. This dissolves the manganous or manganic hydroxide. Trivalent manganese is very unstable in an acid medium. The existing iodide is oxidized to iodine and manganic is reduced to manganous. This reaction as well is quantitative above pH 1·5.

The liberated iodine takes up the excess iodide and forms the complex compound I_3. In the absence of excess iodine a considerable quantity of iodine would evaporate during the titration.

With the help of a thiosulphate solution the iodine is reduced to iodide and the thiosulphate is oxidised to tetrathionate. (The solution should not be too acid, as this would allow the formation of undissociated, unstable thiosulphuric acid, which breaks down into sulfurous acid and sulphur).

The end point of the titration is indicated by starch solution. The starch molecules form an addition compound with the iodine, which has a strong blue colour resulting from the polarization of the electron shell of the iodine. The end point of the titration is best observed in a good light against a light yellow background.

The thiosulphate solution is calibrated by means of a standard solution of potassium hydrogen bi-iodate (titrimetric standard), which has an equivalent proportion of iodide and releases an equivalent quantity of iodine.

$$Mn^{2+} + OH^- \rightarrow Mn(OH)_2$$
$$2Mn(OH)_2 + \tfrac{1}{2}O_2 + H_2O \rightarrow 2Mn(OH)_3$$
$$2Mn(OH)_3 + 3I^- + 6H^+ \rightarrow 2Mn^{2+} + I_3^- + 6H_2O$$
$$2S_2O_3^{2-} + I_3^- \leftrightharpoons S_4O_6^{2-} + 3I^-$$
$$8I^- + O_3^- + 6H^+ \leftrightharpoons 3I_3^- + 3H_2O$$

Range
0 to 10 ml O_2/l

TABLE 3
Oxygen saturation values, 100% rel. atmospheric moisture in ml/l
(according to E. J. Green, 1967)

°C	0	5	10	15	20	30	40 S°/$_{00}$
0	10·31	9·95	9·60	9·27	8·94	8·33	7·76
5	9·01	8·72	8·43	8·15	7·89	7·38	6·90
10	7·96	7·71	7·47	7·24	7·01	6·58	6·17
15	7·09	6·88	6·67	6·47	6·28	6·03	5·56
20	6·37	6·19	6·01	5·84	5·63	5·34	5·04
25	5·78	5·61	5·45	5·30	5·12	4·86	4·59
30	5·27	5·13	4·98	4·84	4·68	4·45	4·20
35	4·85	4·69	4·58	4·45	4·31	4·09	3·87

Accuracy of the Determination
When a 50 ml water sample is used it is possible to determine the oxygen content with an accuracy of ±0·02 ml. A systematic error

in the titration is far less significant than a possible error during the sampling procedure. An important error can be avoided largely only when entirely clean buckets are used for the sampling procedure and siphoning is carried out with the greatest care immediately after sampling. Especially when the sample has a small oxygen content the error during siphoning can be very considerable.

Reagents

Manganese (II) chloride tetrahydrate: dissolve 40 g manganous chloride tetrahydrate in 100 ml distilled water.

Alkaline iodide solution: dissolve 60 g potassium iodide and 30 g potassium hydroxide in 100 ml water.

Sulphuric acid: add 100 ml conc. sulphuric acid to 100 ml water.

Standard solution: dissolve exactly 325 mg potassium hydrogen bi-iodate dried at 105°C and make up to 1 l with distilled water.

Sodium thiosulphate stock solution: dissolve 49·5 g sodium thiosulphate pentahydrate in 1 l water. The stock solution is approximately 0·2 N and is diluted 1:10 for use.

Starch solution: dissolve 1 g soluble starch in 100 ml distilled water and heat. The solution is frequently prepared afresh. It can be sterilized with 1 ml phenol.

Sampling Procedure and Sample Storage

The sampling procedure is the most important part of the oxygen determination! The oxygen sample must always be drawn off first from the bucket. Some 10 to 20 ml is poured into the 50 ml specimen bottle, the bottle is stoppered and agitated. The bottle is then drained and *immediately* filled with the sample. The tube is always kept beneath the surface of the water in the bottle and filling should not be too rapid, or turbulence and air bubbles will form and will be flushed into the sample. Approximately 50 to 100 ml of the sample should be allowed to overflow before stoppering the bottle. No air bubbles should be included! The sample should be analysed at once. If necessary, the sample can be kept in the dark for approximately 10 hours after fixation.

Determination Procedure

Immediately after drawing off the sample add 0·5 ml each of the manganese (II) chloride solution and the alkaline iodide solution. The high specific weight of the concentrated reagents causes these to settle below the water sample. Stopper the bottles. The excess water trapped in the neck of the bottle is pressed out of the bottle.

Now shake the samples vigorously for approximately one minute with short, jerky movements and then allow them to stand in the dark until the precipitate has settled.

After settling (maximum of 10 hours) add 1 ml sulphuric acid, re-stopper and shake again. The sulphuric acid is added in such a manner that the precipitate is not whirled to the surface of the sample. Transfer the sample into a 100 ml beaker and rinse out with 20 ml distilled water.

The liberated iodine is titrated with 0·02 N thiosulphate in very good light. Shortly before the disappearance of the yellow iodine colour add 1 ml starch solution and titrate to the end point.

Calibration of the Thiosulphate Solution

Add 1 ml sulphuric acid, 0·5 ml alkaline iodide solution and 0·5 ml manganese (II) chloride solution to 50 ml water. Mix very thoroughly after adding the reagents. Add exactly 10 ml of the standard iodate solution to the solution from a calibrated pipette. Finally titrate the free iodine as described above.

Calculations

$$1 \text{ equivalent thiosulphate:} \qquad\qquad = 8 g O_2$$
$$1 \text{ ml } 0·02 \text{ N thiosulphate solution:} = 0·16 \text{ mg } O_2$$
$$= 0·112 \text{ mg } O_2 *$$

Consumption of thiosulphate: a ml
Contents of bottle   b ml
Factor of the thiosulphate solution: f
Addition of reagents: 1 ml
Quantity of oxygen contained in the bottle:

$$\frac{a \cdot f \cdot 8}{50} = \text{mg } O_2 \text{ or } 0·112 \cdot a \cdot f \text{ ml } O_2/L$$

Converted to 1 l:

$$\frac{a \cdot f \cdot 8}{50(b-1)} \text{ mg } O_2/l \text{ or } \frac{112 \cdot a \cdot f}{(b-1)} \text{ ml } O_2/l$$

Note

The divalent manganese solution and the alkaline iodide solution contain (little) oxygen. Moreover, if air enters the alkaline iodide solution, iodate may develop, which reacts with iodide during acidification and produces iodine. (1 molecule iodate oxidizes 5 molecules iodide!) More thiosulphate is therefore consumed during the determination than corresponds to the oxygen content of the sample.

*(1 mg O_2 = 0·70 ml O_2 at NTP)

A total blank value can be determined as follows: fill 4 bottles with a random water sample and add 1 x reagents to the first bottle, 2 x reagents to the second, 3 x reagents to the third and 4 x reagents to the fourth bottle. Then continue as above. To dissolve the precipitate add 1 x to 4 x acid. The 'oxygen content' of the samples is calculated, taking into account the increased quantity of reagents, and presented in graph form. A straight line is obtained. The rise of the straight line indicates the reagent blank value. (The frequently described determination of the blank value with the reversed addition of the reagents only determines the proportion of oxidized iodide, but not the total blank value.)

If, in certain special studies, a sufficient quantity of the sample is not available in order to use the above method, oxygen determination can be carried out by a Winkler micro-method, as for example the method elaborated by Fox and Wingfield. This method requires only 1 to 2 ml sample, but nevertheless an accuracy of 2% can be attained. It is also not possible to make use of the normal Winkler determination method in the presence of fairly large quantities of organic substances. The determination must then be based on the iodine differentiation method (W. Ohle) or on the method of Alsterberg (see H. Barnes).

4. Determination of the Hydrogen Ion Activity

Principle of the Measurement

The pH of a solution is defined by the equation $pH = -\log a_H+$. The activity of an ion is understood to be the apparent or actual effective concentration of the particular ion. The activity is linked with the concentration by the equation:

$$A = f \cdot C$$

The 'f' factor is normally <1 and is determined by the solution components, the solvent and the chemical properties of the ion type in question.

The hydrogen ion activity is measured by means of the glass electrode. The glass electrode consists of a small flask made of special glass, filled with a buffer solution. The small flask contains the discharge electrode, an electrode of 'secondary type'. These electrodes have a constant potential and do not lend themselves to polarization. Calomel, silver, silver chloride or thalium amalgam electrodes are used for the discharge of the glass electrodes. The internal resistance of the glass electrodes consists of 10 to 500 MΩ, depending on the type of glass and the temperature. If the hydrogen ion activity in the solution differs from that within the electrode, differences in

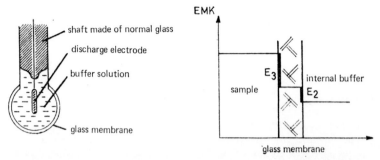

Figure 2. Diagram of a glass electrode and the single potentials between the sample and the internal buffer solution

potential occur at the solution boundary or the glass boundary (see Fig. 2). These differences in potential can be measured when a reference electrode (also of the non-polarizing type) is chosen as the standard in the water sample. A potential difference is formed between the glass electrode and the reference electrode, composed of the following individual potentials:

$$EMK = E_1 + E_2 + E_3 + E_4 + E_5$$

where

E_1 = potential of the discharge in the glass electrode
E_2 = potential between the internal buffer and the glass
E_3 = potential between the glass and the solution
E_4 = diffusion potential during conveyance to the reference electrode (only approximately constant with various solutions in the outer space!)
E_5 = potential of the reference electrode.

As can be shown, under certain conditions the total potential is proportional to the log of the hydrogen ion activity in the outer space. The potential jump at the glass membrane is related to the hydrogen ion activity inside and outside through the Nernst equation.

$$E_2 - E_3 = \frac{RT}{n \cdot F} \cdot \ln \frac{H_{(2)}^+}{H_{(3)}^+} = 0 \cdot 058 \, pH_{(2)} - pH_{(3)} \text{ (at 20°C)}$$

The potential of the reference electrode and the 'f' factor are temperature-determined; the measuring temperature must therefore be known.

Since the internal resistance of the glass electrode is high, heavy demands must be made on the cable isolation and protection, as well as on measurement of the potential.

Accuracy of the Determination

The accuracy of the determination depends on four factors: the calibration, the quality of the pH meter used, the glass electrodes employed and the uniformity of the diffusion potential. It is possible to attain a reproductability of $\pm 0 \cdot 005$ pH and an accuracy of $\pm 0 \cdot 01$ pH if the greatest care is exercised during sampling and calibration.

Sampling Procedure and Sample Storage

The sample for the pH determination is poured into a well-stoppered bottle directly from the bucket. The sample is brought to room temperature and measured as soon as possible after collection. A thermostatic measuring cell is preferable.

Reagents (see Table 4)

Standard solution 1: $0 \cdot 05$ m potassium hydrogen phthalate solution. Dissolve $10 \cdot 211$ g in 1 l distilled water; pH approximately $4 \cdot 0$.

Standard solution 2: $0 \cdot 025$ m potassium dihydrogen phosphate and $0 \cdot 25$ m disodium hydrogen phosphate. Dissolve $3 \cdot 402$ g KH_2PO_4 and $4 \cdot 450$ g Na_2HPO_4 in 1 l distilled water; pH approximately $6 \cdot 9$.

Standard solution 3: dissolve $3 \cdot 814$ g sodium tetraborate in 1 l water; pH approximately $9 \cdot 2$.

All three solutions can be kept for approximately four weeks in glass bottles if well stoppered and kept cool.

TABLE 4
pH standard values in relation to temperature
(according to Bates, 1961)

°C	Potassium hydrogen phthalate $0 \cdot 05$ m	Phosphate $0 \cdot 025$ m KH_2PO_4 $0 \cdot 025$ m Na_2HPO_4	Borax $0 \cdot 01$ m
0	$4 \cdot 003$	$6 \cdot 984$	$9 \cdot 464$
5	$3 \cdot 999$	$6 \cdot 951$	$9 \cdot 395$
10	$3 \cdot 998$	$6 \cdot 923$	$9 \cdot 332$
15	$3 \cdot 999$	$6 \cdot 900$	$9 \cdot 275$
20	$4 \cdot 002$	$6 \cdot 881$	$9 \cdot 225$
25	$4 \cdot 008$	$6 \cdot 865$	$9 \cdot 180$
30	$4 \cdot 015$	$6 \cdot 853$	$9 \cdot 139$
35	$4 \cdot 024$	$6 \cdot 844$	$9 \cdot 102$
40	$4 \cdot 035$	$6 \cdot 338$	$9 \cdot 068$

Calibration

Instrument calibration is carried out in accordance with the directions for the measuring apparatus involved, using two standard solutions. At the same time the temperature is measured and adapted to the compensator. The standard solutions should, if possible, include the pH range of the samples. If good glass electrodes are used, calibration needs only be checked once daily at most. The calibration temperature and the measurement temperature should not differ by more than 3°C, since the temperature compensation does not produce complete equalization. However, a thermostatic measuring cell is preferable.

Measurement of the Samples

After temperature equalization the sample is placed in the cell, the temperature is measured and adjusted on the instrument (not necessary with the thermostatic measuring cell) and the pH is recorded after rectification of the electrodes. The next sample is placed in the cell after intermediate rinsing.

Calculation of the pH *in situ*

If the measuring temperature is $t_m°C$, the measured pH is pH_m and the *in situ* temperature at a depth of 'd' metres is 5°C, then the correct pH is:

$$pH_s = pH_m - \alpha (t - tm) \text{ and}$$

taking the depth into consideration:

$$pH_d = pH_s - \beta.d.$$

The values for α and β are tabulated by Strickland (1965).

5. DETERMINATION OF THE TOTAL ALKALINITY

Principle of the Measurement

The total alkalinity is determined by indirect titration, using a modified procedure by H. Wattenberg and S. Griepenberg. An excess of hydrochloric acid is added to a carefully measured portion of the sample, the carbon dioxide is driven out and the excess hydrochloric acid is titrated back with caustic soda with exclusion of carbon dioxide. The alkalinity is expressed in milli-equivalents per litre.

Accuracy of the Determination

The accuracy of the determination depends to great extent on the reliability of the factor determination of the acid and alkalis and on the quality of the burettes used. The optimal error is ± 0.002 mEquiv./l.

Reagents

Bromothymol blue: triturate 100 mg of the indicator in an agate mortar with 1·6 ml 0·1 N caustic soda and dissolve in 100 ml distilled water.

Methyl red: triturate 100 mg of the stain as described with 3·7 ml 0·1 N caustic soda and dissolve in 100 ml water.

Sodium hydroxide: dissolve 20 g analytically pure sodium hydroxide in 20 ml water, keep it in a closed bottle for several days, then centrifugate in a closed glass bottle. Transfer 2 ml of the clear solution into 2 l water containing no carbon dioxide. The solution is approximately 0·015 N and is kept free of carbon dioxide. The titre of the solution is adjusted as described for the sample, 100 ml distilled water being used instead of 100 ml of the water sample.

Hydrochloric acid: a solution of approximately 0·013 N is produced from Fixanal or concentrated acid and the titre is carefully adjusted as follows (Kolthoff): a quantity of borax adapted to the burette volume (equivalent weight 190·607) is weighed and titrated, together with the acid, against methyl red.

For comparison purposes we use a solution which contains exactly the same quantity of boric acid, sodium chloride, indicator and water as the titred sample. For this purpose 0·105 ml of an 0·1 N boric acid solution containing 0·05 m of sodium chloride for each mg borax is required. The borax is recrystallized below 50°C and kept over a saturated sugar solution.

Determination Procedure

Exactly 100 ml seawater is placed in a 250 ml narrow-neck Erlenmeyer flask; 15 ml hydrochloric acid is then added with a Knudsen pipette, plus 0·4 ml of the BTB indicator. An airjet containing no carbon dioxide is directed into the sample through a frit for 30 minutes. The airjet continues during the back titration. The caustic soda is added from a 5 ml Deron burette with subdivisions of 0·005 ml. The comparison solution is a seawater sample with indicator, adjusted to 6·60 by means of a pH meter.

Calculation

V_p = sample volume
N = normality of the hydrochloric acid to 4 decimals
V = volume of the added acid
$V \triangleq a$ ml caustic soda
b = quantity of caustic soda required for the back titration

$$A = 1000 \frac{N \cdot V}{V_p} \cdot (1 - \frac{b}{a})[\text{mEquiv./l}]$$

Note

All the glassware must be calibrated!

6. DETERMINATION OF DISSOLVED INORGANIC PHOSPHATE

Range

0 to 10 μg—at $PO_4^{--} - P/l$.

Principle of the Determination

Molybdate forms hexamolybdic acid in an acid solution. Under suitable conditions the latter reacts with orthphosphoric acid and forms a heteropoly acid. The composition of this compound is $H_3(PMo_{12}O_{40}) - x\ H_2O$. The 1:12 phospho molybdic acid is yellow. Its optimum development occurs at pH 1·4. Its hydrolysis stability is less than that of molybdate silicic acid. The phospho molybdic acid is reduced with ascorbic acid to a blue compound with the following composition: $H_3[PMo_{12}O_{38}(OH)_2]$. The extinction of this heteropoly acid is measured. The reduced phospho molybdic acid is more stable than the unreduced form. Antimony salts accelerate the formation of the heteropoly acid, forming a compound which contains antimony in a 1:1 ratio with phosphate.

Since the silicic acid reacts with hexamolybdic acid under very similar conditions to those for phosphoric acid, the former would react with molybdate in the presence of a pH which is lower than the optimum. The present specifications demand a pH of approximately 0·8 for the solution. Under these conditions *no* further silico-molybdic acid is formed.

In other specifications for the determination of phosphate in seawater, stannous chloride is used instead of ascorbic acid for the reduction. This increases the sensitivity by a factor of 2. The disadvantage of using stannic chloride is the dependence of the reaction on time, temperature and reagent.

Accuracy of the Determination

If all the sources of error are carefully taken into account, the error of the individual determination is $\pm 0·02\ μg - at\ PO_4^{--} - P/l$.

Sample Collection and Sample Storage

The water sample should be analysed as soon as possible after collection. If this is not possible, the filtered sample can be poured into glass bottles and can be stored for some time after being treated with a little chloroform. All bottles and equipment must be completely clean and free of contamination with phosphate. (Detergents and cleaning agents contain phosphate!)

Reagents

5-N sulphuric acid: dilute 70 ml concentrated sulphuric acid to 500 ml.

4% ammonium molybdate: dissolve 20 g analytically pure ammonium molybdate in distilled water and make up to 500 ml. The solution remains stable in polyethylene bottles.

Ascorbic acid: dissolve $1 \cdot 76$ g ascorbic acid in 100 ml distilled water. The solution can be kept for approximately 30 days in a refrigerator.

Potassium antimony tartrate: dissolve $0 \cdot 274$ g in 100 ml water.

Mixed reagent: mix thoroughly 125 ml of sulphuric acid with $37 \cdot 5$ ml of the ammonium molybdate solution, and add 75 ml of ascorbic acid and $12 \cdot 5$ ml of the antimony solution. The reagent is unstable and should always be prepared shortly before use.

Standard solution: dissolve $1 \cdot 361$ g potassium dihydrogen phosphate in distilled water and make up to 1 l. The solution contains 10 µg – at PO_4^{--} — P/ml and must be stored in a dark glass bottle in a cool place. Diluted standard solutions are prepared shortly before use.

Cleaning of the Equipment

Before use, all the glassware must be left to stand overnight after filling with chromic sulphuric acid; it is then rinsed with distilled water and kept well closed. After use the bottles are immediately rinsed and stored closed. The equipment should be acid-cleaned repeatedly. Contamination with impurities containing phosphate can only be avoided if great care is taken.

Determination Procedure

Pipette 50 ml of the water sample into 100 ml glass bottles, add 10 ml of the freshly prepared mixed reagent and close the bottles. Mix thoroughly and allow to stand for 10 minutes to 2 hours at room temperature. The extinction of the solution is measured in 5 cm cells at 882 nm against distilled water. The reagent blank value is determined in the same manner. Double-distilled water is used instead of water from the sample. The reagent blank value must be redetermined for each batch of reagents. If the water is cloudy the sample is not measured against distilled water, but against water from the sample, to which $0 \cdot 5$ ml 5N sulphuric acid has been added. The reagent blank value is subtracted from the measured extinction and then multiplied with the proportionality factor obtained from the calibration curve.

Establishment of the Calibration Curve

Various quantities of the diluted standard phosphate solution $(0.05$ to 0.5 µg – at P) are placed in 100 ml measurement flasks, which are then filled with distilled water; 50 ml is taken from each flask by pipette and placed in 100 ml glass bottles; the rest of the procedure is as above.

7. DETERMINATION OF NITRATE

Range

0 to 40 µg – at $NO_3 -/l$

Principle of the Determination

The most sensitive determination method for nitrate uses nitrite since the latter can be detected without interference and with high accuracy even in very low concentrations by the formation of an azo dye. The quantitative determination of nitrite in seawater by the procedure of K. Bendschneider and D. J. Robinson is simple, sensitive and has been proved to be good.

The difficulty inherent in the determination of smaller quantities of nitrate by nitrite is related to the controlled reduction of the nitrate. The procedure which uses hydrazine as reduction agent in the presence of copper ions as catalyser is time-consuming, not quantitative, and depends on external circumstances to a great extent. Analysis by this method in the ship's laboratory is not very reliable.

When nitrate is reduced in the presence of ammonium ions and ammonia, i.e. in a very weak alkaline solution, the solution is buffered, and in addition the reduction effect of the cadmium is increased through complex formation. These two effects set up a stable potential for the gross reaction:

$$NO_3^- + (Cd)_s + 2 NH_4^+ + 2 NH_3 \leftrightharpoons NO_2^- + Cd (NH_3)_4^{++} + H_2O$$

Equipment and Reagents

The reducer consists of a glass tube of approximately 40 cm length and 8-10 mm internal diameter, with a goose neck (see Fig. 3). At the top end the tube widens out into a cylinder-shaped vessel containing approximately 100 ml. A frit has been set in at the bottom. The goose-neck has been shaped in such a manner that the amalgam never runs dry.

Cadmium amalgam: approximately 100 g cadmium, coarsely pulverized for analysis (Merck), is cleaned with diluted hydrochloric acid and treated with 100 ml of a 1% mercurous chloride solution

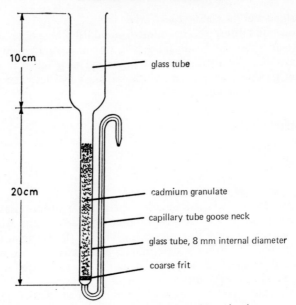

Figure 3. Reducer for the nitrate determination

while stirring vigorously. The amalgam is washed with distilled water which is continuously decanted, and may be conserved for a fairly long time if stored underwater. Before washing the amalgam into the reductor it is again briefly treated, first with diluted hydrochloric acid and then with distilled water.

Ammonium chloride-ammoniac solution: 250 g purest ammonium chloride and 100 ml 6% ammonia solution are made up to 1 l with water. The solution can be stored indefinitely.

Sulphanilamide solution: 5 g sulphanilamide is dissolved in a mixture of 65 ml concentrated hydrochloric acid and 300 ml water, and made up to 500 ml. The solution may be stored for several months.

n - (1 - naphthyl) - ethylenediaminedihydrochloride solution: 0·5 g of the hydrochloride is dissolved in 500 ml water. The solution may be stored for approximately one month in a brown glass bottle.

Filling of the reducer
The reducer is filled with water and the required quantity of cadmium amalgam is poured in all at once. The filling level should be approximately 30 cm. After filling, wash again with at least 40 ml distilled water. The flow rate of the samples should be approximately 50 ml in 15 minutes. During the operating pauses the reducer should

remain filled with water to which some ammoniac has been added. From time to time the amalgam is poured out of the reducer, briefly washed with moderately concentrated hydrochloric acid and freshly amalgamated. If the filling settles, the reducer is filled with diluted hydrochloric acid and allowed to stand overnight.

Determination Procedure

Add 2 ml of the ammonium chloride-ammonia reagent to 100 ml of the filtered water sample and fill the reducer with approximately 40 ml of this mixture. After a preliminary run of approximately 40 ml, fill up again with some 30 ml of the sample and collect the effluent in a 50 ml Erlenmeyer flask. Adjust the volume to 25 ml (calibration mark) by draining off with a needle. Add $0 \cdot 5$ ml of the sulphanilamide solution and after 3 to 5 minutes add $0 \cdot 5$ ml ethylene diamine solution to the reduced sample, mixing thoroughly after each addition. After 30 minutes (to 2 hrs) measure the extinction of the sample at a wavelength of 543 nm or by means of a suitable filter. For a comparison sample use the filtered water sample with the reagent, but without reduction. In this manner the interference caused by the nitrite present in the water sample is eliminated. (Nitrite passes through the reductor uninfluenced). The comparison sample can be used simultaneously for the nitrite determination.

Establishment of the Calibration Curve

To establish the calibration curve known quantities of nitrate are added to seawater with a low nitrate content and the samples are treated as described above. A water sample without added nitrate is used for the comparison sample. The extinction follows the Lambert-Beer law until a content of 20 µg-at nitrate-nitrogen/l is reached. When the nitrate content is greater than 20 µg-at/l, the seawater sample must be suitably diluted with nitrate-free water.

Interference with the Determination

The procedure for the determination of nitrite is very specific and is not disturbed by the ions which occur in sea or fresh water. Small quantities of hydrogen sulphide do not affect the nitrate determination either. The effectiveness of the reducers should be checked frequently. The reducers can be re-activated by rinsing with 1 % EDTA solution.

8. DETERMINATION OF SULPHATE

Principle of the Determination

Although a whole series of volumetric, polarographic and nephelometric methods has been proposed for the determination of sulphate in seawater, none is more specific than the classical gravimetric method of precipitation as barium sulphate. Since potassium and sodium ions are easily co-precipitated when this method is used in seawater, certain conditions must be carefully observed. The co-precipitation can be reduced by the presence of picric acid. The precipitation of calcium sulphate is prevented by acidification of the solution.

Reagents

Picric acid: a saturated solution in water is used.

Barium chloride: dissolve 10 g barium chloride dihydrate in 100 ml water.

Determination Procedure

Using a weighing burette, place 50 ml of the filtered seawater samples in a 600 ml beaker and add 235 ml distilled water, 10 ml of the picric acid solution and 5 ml concentrated hydrochloric acid. Heat the solution in a water bath almost to boiling point, then add 10 ml barium chloride solution while stirring vigorously. The precipitate is allowed to settle for 1 hour, the sample being kept hot during this time. The supernatant solution is decanted through a previously calcined and weighed filtering crucible and the precipitate is rinsed in the crucible with a jet of hot water. The precipitate is then washed with hot water until the filtrate does not show any opalescence with silver ions. The crucible is dried for 30 minutes at 110°C and calcined at 800°C in the oven until a constant weight is attained. All weights must be reduced to the vacuum weight. An accuracy of $\pm 0 \cdot 5\%$ is easily obtainable if the procedure is carried out carefully.

9. Flame-photometric Determination of Sodium and Potassium Ions

Principle of the Measurement

The test solution is sprayed into an oxy-hydrogen flame or an aerosol is first made up and then sprayed into the flame. The drops of the sample evaporate in the flame. A balance $Me \leftrightarrows Me^+ + e^-$ is formed. Energy is conveyed to the atoms through the high temperature of the flame, which is sufficient – particularly with alkali ions – to lift valence electrons to higher energy levels. When the activated electrons return to the level state, light with distinct wavelengths is emitted. Under certain conditions the intensity of the emitted light is proportional to the concentration of the ions to be determined in the sample solution. However, in addition to the form and temperature of the flame and the quantity of the sample added to the flame within the time unit, the intensity of the emitted radiation depends on the size of the drops, etc. Standard values must therefore be taken continually, together with the samples.

Complexes, certain anions and a series of cations disturb the determination. The composition of samples for testing must therefore correspond to the sample solution.

Special filter-flame photometers are commercially available for the determination of alkaline and alkaline-earth metals. A spectrophotometer with the addition of a flame can always be used. For more exact determination a recording flame photometer is preferable.

An accuracy level of $\pm 1\%$ can be obtained under optimal conditions.

10. Determination of Calcium and Magnesium

Principle of the Measurement

Calcium and magnesium ions form relatively stable complexes with ethylene diamine tetraacetic acid; these are called chelates. (log $K_{Ca} = 10\cdot70$; log $K_{Mg} = 8\cdot69$). Under suitable conditions complex formation is a stoichiometric reaction. Its development is rapid and quantitative and thus meets the requirements for a titrimetric determination. The end point of the titration is recognized by the fact that particularly suitable organic compounds (metal indicators) are added (calcon-carboxylic acid for Ca and eriochrome black T for Ca + Mg), which turn to another colour after chelate formation of the Ca or Mg with EDTA. In one assay the sum of the calcium and magnesium ions is determined, and in another the calcium after precipitation of the magnesium as hydroxide is determined separately. Magnesium hydroxide has a smaller solubility product than the chelate under the specified pH conditions.

Accuracy of the Determination

The calcium determination can attain an accuracy of $\pm 0\cdot3\%$. In the magnesium determination the uncertainty produced by the addition of errors during subtraction can amount to approximately $\pm 0\cdot6\%$.

Reagents

Ethylene diamine tetraacetic acid disodium salt: dissolve $14\cdot890$ g in 1 l water. The solution is $0\cdot04$ molar. 1 ml is equivalent to $1\cdot603$ mg Ca. (The factor of the solution must be carefully standardized with calcium carbonate or cadmium or zinc ions.)

Alkaline potassium cyanide solution: dissolve 80 g NaOH and 10 g KCN in 1 l water.

Indicator: triturate 0·5 g eriochrome black or calcon with 50 g sodium sulphate or sodium chloride in the mortar.

Buffer solution pH 10: dissolve 70 g ammonium chloride in 570 ml ammoniacal solution (specific weight 0·90) and make up to 1 l.

Determination Procedure

Mg + Ca: pipette 50 ml of the seawater sample into a 250 ml Erlenmeyer flask; add 10 ml buffer solution pH 10 and a spatula tip of eriochrome black; heat to 60°C and titrate with EDTA solution until the colour changes from red to blue. The titration should not be carried out too rapidly.

Ca: pipette 50 ml of the sample into a 250 ml Erlenmeyer flask, add sufficient EDTA solution to bind almost all of the calcium (the quantity required is established in a preliminary titration), then add 10 ml of the alkaline base solution and one spatula tip of calcon carboxylic acid; after precipitation of the magnesium hydroxide, titrate until the colour changes from wine red to blue. In this way co-precipitation of calcium together with the magnesium hydroxide is avoided.

Note

Strontium and barium ions are detected together with the calcium.

REFERENCES

Books

BARNES, H. (1959). *Apparatus and methods of seawater analysis.* G. Allen & Unwin Ltd., London.

BATES, R. G. (1954), *Electrometric pH determinations.* J. Wiley, New York.

HARVEY, H. W. (1955). *The chemistry and fertility of seawaters.* University Press, Cambridge.

HERRMANN, R., and C. Th.G. AKLEMACH (1960). *Flammenphotometrie.* 2. Aufl., Springer-Verlag, Göttingen.

RILEY, I. P., and G. SKIRROW (1965). *Chemical Oceanography,* I + II. Academic Press, London and New York.

STRICKLAND, I. D. H., and T. R. PARSONS (1965). *A manual of seawater analysis.* Fish. Res. Board of Canada, Bull. No. 125.

Papers

ANDERSON, D. H., and I. R. ROBINSON. (1946). 'Rapid electrometric determination of the alkalinity of seawater, using a glass electrode.' *Indust. Eng. Chem. Anal. Ed.* **18**, 767–773.

BATES, R. G. (1962). 'Revised Standard values for pH measurements from 0 to 95°C.' *J. Res. Natl. Bur. Std.* 66A, 179.

BATHER, J. M., and J. P. RILEY. (1954). 'The chemistry of the Irish Sea. I. The sulphate chlorinity ratio.' *J. du Cons.* **20**, 145.

BUCH, K., and S. GRIPENBERG. (1932). 'Über den Einfluß des Wasserdruckes auf das pH und das Kohlensäuregleichgewicht in größeren Meßtiefen.' *J. Cons. Expl. Mer.* **7**, 232–245.

GRASSHOFF, K. (1962). 'Zur Bestimmung von Sauerstoff in Meerwasser.' *Kieler Meeresforsch.* **18**, 42–50.

(1964). 'Über die Bestimmung von Nitrat in Meer- und Trinkwasser.' *Kieler Meeresforsch.* **20**, 5.

GREEN, E. J., und D. E. CARRITT. (1967). 'New tables for oxygen saturation of seawater.' *J. Marine. Res.*, **24**, No. 2.

FOX, H. M., and C. A. WINGFIELD. (1938). 'A portable apparatus for the determination of oxygen dissolved in a small volume of water.' *J. exp. Biol.* **15**, 437–445.

JACOBSEN, J. P., und M. KNUDSDEN. (1940). 'Urnormal 1937.' *Union. Geod. et Geophys. Intern. Ass. d'Oceanogr.*, Pub. sc. no. **7**, 38, Liverpool 1960.

KALLE, K. (1951). 'Einige Vereinfachungen der Chloridtitration für biologische und wasserbaukundliche Zwecke in Küstengewässern.' *DHZ* **4**, 21.

KNUDSEN, M. (1901). *Hydrographische Tabellen.* G. E. C. Gad, Kopenhagen.

MURPHY, J., and J. P. RILEY. (1962). 'A modified single solution method for the determination of phosphate in natural waters.' *Anal. Chim. acta*, **27**, 31.

OXNER, M. (1962). 'The determination of chlorinity by the KNUDSEN Method.' (Aus.: Reprint of the Hydrographical Tables.) G. M. Manuf. Co. New York.

RAKESTRAW, N. W. (1949). 'The conception of alkalinity or excess base of seawater.' *J. Marine Res.* **8**, 14–20.

WATTENBERG, H. (1933). 'Über die Titrationsalkalinität und den Kalziumkarbonatgehalt des Meerwassers.' *Dtsch. Atl. Exp.*, 'Meteor.' 1925–1927. Wiss. Erg., Bd. **8**, 2, 122–231.

B. Physical Methods

G. SIEDLER

The measurement of physical parameters in the sea plays an important part in many research problems relating to marine biology. In general the biologist will not need to carry out these physical measurements himself, but in order to be able to judge the relevance of certain measurements and the limits of error, the marine biologist should be familiar with the principles of the physical methods which are important for his work. We therefore have attempted to provide an introduction to these methods in the following section; with the exception of the procedures of measurement which are occasionally used by the marine biologist himself, the principle but not the procedure will be discussed. A more extensive description is not possible within the framework of the present book. More detailed information can be found in the literature listed in the appendix and in the directions accompanying the measuring instruments used.

1. WATER BOTTLE

For the study of the physical and chemical properties of seawater and the content of suspended organic components, water samples from certain depths are required if no *in situ* measurement method is available. In general these samples are obtained from a stationary boat with the help of special water bottles at the following internationally defined 'standard depths': 0, 10, 20, 30, 50, 75, 100, 125, 150, 200, 250, 300, 400, 500, 600, 700, 800, 900, 1000, 1100, 1200, 1300, 1400, 1500, 1750, 2000, 2500, 3000, 4000, 5000, 6000, 7000, 8000, 9000 m.

Water from the immediate vicinity of the ocean surface is obtained very simply by scoop buckets; when using these special care must be taken to see that the samples are collected from an area which is not exposed to pollution from the ship. Depth samples are generally obtained with a series of bottles which are suspended one above the other from a cable and are thus lowered into the depths while open; their closing mechanism is activated by a messenger which slips along the wire by propeller movements as raising begins, or by hydrostatic pressure. The bottles are generally made of brass or plastic, and for bacteriological investigations glass, rubber, neoprene or platinum are also used. Both plug valves and flap valves are in use.

We will describe here in some detail the water bottle most commonly used in oceanography, namely the Nansen tipping water bottle (Fig. 4). The bottle consists of a cylindrical container with a capacity of 750-1250 cc, to which a bracket for a reversing thermometer (see I.B.3, p. 32) is attached. The water bottle is provided

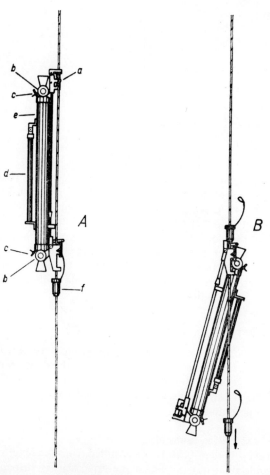

Figure 4. Nansen tipping water bottle (A) before, and (B) after tipping: *a* – releasing device, *b* – stopcock valve for filling, *c* – stopcock valve for drainage, *d* – thermometer bracket, *e* – bottle container, *f* – messenger

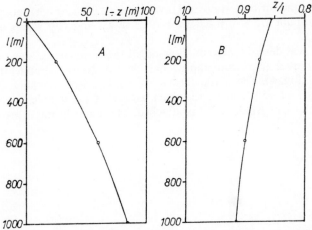

Figure 5. Diagram for depth correction by determination of the differences 1 – z (A) and quotients z/1 (B)

with plug valves at top and bottom. The instrument is attached to the cable of a hydrographic winch by a clamp, the release device of the bottle being at the top. When the desired depth has been reached a messenger is sent down the cable and hits the release device. This releases the top catch and the mounting for another messenger previously set below the water bottle. The bottle tips over by approximately 180° and the plug valves are simultaneously closed and arrested. The lower messenger which has been released travels down the cable in order to close the next water bottle in the series in the same manner. The water sample can then be pulled up on to the ship in a closed condition.

2. DEPTH DETERMINATION

Every sampling or *in situ* oceanographic measurement is accompanied by a determination of the depth or of the depth range of the measurement position, and generally also by a determination of the depth of the ocean floor. Various physical methods are used for these depth determinations.

(a) Cable Length Measurement

The measuring depth can be determined from the length (l) of the wire or cable paid out on which the apparatus is suspended; this requires a straight cable and a known cable angle (α) in comparison to the vertical:

$$z = 1 \cdot \cos\alpha$$

If the inclination of the cable varies because of alterations in depth, the depth at certain points must be determined by a pressure measurement (see next section). The depth data are then corrected as shown in one of the two diagrams in Fig. 5.

(b) Pressure Measurement

At a constant density of seawater, the hydrostatic pressure is closely proportional to the depth. We find that within an accuracy of 1-3 % of the measuring data, the pressure in decibars is numerically equal to the depth in metres. For more accurate measurement the particular density distribution must be taken into account. For practical purposes a distinction is made between single measurements with water bottles and continuous measurements:

Thermometric pressure measurement: the temperature at a fixed depth is measured by means of two deep sea tipping thermometers (see I.*B*.3. p. 32), one of which is protected from the pressure and the other exposed to the pressure. The unprotected thermometer will show a higher temperature than the protected thermometer because of the compression of the system of the thermometer; an increase of approximately $0 \cdot 01°C$ per decibar is generally found.

From the corrected temperature difference ΔT of the two thermometers and from the pressure coefficient $Q = g \dfrac{\Delta T}{P - P_0}$ ($g =$ gravitational acceleration, P or P_0 = pressure *in situ* or at the surface) we obtain with the average density ϱ_m, the observation depth z:

$$z = \frac{\Delta T}{Q \times \varrho_m}$$

The procedure is very accurate, but only yields single measurements at certain depth levels.

Continuous pressure measurement: for continuous depth measurements with immediate tracing or recording on electrical recorders we use measuring elements which yield information by means of the mechanical deformation of especially formed metal surfaces as a result of pressure. Pressure recorders of this type are used to measure the depth of vertically operated apparatus at any time from an anchored vessel, to record the depth of towed equipment and, if necessary, to adjust it suitably by changing the length of the cable or the ship's speed, and to record the pressure fluctuations produced by tidal waves at high sea levels in the deep ocean with instruments placed on the ocean floor. For the latter purpose we also use measuring apparatus which yields information on pressure not through the deformation of metal surfaces, but through the volume change of an air cushion. The principal pressure indicators are the Bourdon tube, the wave tube and the membrane box.

(c) Acoustical Measurements

If the acoustical rate in a certain ocean area is known, the distance traversed may be calculated from the running time of a sound impulse in the water. A time measurement such as this can be used to determine the depth of the ocean floor from shipboard or to establish the depth of a measuring instrument (Fig. 6A), to find the distance of an instrument, such as a camera, from the ocean floor by means of the time difference between the direct and the reflected beam (Fig. 6B), or to check the depth of a towed instrument (Fig. 6C). In general magnetostrictive oscillators are used for sending and receiving the signals within the audiofrequency range or in the supersonic range.

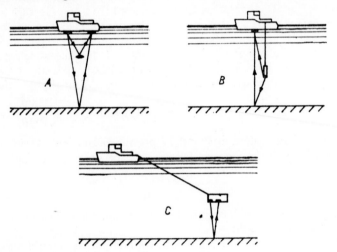

Figure 6. Applications of the acoustical measurements: echosound and fishlens (A), ping (B), and net sound (C)

3. TEMPERATURE DETERMINATION

The temperature of seawater at the various depth zones of an ocean region is generally determined *in situ* with contact thermometers of various types. One exception is the radiation thermometer, borrowed from low flying aircraft for use for surface temperature measurements. Below we describe the principal procedures for contact measurements. Fluid thermometers and electric thermometers, and occasionally bi-metal thermometers as well, are used.

(a) Fluid Thermometers

Mercury Thermometers: for surface temperature determinations a normal mercury thermometer with 1/5 or 1/10°C divisions is used. A sample obtained with the scoop bucket or with an isolation water bottle is used for the measurement. To obtain the temperature at selected oceanic depths we use the Negretti and Zambra or the Richter and Nansen tipping thermometer (Fig. 7), attached to a suitable water bottle (see I, p. 27). After the water bottle has attained the desired depth and an adjustment period of 5-10 minutes has passed, a messenger sliding down the cable releases a stop device, simultaneously causing the thermometer to reverse and the water bottle to close. As the thermometer reverses the mercury thread of the main thermometer breaks at the constriction point on the dead arm of the capillary and runs down. If the temperature later increases, any mercury which follows from above is trapped in the receiving loop. After an adjustment period for the thermometer system on board, the main and auxiliary thermometers are read with a magnifying glass. In addition to a calibration correction, the value shown by the main thermometer must be corrected with regard to the temperature at the time of the reading, which is given by the auxiliary thermometer, in order to equalize the changes in volume shown by the broken mercury thread as a result of the alteration in the surrounding temperature.

The following formula is used:

T_g = Temperature reading of the main thermometer, with calibration correction

t_g = temperature reading of the auxiliary thermometer, with calibration correction

K = Heat expansion coefficient of the thermometer system

V_0 = Hg volume below the 0°C mark (in equivalent °C)

$$T_w = T_g + C_g \text{ with } C_g = \frac{(T_g + V_0)(T_g - t_g)}{K - 100}$$

When all the corrections have been carried out an optimal measurement accuracy of $\Delta T = \pm 0\cdot 01°C$ can be obtained.

For the unprotected thermometer (see 2, p. 32) the following formula is used:

T_u = temperature reading of the main thermometer, with calibration correction

t_u = temperature reading of the auxiliary thermometer, with calibration correction.

T_{wu} = temperature reading *in situ*

$$T_{wu} = T_u + C_u \text{ with } C_u = \frac{(T_g + V_0)(T_g - t_g)}{K - 100}$$

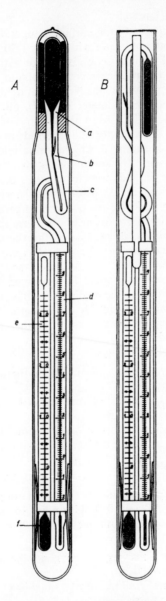

Figure 7. Deep sea tipping thermometer after reversing, shown in protected (A) and unprotected (B) form: *a, f* – mercury container, *b* – constriction point, *c* – collecting loop, *d* – scale of the main thermometer, *e* – scale of the auxiliary thermometer

Bathythermograph: for continuous recording of the temperature as a function of the depth with less accuracy up to depths of 270 m we use the Spilhaus bathythermograph (Fig. 8). A pressure-responsive corrugated tube moves a smoked or gold-coated glass plate on which a stylus scratches a tracing in response to the temperature. For the temperature measurement we use an organic fluid which, under the influence of heat, expands in the spiral-shaped closed thermometer tube and thus produces an increase in pressure. The

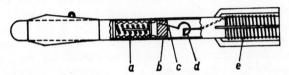

Figure 8. Bathythermograph: *a* – corrugated tube for pressure measurement, *b* – recording plate, *c* – stylus, *d* – Bourdon tube, *e* – thermometer capillary

Bourdon tube connected to the thermometer tube reacts to this increase in pressure and thus shows the temperature elevation. Important advantages of this instrument are its simple, robust construction and the fact that it can be used not only from an anchored ship, but also from a moving ship.

(b) Electrical Thermometer

Metal resistance thermometer: metals have the ability of increasing their electrical resistance when heated. For measurement purposes we generally use platinum, and occasionally, when less accuracy is required, nickel. Platinum thermometers are used particularly in those cases where a high degree of accuracy and exchangeable probes are desired. The relative alteration in the electrical resistance is almost constant and consists of approximately 0.4% per °C. The optimal accuracy obtainable in measurements at sea consists of $\Delta T \approx \pm 0.01$°C. Platinum resistance thermometers are generally used in the Wheatstone bridge circuit together with a multicore cable.

Thermistors: certain semi-conductors can react to heat with a strong fall in resistance, which can be approximated by an exponential law. These 'thermistors', which are also described as 'NTC resistances' or 'hot conductors' have the advantage of a very high sensitivity of approximately 3-6 % per°C (at 20°C). Electrical measurement is thus simplified in comparison with platinum thermometers. However, the strong non-linear curve and drift phenonema produced by ageing processes present a disadvantage. Thermistors are also frequently activated by a Wheatstone bridge, as for example in recording thermometers for the measurement of surface water temperatures from moving ships, in anchored automatic recording thermographs and in vertical temperature sounds for a single use from a moving ship (expendable BT). Two measurement systems with thermistors have been developed for continuous recording of the total water temperature stratification close to the ocean surface. The Richardson and Hubbard thermistor chain consists of a heavy chain weighted with fins, provided with 23 thermistors and a Bourdon tube pressure recorder for depth checks, which is towed by a moving ship. The thermistors are scanned in repeated cycles, and on board a recording of the isotherm is obtained directly through a special transformer system. A less expensive system is the 'Delphin' of Joseph and Weidemann, in which a streamlined measuring device is automatically steered up and down behind the moving ship. Direct recording of the isotherm is also possible in this case.

4. Salinity Determinations by Physical Methods

Salinity, as defined by Forch, Knudsen and Sörensen (see I.A, p. 3), can be determined indirectly by means of physical methods, using various values whose correlation with the salinity is known through empirical data. The salinity can be expressed in the form of the following functions:

$$S = f_1 (\zeta, T, P_0)$$
$$S = f_2 (n, T, P_0)$$
$$S = f_3 (L, T, P)$$
$$S = f_4(Cl)$$
$$S = f_5 (v, T, P)$$

where

S = density, T = Temperature, P = pressure (P_0 = atmospheric pressure), n = refractive index, L = electrical conductivity, v = velocity of sound, Cl = chlorinity, ζ = density.

Functions f_1 to f_4 are used for the salinity determinations.

(a) Salinity Determination with the Hydrometer

In accordance with the above-mentioned function f_1, the salinity may be calculated from the density of the seawater at a known temperature. Hydrometers in the form of calibrated glass floats are used to measure the density. Two types are employed, namely, columnar hydrometers and weight hydrometers.

When using the weight hydrometer one must be careful to see that the float is submerged up to a fixed point by putting on weights. The size of the weight added is a measurement of the density of the seawater sample. With the more commonly used columnar hydrometer (Fig. 9) the floating part has a scale which is directly calibrated in

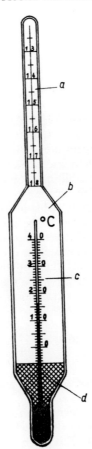

Figure 9. Columnar hydrometer for salinity determination: *a* – density scale with reference to δ = (Q − 1) 10³, *b* – buoyancy body, *c* – thermometer, *d* – metal weight

density units. The water sample is poured into a standard cylinder, allowed to adjust to the environmental temperature, and the hydrometer is inserted. Following the free balancing of the floating part the density value is read from below, through the water, from the point where the spindle emerges from the water surface. The value indicated is converted into the density at 17·5°C and the salinity is obtained from Knudsen's Hydrographic Tables.

The optimal precision attainable is $\Delta S \approx \pm 0 \cdot 05‰$; for this the temperature error must be smaller than $\Delta T = \pm 0 \cdot 1°C$. The precision of the procedure is much lower when the measurements are carried out on shipboard.

For more accurate measurements the Stott procedure – in which a float is brought into suspension by increasing the temperature of the seawater – may be used, or Cox's procedure in which suspension is attained by increasing the air pressure above the sample.

(b) Salinity Determination with the Refractometer

It is possible to determine the salinity by measuring the refractive index at a constant temperature in accordance with the above-mentioned function f_2. The refractive index of seawater increases as the salinity increases, since additional ion hydrates raise the dielectric constant of the water. The increase per $\Delta S = 0 \cdot 1\%$ consists of $\Delta n \approx 2 \cdot 10^{-5}$. A temperature reduction of $\Delta T = 1°C$ raises the refractive index by approximately the same amount. For practical measurements we use a Pulfrich immersion refractometer, in which the critical angle of the total reflection is determined when a monochromatic light beam strikes the boundary surface between the seawater and optically denser glass. For this critical angle α_g (against the perpendicular) of the total reflection the following formula is used:

$$\sin \alpha_g = \frac{n_w}{n_G}$$

where

n_w = refractive index of the seawater, and n_G = refractive index of the glass.

Fig. 10 shows the measurement arrangement. A measuring prism is immersed in a circulation cell, which in turn is located in the water bath of a thermostat. The prism is illuminated from below, through a mirror, with a monochromatic light beam. Rays with $\alpha < \alpha_g$ enter into the prism, rays with $\alpha > \alpha_g$ are totally reflected. When the prism is observed through a telescope we find a light-dark boundary on the scale. The scale value shown is converted into the normal temperature and the salinity is obtained from a table.

The optimal precision attainable is somewhat better than $\Delta S = \pm 0 \cdot 1\%$. This precision can also be obtained on board ship.

(c) Salinity Determination by Conductivity Measurements

In accordance with the above-mentioned function f_3, conductivity measurement can be used for the salinity determination if the temperature of a drawn sample is kept very constant or compensation is made for the influence of the temperature, since the specific electrical conductivity of seawater rises by $\Delta L \approx 0 \cdot 01$ mS/cm at

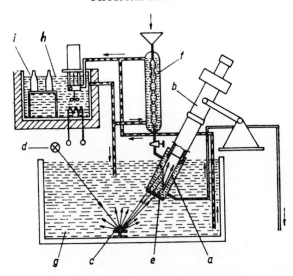

Figure 10. Immersion refractometer for salinity determination: *a* – prism, *b* – telescope, *c* – mirror, *d* – lamp, *e* – container for sample, *f* – circulation cooler, *g* – water bath, *h* – thermostat, *i* – bottle with sample

a rise in salinity of $\Delta S = 0\cdot01\%_0$, of temperature by $\Delta T = 0\cdot01°C$ and of pressure by $\Delta p = 20$ decibars. When carrying out *in situ* measurements it is necessary to take a very accurate temperature measurement and a pressure measurement simultaneously with the conductivity measurement. Devices with two or three electrodes or inductive measuring cells can be used as measurement elements.

Whereas in electrode cells we measure the alternating current resistance of a water column between two or three electrodes, in inductive cells the conductivity of the seawater controls the linkage of two transformers. Fig. 11 shows examples of simplified measuring arrangements. In the electrode measurement cell (Fig. 11A) the resistance R_W of the seawater, which depends on the salinity, is the component of an alternating current measurement bridge, which can be calibrated by means of the potentiometer R_P and the zero indicator G when the fixed resistances R_1 and R_2 are used. The resistance R_W can be calculated according to

$$\frac{R_W}{R_P} = \frac{R_1}{R_2}.$$

An alternating current must be used in order to avoid polarization effects. The advantage of the electrode measurement cell over the inductive measurement cell is the low expense when constructing the

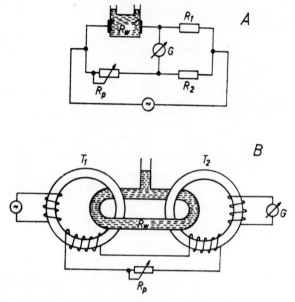

Figure 11. Principles for the procedure to measure the specific electrical conductivity with electrodes (A) inductively (B)

apparatus; its principle disadvantage is the poor long-term constancy as a result of changes on the electrode surfaces.

In the inductive measurement cell (Fig. 11B) the salinity-dependent resistance controls the current, through which two toroid transformer coils are coupled. The alternating current induced through T_1 and the seawater loop in T_2 are compensated by a current in the opposite direction, which can be set up through the potentiometer R_P. After calibration to zero through the indicator G, the position of the potentiometer indicates the salinity. Instruments for the salinity determination through electrical conductivity are frequently called 'salinometers'.

Laboratory conductivity measurement apparatus. Two types of apparatus are now in use for precision determinations. In the Cox salinometer, the construction of which is basically identical with the previous-developed models of Wenner, Smith, Soule, and Schleicher as well as of Bradshaw, eight electrode measurement cells whose cell constants have been made of identical size by series resistances and parallel resistances, are kept very exactly at a temperature of 15°C by a thermostatic fluid. The measurement is carried out in a transformer bridge which can be used stepwise;

zero calibration can be carried out by means of a sign with a cathode ray oscillograph and an indicator with a phase-dependent rectifier. For calibration purposes, normal water measurements are always undertaken between the measurements of the unknown seawater samples. With this procedure a reproducibility of $\Delta S = \pm 0 \cdot 002‰$ and a precision of $\Delta S = \pm 0 \cdot 01‰$ can be attained. The measurement devices show the relationship of the conductivity of the unknown seawater to that of the normal water. The salinity is then obtained by means of tables.

Brown and Hamon attained approximately the same degree of accuracy with less expense by omitting the thermostat and using a temperature compensation over a certain temperature range by means of a thermistor. The measurement was carried out with an inductive measurement cell as follows:

The water samples and the standard or sub-standard water are brought to room temperature (maximum permissible deviation: approximately $\pm 2°C$) and the salinometer is set up for the appropriate temperature range. The measurement cell is filled by suction of the standard water sample and calibration is carried out. After this the unknown seawater samples are poured in one after the other and their resistance in relation to the standard water is determined. After 10 to 20 measurements a new measurement with the standard water is carried out, which can be used for correction if there is any indicator drift. Disturbances may be produced above all by the formation of gas bubbles. Each salinity value is again obtained through tables. The precision of this measurement method is $\Delta S = \pm 0 \cdot 003‰$ for ocean water.

In-situ Conductivity measuring apparatus. For measurements in shallow coastal water areas where the precision requirements are not particularly high, *in situ* measurement apparatus with electrodes or with inductive sounds is used for the salinity determinations. An example is the portable salinometer of J. Williams. As measurement probe an inductive receiver is used for the conductivity and two thermistors connected in parallel are used for the temperature. The measurement sound is placed at a specific depth in each case. The temperature is measured by means of a normal Wheatstone bridge; the thermistors are then connected to a second bridge together with a potentiometer, which is adjusted through the conductivity sound. By appropriate selection of the bridge resistances it is possible to obtain a compensation of the temperature effect on the electrical conductivity and thus to attain a direct indication of the salinity with a precision of $\Delta S = \pm 0 \cdot 3‰$. The bridge calibration occurs automatically with this apparatus.

Apart from simple equipment of the type mentioned, there is in use *in situ* measurement apparatus for the open ocean with which a measurement precision of $\Delta S = \pm 0 \cdot 03‰$ can be attained.

5. LIGHT MEASUREMENTS

The influence of solar radiation plays a significant part in marine biological research because of its importance in photosynthetic processes as the basis of all marine life. Within the framework of this introduction we will discuss several measurement methods which enable us to determine some values of significance for the description of the radiation balance. These are the physical coefficients which characterize the absorption and dispersion within the sea.

Only part of the solar radiation falling on the surface of the sea penetrates into the seawater, since a certain fraction is reflected. Again, only a part reaches the lower depths of the water; the remaining rays are absorbed or are scattered. Absorption differs at various wavelengths, and therefore also for the spectrum colours of visible light. In pure seawater its minimum is in blue; however, the content of dissolved substances may cause the position of the minimum to shift. The scatter is also dependent on wavelength (Rayleigh scatter), particularly in regard to the particles suspended in the water. Part of the scattered light leaves the sea again through the water surface and thus essentially determines the impression of colour for the observer. By comparison with several standard colours it is possible to establish this seawater colour and to draw gross conclusions from this in regard to the content of dissolved and suspended susbtances.

For the description of the effect of absorption and scatter in seawater we use a number of physical coefficients which describe the extent of the radiation intensity (I_0) before and (I) after traversing a distance (z) by the following law:

$$I = I_0 e^{-kz}$$

(e = basis of the natural logarithm, k = fading coefficient).

We must differentiate between the coefficient for a beam and the coefficient for the total diffuse radiation. To describe the process relating to the beam we use the coefficients of forwards and backwards scatter and their sum, the dispersion coefficient, as well as the physical distinction coefficient, which indicates the total loss of a beam through absorption and dispersion. For diffuse radiation we use the vertical extinction coefficient, which indicates the attenuation of the diffuse, downwards-directed radiation, and the diffuse reflection coefficient, which gives the ratio of upper light to under light. For the complete description it is sufficient to measure three of these coefficients; the remainder may then be calculated from them.

We give below some measurement procedures for the determination of these coefficients. As radiation receptors we use alkali photocells, selenium photocells or cadmium sulphide photoresistances; glass filters are used to determine the spectral dependence.

(a) Determination of the Physical Extinction Coefficient in the Laboratory

A cell filled with seawater is placed in the beam of a Pulfrich photometer. By comparing the intensity of a light beam which has gone through the cell and that of a beam which has traversed a comparison distance we obtain, behind a monochromator, the ratio $1/1_0$ and thus the physical extinction coefficient. An advantage of the method is that the measurement is easily carried out for many wavelengths by means of selective filters; a disadvantage is the error which may occur because of the additional scattering resulting from the formation of gas bubbles in the scooped water sample.

(b) Determination of the Physical Extinction Coefficient *in situ*

A frequently used instrument is the Joseph visibility measurement instrument (Fig. 12). A parallel light beam runs through a measurement distance of 1 or 2 m, thus undergoing attenuation. The remaining light intensity is measured through a red or blue filter with a selenium cell. The resulting voltage is conveyed through a multicore

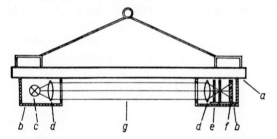

Figure 12. Visibility measuring instrument by Joseph. a – frame, b – casing, c – lamp, d – lens, e – filter, f – photocell, g – measurement stretch

cable together with a temperature measurement signal to a recording instrument on board. The maximum possible depth with this instrument is approximately 600 m. Recently measurement instruments of this type which can be used up to a depth of 6,000 m have been developed.

(c) Determination of the Vertical Extinction Coefficient

The dependence of the intensity of the upper light on the depth is measured by a simple process using Secchi discs. A round white disc of approximately 50 cm diameter is lowered into the water until it disappears from view. According to Joseph, the correlation between the depth of visibility (s) measured in this manner and the vertical extinction coefficient (b) is:

$$b \cdot s \approx 0 \cdot 95$$

Another procedure for measuring the dependence of the depth on b uses a lowered selenium photocell directed upwards, and filters which may be inserted selectively. The signals are conveyed to the ship through a multicore cable and compared at the surface by means of the potential of a photocell.

(d) Determination of the Diffuse Reflection Coefficient

If, in addition to the downwards-turned photocell, a photocell turned upwards is simultaneously used in the latter method, this will yield the ratio of underlight to upper light and thus the diffuse reflection coefficient.

6. CURRENT MEASUREMENTS

Horizontal current distributions can be obtained essentially by two different – and in principle equivalent – methods: determination of the path lines of individual water particles through drift measurements, or by recording the flow lines shown by the current at several fixed points.

Drift measurements in the surface current. The principal basis of present-day surface current maps is the data on ship displacements collected at hydrographic stations. In this method the entire ship is

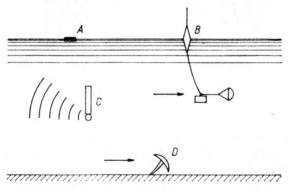

Figure 13. Methods of drift measurement: drift mat (A), parachute buoy (B), Swallow buoy (C), and Woodhead drifter (D)

considered as a drifting object and the deviation from the planned course yields information on the strength and direction of the surface current. One problem associated with this is the interference created by displacement resulting from the direct action of wind. For special ocean areas bottle post investigations are carried out with drift bottles or drift mats enclosed in plastic bags (Fig. 13A). Many hundreds or thousands of such objects are brought out to selected sites. Those who find them on the coast record the time and place where found on the enclosed map and return them for a reward to the sender.

(a) Determination of Current Direction

Drift measurements in moderate water depths. Two types of apparatus are used. In the case of the parachute buoy (Fig. 13B), a small surface buoy with markings, radar reflectors or a radio transmitter has a parachute or kite-like structure with high current resistance attached to it by means of a thin wire with a weight. The whole arrangement moves with the current at the depth of the parachute, which can be tracked by observing the surface buoy. The swallow buoy (Fig. 13C) is a tubular structure with an automatic sound-generating device which is suspended at a depth of a few hundred metres; it is tracked and followed by listening simultaneously with two hydrophones on board ship.

Drift measurements of the bottom current. Mushroom-shaped drift bodies as described by Woodhead (Fig. 13D), which can easily be set on the ocean floor by means of the foot, are put out in large numbers at specified sites and are then carried along with the bottom current. When found again in bottom trawl nets the drift can be established.

(b) Determination of Current Velocity

Current measurements from the anchored ship. Propellers or Savonius rotors are used as measurement elements for the determination of the current velocity; the current direction is shown by the movement of a current vane or of the total casing in relation to a compass needle or to the direction of the ship's longitudinal axis. The ship must be anchored.

The previously widely used Ekman current meter is dropped by cable to the desired depth from an anchored vessel; recording begins when the propeller is released by a messenger. The propeller activates a counting device and after a certain number of turns simultaneously leaves the way free for a bronze ball which then rolls along a groove on the compass needle and drops into a compartmental arrangement. A second messenger releases the propeller again, generally after ten minutes. When the instrument has been brought to the surface the number of turns can be read from the counting device and the current velocity can then be obtained from a standard table. The frequency distribution of the balls in the compartment is a measure of the direction.

When measurements are carried out close to the surface the ship's magnetism disturbs the compass; another type of direction measurement must therefore be used. The frame of the Rauschelbach bifilar current meter is lowered by means of two cables to a maximum depth of 50 m. The frame is always suspended perpendicularly to the longitudinal axis of the ship, and the current direction can therefore be obtained from the turning of a direction vane in relation to the frame, with the help of the ship's compass. Measurement of the current velocity is carried out by means of a propeller. The direction and velocity values are conveyed to the ship by electricity through a multicore cable and are read and recorded on board ship. For continuous measurements at great depths we use instruments which are provided with Savonius rotors for velocity measurements and a current vane for direction measurement. Recording is carried out by a strip recorder in the underwater instrument or after the measurement values have been conveyed to the ship by means of a cable.

Current measurements with moored automatic recording instruments. The anchoring of automatically recording current meters at selected depth horizons allows recording over several weeks. From the large number of instruments used we will select two for description here.

In the Rauschelbach paddle-wheel current meter or its further development, the propeller current meter, a paddle wheel or a propeller moves a counting device in a water-tight casing which is adjusted in the current direction. The scales of the counting device and a compass built into the casing are photographed every five minutes. When the instrument has been recovered the recording film is read and the current velocity is calculated with the aid of standard tables.

In the Richardson rotor current meter a Savonius rotor, a current vane and a compass are scanned at certain time intervals with the help of special coding arrangements in such a manner that the information is recorded in digit form as a pattern of dots on a photographic film or as a series of impulses on a magnetic band. The film or magnetic band is interpreted with the help of automatic reading and tracing instruments.

For the measurement of low current velocities various other procedures use pendulums which provide a measurement of the current velocity by means of their angle of inclination in the current.

7. CONCLUDING REMARKS

In general, the measurement methods of physical oceanography must meet the following requirements: lack of sensitivity to the ship's movements or to the movements of the instruments, and particularly to shocks; protection against great pressure; resistance to seawater; high precision in measuring. Other considerations must be added at present. The direction taken by the development of physical measurement methods in oceanography today is determined by recognition of the fact that even in the depths the temporal changes in the stratification and flow of many ocean areas are considerably greater than was first assumed. In the future increasing importance will be given especially to those measurement procedures which allow measurements at close time and distance intervals and automatic processing of most of the measurement data obtained.

REFERENCES

Comprehensive Works

VON ARX, W. S. (1962). *An Introduction to Physical Oceanography*. Reading.

DEFANT, A. (1961). *Physical Oceanography*. Bd. 1, 32-63; Bd. 2, 31–65. Berlin.

DIETRICH, G., und K. KALLE. (1957). *Allgemeine Meereskunde*. Berlin.

DIETRICH, G. (1964). *Ozeanographie*. Das Geographische Seminar. 2. Aufl., Braunschweig.

JERLOV, N. G. (1963). 'Optical Oceanography'. *Oceanogr. Mar. Biol. Ann. Rev.* 1, 89–114.

RILEY, J. P., and G. SKIRROW. (1965). *Chemical Oceanography*. Bd. 1, 73–120. London.

SVERDRUP, H. N., M. W. JOHNSON, and R. M. FLEMING. (1954). *The Oceans*. 5. Aufl., 331–388. New York.

U.S. Hydrographic Office. (1959). *Instruction Manual for Oceanographic Observations*. 2. Aufl., Washington.

U.S. Hydrographic Office. (1957). *Processing Oceanographic Data*. 2. Aufl., 114 Washington.

WUEST, G., G. BOEHNECKE, und H. M. F. MEYER. (1932). *Ergebn. Dtsch. Atlant. Exp.*, 'Meteor.' 1925–1927. Bd. 4 (1), 7–76 und 178–261. Berlin.

Section II

ANALYSIS OF THE STOCK

A. Plankton

J. LENZ

The word 'plankton', which originates from the Greek (literally: that which wanders about), is a collective term for all the organisms which float in the water and do not execute individual movements of any importance.

This biological community ranges from tiny one-celled organisms measuring only a few micrometers to jellyfish measuring 2 metres. We distinguish between the (plant) phytoplankton and the (animal) zooplankton. The zooplankton in turn consist of two groups, the holoplankton, which spend their entire life floating in the water, and the meroplankton, which are only present in the plankton during certain phases of their development. To the latter belong the juvenile stages of many bottom-dwelling animals (such as molluscs, polychaetes, echinoderms) and fish larvae.

The wide variation in the size of the plankton organisms makes a classification by size necessary (Fig. 14). The net plankton contrasts with the nannoplankton, which cannot be caught completely with even the finest plankton net because of its small size, and therefore requires other methods.

Direct observation of plankton in the water is unfortunately generally limited to a few large animals and requires favourable circumstances. Hydromedusae, jellyfish and colonies of salps frequently lend themselves to good observation in surface waters. Wider possibilities are offered by snorkeling and diving as a sport. Attempts are made to investigate the vertical distribution of the larger plankton animals from the diving boat. Large copepods probably represent the lower limit of identification by this method.

The mass occurrence of plankton organisms is frequently indicated by a change in the colour of the water. Well-known examples are the reddish 'krill' of the whaler, which consists of swarms of Euphausiacea, the green-coloured phytoplankton blooms, and the

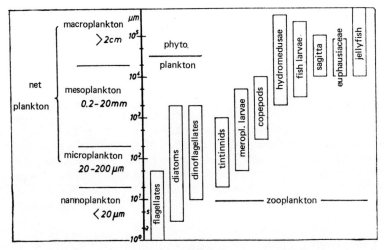

Figure 14. Classification of the plankton according to size

so-called 'red tides' produced by a massive development of certain dinoflagellates. This category also includes phosphorescence of the sea, which is generally produced by *Noctiluca* and other dinoflagellates.

Indirect observations relating to the presence of zooplankton are made with the help of the echo-sounder. However, such observations require much experience, as it is often very difficult or even impossible to differentiate between fish and plankton accumulations from the indications on the echograph. This question plays an important part in the study of the deep scattering layers distributed throughout the world. Diffuse signs indicate plankton collections. According to our present experiences, only macroplankton act as echo-senders. In order to interpret echo-sound signals the supersonic frequency involved must be known.

Transparency or turbidity measurements offer another aid for the determination of plankton concentrations, and particularly for phytoplankton concentrations. A photometer, or at its simplest a Secchi disc (see p. 41), is used to measure the decrease of daylight in the water, i.e. the vertical extinction. The use of transparency measuring equipment – abbreviated as T-equipment – does not depend on the time of day and the depth. The physical extinction of the water is measured by using a uniform light source and distance.

However, apart from the exceptional ideal case, these two methods yield only approximate data relating to size, since the extinction greatly depends on the size of the light-scattering particles and the water near phytoplankton cells nearly always also contains a great number of detritus particles. Nevertheless these measurements are certainly useful for the supplementation of other methods. Only in monocultures is it possible to use the light extinction as a direct measurement for the concentration of the suspended particles.

1. Obtaining the Material

When obtaining the material one must set out with the understanding that it is virtually impossible to collect all members of a plankton population at one time in a single haul. The obstacle is not only the variation in size of the organisms, but also the very great differences in the size of their living space in the water. Thus, a few millimetres of a water sample are generally sufficient for a representative nanno-plankton sample, but for a macroplankton catch of corresponding size the net must go through a water volume of many cubic metres. Consequently, various types of equipment are required for collection of the plankton, the choice depending on the particular problem involved. Essentially one must choose between water bottles and plankton nets, and suitable mesh sizes must also be selected.

The principal environmental factors which determine the presence of individual plankton species are the temperature and the salinity. These two factors characterize the various water masses of the sea, together with their respective plankton communities. Certain plankton species are very typical for specific water masses. For example, the inflow of Atlantic water into the North Sea can be identified by the occurrence of such 'indicator' species.

Particularly in ecological and zoogeographical studies it is important to measure these environmental factors as well. They include the oxygen content of the water and – in connection with questions relating to biological production – the concentration of nutrient salts (see I.A., pp. 1-25).

(a) Plankton Nets

The plankton net has always been the preferred collection device in plankton work. Various types exist, but only a few have achieved general use and have been developed into standard types.

The net gauze, which was previously made of silk, is now usually made of nylon, perlon or similar synthetics, which have the advantage of greater endurance. A criterion for the quality of the gauze is that the mesh must have a defined and uniform size and does not become distorted even when subjected to fairly large tensile stress. Also, the ratio of the mesh apertures to the mesh frame, which determines the filtration surface of the gauze, should be as large as possible; the filtration surface of a gauze is given as a percentage of the total surface; the finer the gauze the smaller its filtration surface. Thus, for example, with a mesh width of 10 μm the finest gauze now available has a filtration surface of only 4%.

The mesh width of the net gauze is the most important characteristic of the plankton net because it determines the size groups of plankton which can be collected with the net. The mesh width must be in the correct ratio to the size of the plankton. It is evident that if the mesh of the net is too large, no small plankton forms will be caught. But the inverse is almost as unsuccessful. A fine-mesh net has too small a filtering rate to collect large, rapidly moving zooplankton. It cannot be towed through the water fast enough without producing strong water resistance in front of the net opening, which drives the animals away. The filtering rate thus determines the favourable towing or raising rate when collecting.

Various types of gauze were previously classified according to a certain numerical system (Table 5). Today the mesh width itself in micrometers is indicated as the distinguishing characteristic.

Table 5
The most common types of gauze

Gauze no.	Average mesh width in μm
0	570
3	335
4	315
8	200
14	100
20	75
25	55

The simplest type of plankton net consists of a conical net pouch which is attached to a ring and provided with a net bucket at the bottom. This type of net, which may be considered the prototype of all plankton nets, is particularly suitable when a qualitative survey of the composition of the plankton is desired. Fig. 15 shows a qualitative net of this type, which is also called a *standard net*. It is used in all sizes and with a very wide variety of mesh widths. A large form with coarse gauze (300 or 500 μm) is the so-called *ring trawl*, which is used to collect large zooplankton and fish larvae.

Brief mention must be made here of another two nets of entirely different appearance, the *Knüppel net* and the *Isaacs-Kidd midwater trawl*. Both are smaller models based on the principle of the fisheries' trawling nets. The Knüppel net is used particularly to catch fish

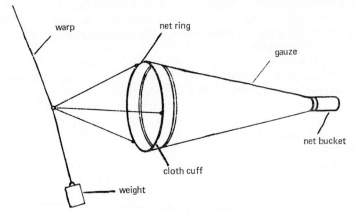

Figure 15. Standard net

spawn. The Isaacs-Kidd midwater trawl has been built especially for the high seas, to make possible the collection of macroplankton and mainly smaller fish at greater depths.

The plankton hauls can be carried out in three different ways: as a vertical catch, as a horizontal catch and as an oblique haul.

In a *vertical haul* the entire water column is filtered through from the bottom to the surface, or only a top part is filtered. The net is lowered to the determined depth from the anchored ship and slowly hauled up again. The hauling speed is determined principally by the mesh width of the net. As a guide we can use the value 1 m/sec for a 300 µm net. A very heavy weight (10 to 25 kg), which is attached to the net ring with long lines in such a manner that it hangs down below the net bucket, helps to lower the net and prevents its displacement by currents.

The *horizontal haul* is used in order to obtain a plankton sample from a particular water – layer or depth layer. A weight holds the net while it is being towed in the depth (see Fig. 15). With bigger nets, such as the ring trawl, an especially shaped shearing body or depressor is generally used. The depth position of the net can be determined very roughly from the cable angle and the length of cable which has been paid out.

The *oblique haul* is a combination of the two other types of haul. The net is lowered from the moving ship and is then drawn in again very slowly and uniformly while being towed. The advantage of this method is that the water column is more intensively filtered in this manner than in a purely vertical haul.

(α) *Quantitative plankton nets.* A very important requirement for quantitative plankton research is the availability of sampling apparatus with precisely defined propertics; only with its help is precise information on the plankton distribution in the water possible.

Two requirements in particular are made of quantitative plankton nets; they should not drive the plankton away and they should filter the water streaming in through the opening of the net completely and

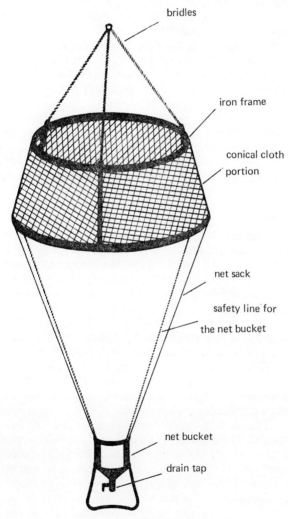

bridles

iron frame

conical cloth
portion

net sack

safety line for
the net bucket

net bucket

drain tap

Figure 16. Heligoland larva net

without creating water congestion. This means that the filtration rate should amount to 100%.

The filtration rate of a net indicates the actual volume of water filtered in comparison with the theoretically possible value (= cross section of the net aperture multiplied by the towing distance). The rate depends on the relationship of the mouth to the filtering surface of the net bag. A favourable correlation between these two can be attained by two means:

In most cases the mouth of the net is reduced by the addition of a conical portion made of linen or some other strong material. Typical examples of this type of net are the *Hensen egg net* (mouth 73 cm, mesh width 330 μm) and the larger net built on the same principle, the *Heligoland larva net* (mouth 143 cm, mesh width 570 μm) (Fig. 16). Both nets were originally built especially for the collection of fish spawn.

The second procedure involves precisely the contrary, i.e. the filtering surface of the net is enlarged. A cylindrical net section which takes over a part of the filtration is added to the conical net bag. This principle has been used in the large '*Indian Ocean* (IOE-) *Standard Net*' (mouth 113 cm = 1 m², mesh width 300 μm), which was used by the international expeditions in the Indian Ocean.

In practice the filtration rate of the above-mentioned nets fluctuates between 65% and 95%. Frequently a partial clogging of the net mesh cannot be avoided. The filtration rate is therefore monitored with the help of small mechanical *flow meters*. These are best attached in the net mouth in such a manner that the distance to the net ring is equivalent to half the radius, as the inflow rate is somewhat reduced at the centre and is therefore not representative. The filtration rate is measured directly by means of a second flow meter which is attached at the same distance externally on the net ring.

A certain avoidance effect of the plankton is exerted by the pressure wave which precedes the net when the latter is towed through the water. This pressure wave is caused in particular by the net suspension, the bridles, in front of the mouth of the net, and by resistance due to a poor filtration rate. Both faults can be avoided with a good net. Water suction should develop instead of water resistance.

The determination of the precise haul depth is a problem which occurs especially with all horizontal hauls. There are mechanical *depth recorders* (see p. 28) which make it possible to reconstruct – after the haul – the route taken by the net in the water. Even better are electrical depth meters (see p. 29) with simultaneous recording on board ship. In this case the trawl line consists of a cable with electrical conduction. The depth position of the net can thus be traced precisely and be guided accordingly.

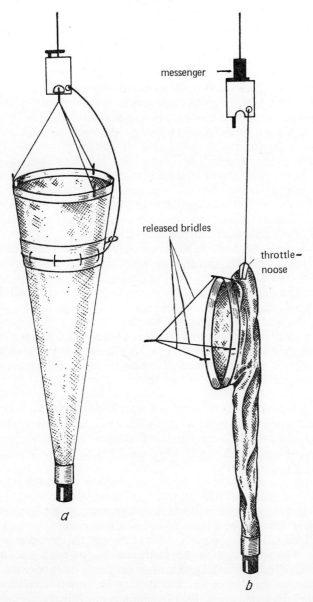

messenger

released bridles

throttle-
noose

a

b

Figure 17. Nansen closing net (from Wickstead 1965).* (It is unfortunately not entirely evident from Figure 17 *b* that the released throttle noose closes the net.)
*We should like to thank the Hutchinson Publishing Group Ltd. for their kind permission to reproduce Figures 17, 18, and 19

A further improvement of the quantitative net consists of an *opening* and *closing mechanism* which allows the collection of discrete plankton samples from specific water layers.

The simplest model of a net of this type is the *Nansen closing net* (Fig. 17). The net can be used for towing and for vertical hauls. The latter usually take the form of so-called stepwise catches, from one depth to the next, in order to determine the vertical distribution of the plankton. The net can be closed at any desired depth, by means of a messenger which releases the throttle noose. The *Apstein* closing net has a closing mechanism with a cover.

Various nets have been developed for stepwise catches in a single haul, in order to save work and ship time. One example is the Bé multi-net. Several collapsible net sacks are attached to a square frame; during the haul these nets open and close in series. A special mechanism which reacts to the water pressure releases the change of net at the predetermined depth (Bé *et al* 1959).

Another method which has been adopted is the exchange of the net bucket. A bucket-exchange device of this type, which is approximately equivalent to the revolving nosepiece of a microscope, is generally actuated electrically from the ship. This has the advantage that the depths at which the exchange takes place can be determined while the haul is still in process.

In quantitative catches the number of organisms found is generally related to a water volume of 1 m^3. If the entire water column has been filtered in a vertical haul, it is possible to relate the number – as for example in fisheries – to 1 m^2 water surface.

(β) *High-speed collection instruments.* Their purpose as such is to catch the especially rapid zooplankton which escape most plankton nets, the older fish larvae. These collecting instruments are generally torpedo-shaped and consist of a solid casing which protects the filtering gauze within.

In some respects the *Clarke-Bumpus sampler* represents the transition from conventional nets to rapid tow instruments (Fig. 18). The solid front section is provided with an opening and closing device which is activated by a messenger, and a flow meter. The net itself is free. Depending on the mesh width of the net, the tow rate is $0.5/3$ knots. The advantages of this instrument are its manageability and the possibility of attaching several on one cable, one above the other. In this manner various water strata may be filtered simultaneously. A small disadvantage is that the mouth is not entirely free from the rods.

A very stable tow instrument is the '*Hai*', the German modification of the American *Gulf-III sampler* (Fig. 19), which can also be pro-

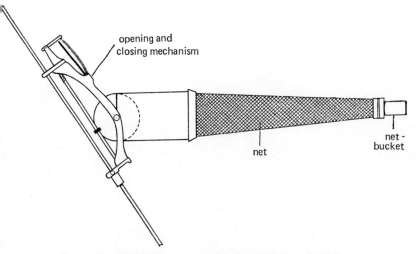

Figure 18. Clarke-Bumpus sampler (from Wickstead 1965)

vided with a closing mechanism. The inside of the net frequently consists of bronze gauze. The towing rate is 6 knots.

The *indicator* is a very small collecting instrument, also of torpedo shape, which can be towed at high speed and is easily worked by hand. It has been developed especially for English and Scottish-herring fishermen, to enable them to use information on plankton abundance as an indicator for the occurrence of fish (Glover 1953). For simultaneous fishing at various depths, several of the indicators can be towed suspended one above the other (Glover 1961).

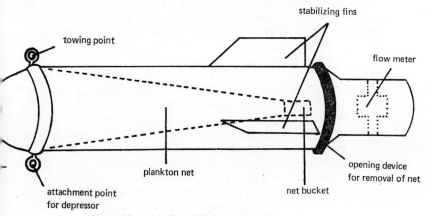

Figure 19. Hai (from Wickstead 1965)

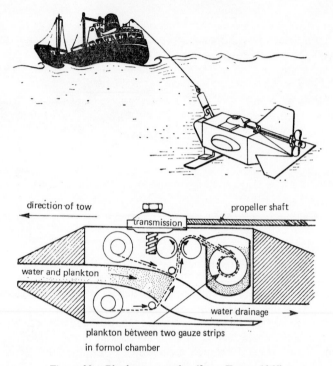

direction of tow

propeller shaft

transmission

water and plankton

water drainage →

plankton between two gauze strips
in formol chamber

Figure 20. Plankton recorder (from Fraser 1965)

(γ) *Plankton recorder.* A special – and in its way unique – construction is the plankton recorder developed by Hardy (Fig. 20). In this instrument the net gauze runs through the water tunnel in the same way as a film in a camera, filtering the plankton which streams through. The covering gauze then covers the collecting gauze and both together are rolled up into a container which is filled with formalin solution in order to preserve the plankton. The strip of gauze is moved by the action of a propeller, which is driven by purely mechanical means as the result of the trawling process.

The rate of transport of the gauze is regulated at a speed which is so small that the advance of 1 cm is equivalent to 1 nautical mile of the trawling stretch. A loaded recorder consequently has a scope of approximately 500 nautical miles. Despite the small mouth, measuring less than 2 cm², the collecting capacity – even for larger organisms – is very good. This is determined by the favourable ratio (1:32) of the mouth to the gauze surface in the water tunnel. The second reason for this is the high trawling speed of 8 to 17 knots.

The gauze has a mesh width of 300 μm. Collection of the zooplankton is therefore roughly quantitative. However, only a small fraction of the phytoplankton is collected; the size of this fraction can be considered fairly uniform as long as no clogging of the net mesh occurs.

The plankton recorder allows truly large-scale detection and observation of the plankton stock in the sea. The low collecting rate equalizes all small-scale differences in the plankton distribution, which are often so marked in bottle and net samples taken at random.

The second aim of construction was the independent use of the recorder from a research ship. During a collecting period it requires virtually no attention and can therefore be used by untrained personnel on any desired ship. The Oceanographic Laboratory in Edinburgh has built up a very extensive recorder service on the basis of this fact. At regular intervals the line ships tow plankton recorders through the North Sea and the North Atlantic along the navigational routes which span the sea like a net (Glover 1962).

The interpretation of the recorder material requires very intensive work, as the winding frequently compresses and deforms the plankton considerably on the gauze.

(δ) *Plankton pump.* Any pump which conveys the seawater with its plankton onto the ship without contamination is suitable as a plankton pump. The type preferred is the centrifugal pump. If rapidly mobile plankton, such as copepods, are to be caught, the cross-section of the tube must be sufficient to permit a capacity of approximately 100 l/min. Screening of the organisms is very simple: the water is poured through one or several plankton nets suspended one above the other. In this way the plankton are immediately available in the desired sizes.

The use of the plankton pump has the following advantages:

1. The most important point is that all of the comparative studies involving the plankton, as well as the measurement of the abiotic environmental factors, are carried out in a single water body of known volume.
2. The spot-type sampling allows examination of the micro-structure of the plankton distribution (patchiness).
3. Continuous sampling from the moving ship is also possible.

A disadvantage of the pump procedure is that the larger and more sensitive plankton forms are often damaged when pumps are used. The second limitation is the limited scope in depth. For technical reasons, such as the coiling of the tube, the plankton pump can only be used for the upper 100 to 200 m.

(b) Nannoplankton

The nannoplankton, which consists predominantly of small flagellates, can only be obtained quantitatively from bottle samples. Bottle samples of several litres are often sufficient for detection of the total phytoplankton content. In eutrophic waters, which are generally very rich in nannoplankton, a large water drop (such as $0 \cdot 1$ ml) is frequently sufficient for detailed study purposes. In the open ocean areas, however, the small organisms must first be concentrated.

This may be achieved by three different methods.

Centrifugation of small water masses (10 to 100 ml) is the best method for live examination of nannoplankton. As the forms involved are mainly very delicate ones, the centrifugation rate must not be too high. The best range is between 3,000 and 5,000 r.p.m. In essence even a hand centrifuge will be adequate. Lohmann (1908) and Hentschel (1932) used the centrifuge for their well-known studies on the significance of the nannoplankton in the sea.

The concentration of nannoplankton by *sedimentation* is carried out principally with fixed water samples (see the Utermöhl technique, discussed further below).

The third method, *filtration* of the seawater through a fine membrane filter, is practicable in a limited way only, since a portion of the nannoplankton are damaged and disintegrate when this method is used.

2. EXAMINATION OF THE MATERIAL

The examination of living plankton is usually only possible immediately after the sample has been obtained. The complete fullness of form which is revealed to the observer under the binocular microscope or the simple microscope, and which also has great aesthetic attraction, is unfortunately of a very temporary nature. The delicate organisms, some of which have already been damaged during collection, can only tolerate the rapid changes in the environmental conditions (such as too high a temperature, excessive light intensity under the microscope and oxygen deficiency) for a short time, and then die. The cooling and ventilation of a catch lengthen its life span somewhat.

Fixation causes many plankton to change or lose their original colours, which must consequently be recorded by notes or sketches while the animal is still alive. The lifetime of the nannoplankton under the microscope is particularly limited. Many disintegrate during fixation or become totally deformed, so that they can only be observed in the living state. Photography is of great help in these cases. Even the rapidly moving organisms can be photographed with a flash arrangement.

Menthol and chloral hydrate are suitable anesthetics, among others. Recently the more effective propylene phenoxetol has been used as a $0 \cdot 1 \%$ solution in seawater.

The microprojection of living plankton has been found very useful for simultaneous observation by several persons and for demonstration purposes. This method allows particularly high magnification of the objects under study. It can also be carried out successfully on board ship.

Many zooplankters are very difficult to conserve in the aquarium for any length of time. The principal problems involved are the renewal of the water and the lack of movement in the water contained in a tank. Many different systems have been evolved in order to eliminate these difficulties. For example, Schumann and Perkins (1966) describe a simple circulation system for the conservation of plankton organisms; this method has also been found successful on board ship.

Fixation of the plankton for later work must be carried out immediately after the catch. The large net catches which have first been drained into a bucket are fixed with a small shot of formalin, and are transferred after some time to a preserving bottle with the help of a plankton strainer. For preservation 40% formalin is added in a quantity sufficient to make a final concentration of approximately 4%.

Despite its unpleasant smell, formalin is the most commonly used preservative for plankton samples. Before use it is saturated with calcium carbonate, borax or hexamine (hexamethylenetetramine), so that the easily formed formic acid, which dissolves the calcareous parts of the organisms, is neutralized. The advantage of hexamine is that it enters directly into a neutral union with the formic acid. Diatom plankton should be kept in glass bottles, as the silicic acid shells will otherwise dissolve in the course of time. For larger zooplankton alcohol is also a good preservative.

Microscopic preparations can be made by means of numerous methods, which are described in the pertinent technical reports (such as Romeis 1948). Diatom preparations require a special technique. In a very simple method which is adequate for many purposes the organisms are embedded in glycerine gelatine, having been kept in an aqueous glycerine solution for some time previously. Even more convenient is the use of polyvinyllactophenol (manufacturer: Gurr Ltd., London), which makes it possible to embed directly. Methylene blue and Bengal pink are stains useful for many purposes. For longer preservation it is advisable to circle the edges of the coverglass with a synthetic resin. Larger animals, such as copepods, are preferably kept in small, labelled test tubes because it is often necessary to remove certain limbs in order to identify the species precisely.

The quantitative analysis of large plankton catches requires a subdivision of the sample. The aliquot should be representative of the total catch. Consequently no selection of specific animals should be carried out when the partial sample is taken. One of the oldest and most proven procedures – although only for smaller plankton – makes use of the Hensen Stempel pipette (Fig. 21 *a*). The catch is evenly distributed in the round container by shaking it, and a fraction of the sample (known volume) is removed by pipette. The Folsom

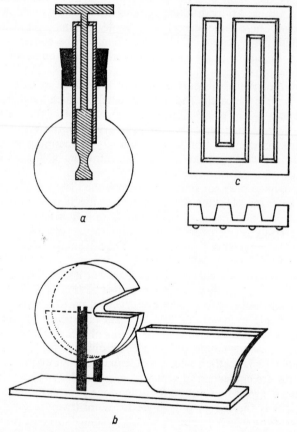

Figure 21. (*a*) Stempel pipette; (*b*) plankton separator according to Folsom; (*c*) counting tray according to Bogorov

plankton separator (Fig. 21 *b*) (McEwen *et al* 1954), on the other hand, requires division of the total catch in half, which can be continued if desired. The determination and counting of the organisms under the binocular microscope is facilitated by the Bogorov counting pan (Fig. 21 *c*). For phytoplankton one may use a small square counting dish which is divided into squares, and for nannoplankton a cell holding 0·5 ml, on a slide.

Utermöhl (1958) developed a special technique for determining the nannoplankton content in the water. Fresh water samples, generally consisting of 100 ml, are fixed with concentrated formalin or with lugol solution in which elementary iodine has been dissolved in KI. The latter stains and weights down the sediments. The fixed water samples are allowed to settle out in tubular counting chambers. They are then examined under the reversed microscope, which is also known as the Utermöhl microscope. In this type of microscope, the lighting units and the lens are in reverse order in comparison with the ordinary type. The light rays fall from above through the counting chamber into the lens. The sedimented organisms are examined from below through the thin glass floor of the chamber. When counting, it is best to divide the chamber floor into diagonal strips, which cancels out the inequalities in sedimentation.

Very recently an electrical measurement method has been used for the counting of small plankton organisms (Coulter counter procedure). The plankton are conveyed into a counting tube which contains a counter-balanced electrical field. The passing particles disturb this field and produce an impulse which is proportional in amplitude to the volume of the particles. This method is suitable in particular for experiments carried out with plankton cultures in the laboratory.

TABLE 6

The accuracy as a function of the number of counted individuals at an approximate probability limit of 95%

Organisms counted	Accuracy in percent	Range
4	±100%	0 – 8
11	± 60%	4 – 18
16	± 50%	8 – 24
25	± 40%	15 – 35
44	± 30%	31 – 57
64	± 25%	48 – 80
100	± 20%	80 – 120
178	± 15%	151 – 205
400	± 10%	360 – 440

When any plankton counting procedure is involved we must remember that the results are subject to statistical scatter. Apparently clear differences are frequently only the expression of the random distribution. Table 6 shows how slowly the accuracy increases with the size of the numerical value.

For a doubling of the accuracy the numerical value must be multiplied by 4. However, differences in the number counted may only be considered as significant when they cannot be collected in a fluctuation range, as for example 4 and 20. It is therefore desirable to count to at least 30 for each species, if possible, in order to improve the statistical significance of the existing frequency differences. In principle, statistical analysis should be carried out only with the original numerical values, but not with their multiples.

REFERENCES

Special Literature

GLOVER, R. S. (1953). 'The Hardy Plankton Indicator and Sampler: a description of the various models in use.' *Bull. Mar. Ecol.* **4**, 7–20.

— (1961). 'The Multi-depth Plankton Indicator.' *Bull. Mar. Ecol.* **5**, 151–164.

— (1962). 'The Continuous Plankton Recorder.' (Contributions to Symposium on Zooplankton Production 1961.) *Rapp. Proc.-verb. Cons. Exp. Mer.* **153**, 8–15.

HENTSCHEL, E. (1932). 'Die biologischen Methoden und das biologische Beobachtungsmaterial der "Meteor"-Expedition.' *Wiss. Ergebn. Dt. Atlant. Exped. 'Meteor'*, Bd. 10.

LOHMANN, H. (1908). 'Untersuchungen zur Feststellung des vollständigen Gehaltes des Meeres an Plankton.' *Wiss. Meeresunters., Abt. Kiel, N. F.*, **10**, 131–370.

MC EWEN, G. F., M. W. JOHNSON, and T. R. FOLSOM (1954). 'A statistical analysis of the performance of the Folsom plankton sample splitter, based upon test observations.' *Arch. Met. Geophys. u. Klimatol.*, Ser. A, **7**, 502–527.

ROMEIS, B. (1948). *Mikroskopische Technik.* Oldenbourg München.

SCHUMANN, G. O., and H. C. PERKINS (1966). 'A Jar for Cultering Small Pelagic Marine Organisms in Running Water during Shipboard or Laboratory Studies.' *J. Cons. perm. int. Explor. mer* **30** (2), 171–176.

UTERMÖHL, H. (1958). 'Zur Vervollkommnung der quantitativen Phytoplankton-Methodik.' *Mitt. int. Ver. Limnol.* **9**, 1–38.

General Introductory Works

BARNES, H. (1956). Oceanography and Marine Biology. *A Book of Techniques.* Allen & Unwin, London.

FRASER, J. (1965). 'Treibende Welt. Eine Naturgeschichte des Meeresplanktons.' *Verständl. Wiss.* **85**, 151 Springer, Berlin-Heidelberg-New York.

HARDY, A. (1958). 'The Open Sea.' *The World of Plankton.* 2. Aufl., Collins, London.

NEWELL, G. E., and R. C. NEWELL (1963). *Marine Plankton. A Practical Guide.* Hutchinson, London.

SCHWOERBEL, J. (1966). *Methoden der Hydrobiologie (Süßwasserbiologie).* Franckh, Stuttgart.

WICKSTEAD, J. H. (1965). *An Introduction to the Study of Tropical Plankton.* 160 Hutchinson, London.

WIMPENNY, R. S. (1966). *The Plankton of the Sea.* 426 S. Faber & Faber, London.

Systematic Identification Studies

BRANDT, K., und C. APSTEIN (1901–1938). *Nordisches Plankton.* 8 Bde. Lipsius & Tischer, Kiel u. Leipzig. (Neudruck bei Asher & Co., Amsterdam 1964.)

GRIMPE, G., und F. WAGLER (1925–1938). *Die Tierwelt der Nord- und Ostsee.* Akad. Verlagsgesellsch., Leipzig.

HUSTEDT, F. (1930–1959). *Die Kieselalgen.* (Rabenhorsts Kryptogamenflora, Bd. VII.) Teil 1. Akad. Verlagsgesellsch., Leipzig 1930 (Neudruck Weinheim 1962). Teil 2. Leipzig 1959.

JESPERSEN, P., F. S. RUSSEL *et al.* (1939). *Fiches d'identification.* Host. Copenhagen, ab 1939 fortlaufend.

TRÉGOUBOFF, G., et M. ROSE (1957). *Manuel de planktonologie Mediterranéene.* Vol. I u. II. Centre Nat. Recherche Scient., Paris.

B. Nekton

R. KÄNDLER

The term 'nekton' ('that which swims') includes all water animals capable of actively moving from place to place. Since only the larger animals possess the required muscular strength for extensive directed locomotion, the great mass of the invertebrates is excluded from the biological community of the nekton; only squid (Cephalopods) and some forms of higher crustaceans (among the Decapods) are included. Nekton consists principally of vertebrates, particularly of *fish*; the abundant number of their species entirely overshadows forms from other classes of vertebrates. It is therefore clear why the statements which follow will relate only to fish as representatives of the nekton.

Analysis of a fish stock. In view of the motility of fish, a quantitative study in a limited sphere – somewhat similar to that covering the total plankton or the bottom-living animals – is scarcely possible since some of the fish always escape from the collecting gear. We must therefore be content with a qualitative stock analysis, although the latter is usually incomplete because the mesh width of the nets determines a selection based on size. Depending on the construction and handling of a net all the fish above a certain size which come into its range will be retained, or a selection will take place. The collecting gear used for practical fishing purposes has differing properties in this regard, which must be taken into account when analysing the catch.

The material brought back by commercial fisheries provides many opportunities for insight into the marine fish fauna and for collecting material for personal studies. However, the composition of this material generally does not correspond to the catch obtained from the fishing site, since only the marketable species and sizes are brought in and the total 'accompanying catch' which is of particular interest to the biologist, is immediately thrown overboard again. In order to study samples from unadulterated, original catches we must refer to scientific catches with research vessels or to the participation of commercial fisheries in collection voyages.

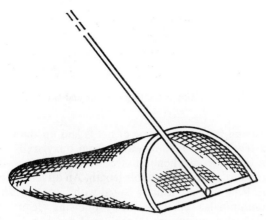

Figure 22. Push net

1. Collection Equipment

For catches directly on the shore in shallow water on a firm sandy bottom we use the *sliding push net* (Fig. 22). This consists of a semi-circular tubular or metal bow with a straight edge, to which a bag made of narrow-mesh net twine is attached. The net is fastened to a rod which is pushed ahead by hand in the water. With this net it is possible to catch all the fauna temporarily living on the bottom, particularly juvenile stages of flat fish and of shrimp. The results of the catch depend on the speed and duration with which the push net is pushed through the water, since the startled animals attempt to escape laterally. When walking through fields of sea grass and sea-weed typical inhabitants of this biotope are also harvested. This type of net cannot be used on ground which has large stones or is rocky, but if these are present one may try one's luck with a simple landing net or fish-hook, which probably does not require description. The possibilities opened by modern diving as a sport can only be mentioned peripherally here. It is of increasing importance for the study of the living creatures of warmer waters and in more northerly coastal areas it can also be expected to further our knowledge some-what, although the poorer visibility produced by strong water turbidity makes all underwater observations of rapidly moving objects more difficult, even with cine- and television cameras, and restricts them sharply.

Another very useful instrument for collecting fish (and the larger crustaceans) in close and shallow coastal areas is the *ring net* (Fig. 23); a simple or double ring is used, made of synthetic netting on a

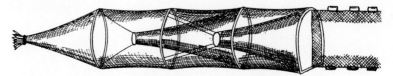

Figure 23. Ring net with guide line and two necks

tubular or metal frame. This type of net is laid on the bottom at a depth of a few metres in calm water, from quay walls or landing bridges, or, better, from a boat into free water; a number are usually used, attached to a cable of suitable length. An attached float marks its location. Ring nets are usually baited with pieces of fish, mollusc flesh, etc; the animals reach the inside by following a lead line through a progressively narrowing funnel (neck) and are then unable to find their way back to the outside.

The sliding push net, ring net, landing net, and fish-hook make up the easily manageable catching equipment. They offer only a small choice of objects for investigation. Much better possibilities are offered by the use of the *trawl*, in the various shapes and sizes which have been developed for commercial fishing. The trawl consists of a top section and a bottom section which, when strapped together, make up a net bag with lateral wings (Fig. 24). The trawl is dragged over the ocean floor by means of two otter boards to which the wings are fastened laterally, generally with long supporting lines. The boards are arranged in such a manner that they keep the net open; the ground rope, which is weighed down with lead or iron chains, glides closely along the ground, while the shorter head line is kept up by floats. The warps, boards and wings drive some of the catch away; the catch collects in the cod end, whose mesh width limits the lower size of the catch. Smaller devices of this kind with narrower net mesh for the collection of young fish can be easily dragged from a motor boat in fairly shallow water; for greater dimensions a larger vessel is required, with stronger motors and a motor winch to

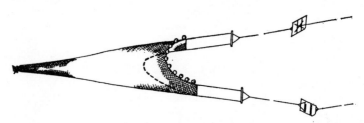

Figure 24. Trawl with otter boards

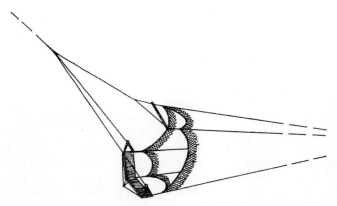

Figure 25. Isaacs–Kidd midwater trawl. After A. V. Brandt from: *Fish Catching Methods of the World.* Fishing News (Books) Ltd. London 1964 (altered)

wind up the cables and the otter boards. The instruments used for the collection of pelagic fish (anchored nets and drift nets, ring trawl, etc.) are only exceptionally used for scientific fishing purposes. The newest development is the pelagic *mid-water trawl*, which opened up the broad pelagic area to trawl fishing. The Isaacs-Kidd midwater trawl (Fig. 25), which can only be mentioned briefly here, is used in oceanic expeditions to collect animals in open waters at fairly great depths. We must also mention the use of ultrasonic devices to locate fish shoals in vertical and horizontal directions (fish lens, echograph, echo-sounder). Certain types of fish produce characteristic signals on the echogram, which yield information on the extent and density of the fish shoals.

2. THE CATCH

A catch does not represent the existing stock in all cases; it often represents a selection, produced by the selectivity of the nets. Furthermore, we must be aware of the fact that *chance* plays a part in sampling, which limits the significance of the sample. All the characteristics of a catch sample, particularly the measurable and calculable characteristics, are affected by a certain error in relation to the true, but unknown, characteristics of the total stock; the size of this error is determined by the laws of chance. The results obtained in this way must be checked statistically before far-reaching conclusions are drawn from them. This necessity must be pointed out at the very beginning of our report.

A data-sheet must be worked out for each catch intended for later scientific investigation; this data sheet must contain all the important data relating to the time, place, water depth, wind, duration of collection, net, etc. If possible, the entire catch should be investigated; if this is not possible, a part should be removed and its relation to the total catch should be established in terms of weight or numbers. However, it is advisable to sort out rare species or particularly large specimens to begin with and to record them separately. The catch is first *listed according to species;* this requires a good knowledge of systematics which may be obtained by study or by means of an identification key. For the beginner and even for the skilled investigator this is in itself a task requiring full attention and a critical approach when catches from previously unknown faunal areas are involved. The 'accompanying catch' of invertebrates and algae also always requires attention and the species and numbers should be recorded, as any additional observation may be useful for the subsequent analysis.

The statements below may serve as a guide for the practical analysis of a fish catch. We will assume the most complete possible *stock analysis of one fish species,* present in a fairly large number, as a task for this purpose.

3. Length Measurements

In order to carry out this process as logically as possible, we use a metal or synthetic *measuring board* provided with a stop at the front, against which the mouth of the fish is laid. We usually measure the total length to the furthest tip of the tail fin when folded back onto the axial line. If this proves impossible because the fin rays are too rigid, we measure up to the intersection of the axis and a line drawn between the two tail tips. It is advisable not to aim at the highest degree of precision, but to measure instead as many specimens as possible. A measuring board of approximately 1 m length will suffice in most cases; its measuring scale is divided into centimetres, the first 20 centimetres being further sub-divided into millimetres; small fish can then be measured in terms of millimetres, and the larger in terms of full centimetres. We can select measurements to the nearest centimetre (for example: 15 cm = 14·5 to 15·4 cm) and to the lower centimetre (15 cm = 15·0 to 15·9 cm). Although the former limitation of an interval is generally the one commonly used for statistical purposes, measurement to the lower centimetre is usually preferred for length measurements of fish, as it can be carried out more rapidly when using a cm scale than estimation of the nearest centimetre, by simply ignoring the part of the tail fin which projects beyond the last cm line. We must remember, however, that in this case the conventional designation of the class variants by whole numbers is not exact, since each mean is higher by 0·5 cm. Very large fish are measured with a *tariff inch rod;* the rod is not laid over the arched body of the fish, but paralled to it. It is important to measure all fish in the same manner, rays lying ventrally, flatfish on the blind side, round fish on the right side, the left hand lightly pressing the tip of the mouth against the stop of the measuring board and the right hand bringing the body and tail of the fish into the correct position. The measurement results are listed in the form of tables, which give the variants observed and the related frequencies. If several catches or stocks with samples of varying extents are to be compared, we calculate the percentage frequency of the length groups and the average length. A graphic representation of the measurements in the form of a *diagram* is very clear. Since the length represents a continuously changing characteristic, which is collected in class variants, the length distribution is best represented in the form of a *polygon;* in a co-ordinate system the means of the length groups are plotted on the abscissa and the frequency values plotted on the ordinate, and the correlated points in the co-ordinate field are connected by straight lines (Fig. 26). Overlapping classes makes it possible to cancel out somewhat the irregularities which are apparently produced by involuntary favouring of uneven or even values.

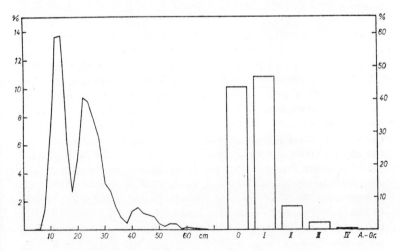

Figure 26. Example of the representation of the length and age distribution (codfish catch in the Western Baltic, November)

This often occurs for example, when two 1-cm groups are combined into one 2-cm group. The average length is recorded as the vertical on the corresponding point of the abscissa. Differences in length distribution in various samples can show up strikingly in this type of graphic representation.

4. WEIGHT DETERMINATIONS

Precise weight measurements on shipboard present considerable difficulties, as the ship's movements create great disturbance. However, on calm days acceptable weight determinations can be obtained with suitable scales (spring and beam balances). On land any type of balance showing the weight on a scale can be used. In general it is entirely sufficient to weigh small fish to within 1 g, and large ones to within 5–10 g, since the variations in condition and moisture make more exact weighing of little significance. Occasionally it is necessary to weigh separately the parts of a fish, such as the gonads, liver or intestinal tract, and possibly also the stomach content.

The *correlation between length and weight* of a fish is very instructive in some regards. Mere observation reveals that the correlation depends on the shape, which may be long and slender or short and stumpy. Obviously the weight/length relationship also depends on the nutritional condition, and particularly on the state of development of the gonads. Investigation of this question must start with an examination of the manner in which the weight of a fish changes as the length increases. According to a mathematical law, the volumes of similar bodies are equivalent to the third power of one of their dimensions. If we consider the various sizes of a fish species as similar bodies and their specific weight as constant, we can calculate the weight of the fish rather than the volume, and can consider this as a function of the third power of the length. In order to obtain relationships near 1, this correlation has been given the formula $W = k \cdot L^3/100$, W (weight) being given in g, and L (length) in cm. Factor k represents the length-weight coefficient; the more slender the fish, the smaller factor k; the stumpier the fish, the larger factor k (both individually within one species and also, to a much greater extent, from species to species). However, it happens that in contrast to the above assumption the growth of fish is by no means isometric, and that even at the same length, the weight throughout the year's cycle is subject to considerable variations because of maturation and emptying of the gonads and accumulation and consumption of reserve substances. Consequently the length-weight coefficient is *not constant* even within one species, but alters very characteristically during the various life phases and within the sexual cycle. It also fluctuates very widely during the maturation period in accordance with the nutritional condition of the fish, and has therefore sometimes been called the 'nutritional coefficient'; however, this only partially does justice to its nature. In order to eliminate these individual fluctuations one should attempt to weigh as many specimens as possible and use the average weight of the length groups or maturation stages for the calculations. As the calculation includes the length to the third power, it must be obtained as accurately as possible (in terms of millimetres). Furthermore, sources of error – such as a fairly full stomach – must be excluded before determining the weight; the picture would be falsified if the voluminous maturing gonads (particularly the ovaries) were removed in order to exclude the influence of the maturation cycle, as they are made up of body substance to a great extent and represent a portion of the latter.

The length-weight coefficients are compared to the appropriate averages of the length groups in a table; they can also be averaged over the total length range, and can then be followed in the course of the year or be classified into the various maturity stages, separately for males and females. The results are shown in graph form, the ordinates representing the length groups, maturity stages or months; the correlated points are connected by straight lines. To show the extent of the fluctuations, the average k-value obtained from all determininations is marked on the diagram as a vertical running parallel to the abscissa. One will note that as the length increases, the length-weight coefficient often becomes larger and rarely becomes smaller; i.e., some fish species develop a more stumpy body shape but more rarely a slender one. In such cases the development is obviously clearly *allometric,* and the formula given does not seem very suitable in order to reflect correctly the functional correlation between the weight and the length.

For this reason this relationship has recently frequently been expressed in terms of the allometric formula $y = a \cdot x^b$ which becomes $W = a \cdot L^b$ in the present case. In place of the one constant in the linear formula it contains two from it, one of them as exponent. To calculate the formula we take the logarithm of this power formula and obtain the linear formula $\log W = b \cdot \log L + \log a$ in this manner. The group averages obtained for W and L are introduced and constants a and b are calculated according to the method of least squares. To establish which formula is preferable, the logarithms of the pair of values are recorded in a coordinate system in which the logarithms of the length are listed on the abscissa and those of the weight on the ordinates. The position and direction of the straight lines which connect the points in the best approximation indicate whether the exponent is substantially different from 3. If necessary the significance of the difference may be tested. The length–weight coefficient k obtained in the linear formula provides more unambiguous information than the two constants of the allometric formula, particularly as relatively slight deviations of exponent b from 3 result in very great changes of a in the opposite direction. For comparative examinations the simple relation between weight and the third power of the length is therefore preferable.

5. Age Determinations

Accumulations of certain *length values* frequently occur in a measurement series, particularly in the low length ranges; these accumulations are clearly separated from each other and can be grouped into individual year's growths. This simple method of growth studies can be used in many animal groups, but does require verification by means of exact age data, which often presents great difficulties. Stock investigations of fish obtain their special character from the very reliable age determinations which can be carried out with the help of the *hard structures* (scales, otoliths, bones). As the fish grows, they also increase in size by the accretion of substance, their increment zones clearly showing the character of *annular rings*. The multiple ways in which these structures are marked can only be alluded to here, and considerable experience is required to identify the zones correctly and to recognize accessory or secondary rings as such and exclude them from the count.

The structures on the upper side of the scale (Fig. 27), which produce the impression of annular rings, may differ greatly according to the system to which the fish belongs. In principle they are based on the fact that the scleritin rows or ridges which are grouped more or less concentrically around the central field (the first scale deposit) show discontinuities – either because they become narrower in the autumn and grow considerably larger again without transition the following spring (salmon, cod types, flatfish); or because the increasingly shorter scleritin ridges are deposited one against the other in cap form until, in spring, a complete ring indicates the beginning of a new growth period (mackerel); or because the scleritic ridges break off in the winter and continue the following spring, frequently in a somewhat different direction, after a distinct absence (herring). Combinations of these various types also occur. In all cases, at low magnification and with suitable lighting, one has the impression of *dark* and *light* zones or lines which delimit

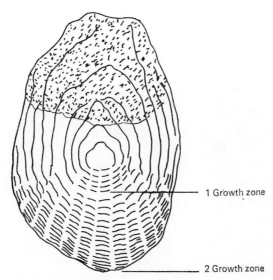

1 Growth zone

2 Growth zone

Figure 27. Fish scale with two annular rings (large sand eel, *Hyperoplus lanceolatus*, A. – Gr. II

the growth for one year, similarly to the annular rings seen in a cross section of a tree trunk.

The most easily readable scales are generally found between the breast and dorsal fin above and below the lateral line, as well as above the after fin; scales from the lateral line are not very satisfactory because of their abnormal form. Before *removal of the scales*, the mucus and any loose adhering extraneous scales are carefully removed from the skin. Using a scalpel and moving in the direction of the head we scrape off a fairly large number of scales which are first stored in small bags marked with the information relating to the fish or with a serial number under which detailed information can be found. The larger scales are removed individually from the scale bags for the investigation; they are carefully cleaned in water or 5% potash lye and examined, wet or dry, under the coverglass at moderate magnification and with suitable lighting.

Of the three *otoliths* on each side in the labyrinth of bony fish, only the *sagitta* generally seen on the right in the sacculus can be used for age determinations (Fig. 28). The other two otoliths are extremely small, apart from those of the Ostariophysi, which live predominantly in fresh water; in these the asteriscus in the lagena is the largest. In the following, 'otolith' is therefore understood to mean the sagitta. This structure, consisting of a collagen matrix

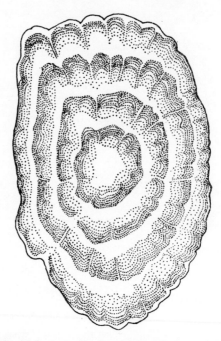

Figure 28.   Otolith with four annular rings (plaice, *Pleuronectes platessa*, **A.** – Gr. IV). According to R. S. Wimpenny from: *The Plaice*, Edward Arnold and Co., London 1953 (altered)

with intercalations of calcium carbonate, shows in its organization a series of growth zones which – at a weak magnification and particularly clearly in a cross section – can be identified as concentric annular rings around a nucleus, the first deposit of the otolith in the larval state. Examination of the otolith in a light from above against a dark background shows the growth by an alternation of white (opaque) and dark (hyaline) zones. One *opaque* and one *hyaline zone* form together one *annular ring;* the differences in the permeability to light are produced by the variations in the arrangement of the calcium needles in the organic matrix. If a penetrating light is used (thin sections) the contrary optical impression is created: the hyaline zones then appear bright and the opaque ones dark. These differences in the development of the growth zones are the result of changes in the metabolic intensity occasioned by variations in temperature and therefore of a seasonal nature; the development can only be affected by other internal and external factors, which may lead to disturbances and the formation of secondary rings, as in the scales. Consequently, the 'reading' of otoliths also requires considerable experience and comparative material for all stages of development.

To *remove the otoliths* we use a strong knife and open the cranium box in the vicinity of the labyrinthine region by a cut which is vertical or slanting towards the back (for flatfish a lateral cut, approximately at the level of the posterior margin of the pre-operculum); we then bend the head down and the otoliths, which are readily attainable if the cut has been correctly made, are removed with pointed tweezers. This requires considerable practice, particularly with small fish. Damage to the otoliths is to be avoided if possible. The annular rings are most clearly seen if the otoliths are examined immediately after removal, while still saturated with endolymph, against a dark background. Usually, however, the otoliths are kept in bags, or, preferably, in small glass tubes for later examination.

After careful rinsing of blood and dried tissue residue, small and flat otoliths which show the annular rings distinctly without preliminary treatment are examined in a small black glass dish with water or another fluid with a suitable refractive index, in which the light and dark zones are distinctly visible (glycerine, xylene, methyl benzoate, etc.). Since an otolith grows in all dimensions, even if the extent of growth varies according to its form, it becomes thicker with the years and the newly-deposited layers mask the nucleus and the earlier growth zones so that these become unrecognizable. In such cases it is advisable to grind them down on one or both sides in order to reveal the internal structure. Large otoliths, such as those of gadoids, are carefully broken in half, having been cut previously with a three-cornered file, and the surface of the fracture is smoothed with sandpaper. The fracture must follow as far as possible the formation centre of the otolith. For the interpretation the tip of one half is stuck into a plastic mass, the fracture surface – which may have been moistened with an illuminant – is lit up with a spot lamp and the otolith is examined with a stereobinocular microscope at low magnification.

Among other *bones*, the vertebrae on the concave, inner surfaces, and the thin, flat elements of the operculum often show distinct annual zones. Spines and thorns, even of cartilaginous fish, can be used for age determination. The preparation of the bones requires considerable care; they must be boiled and carefully de-fatted so that the zone development shows up clearly. Occasionally a comparison of the results obtained from examination of the scales, otoliths, and bones is very instructive.

Apart from species with particularly large scales, the use of the otoliths is generally preferable. They are established during the larval stage, whereas the scales only appear after the fish has reached a certain size. In some species further scales grow later, and lost scales are regenerated. A number of scales must always be examined, therefore.

Age determinations yield information on the *composition of a catch sample in regard to age groups or years*, thus providing further important information on the fish stock at the site of the catch. It has become traditional to consider the age in relation to the classification of the fish within an *age group*, these age groups being designated according to the number of years of life completed; thus, animals in their first year of life belong to group 0, those in their second year of life to group 1, etc. The fixed day selected for determination of one day of life should correspond roughly to the average spawning time of the brood. Since the great majority of commercial fish in moderate zones spawn at the end of winter, the 1st of April has been proposed as the uniform fixed date. However, when the spawning period diverges from this considerably, as occurs with summer or even autumn-spawners, this date makes little sense and in such cases a later fixed date should be selected. The additional recording of the year group and of the catch date will prevent misunderstandings.

As with the length distribution, the *age distribution* is reproduced in tabular form, either in percentages, or in the form of diagrams; the most useful is a *column diagram* (histogram), since slight variants are involved (Fig. 26). The combination of the length determinations and the age determinations makes it possible to list the *length distribution in the individual age groups* in tables and to calculate the average length for each; during this process one must remember the true class representatives when measuring to the lower centimetre.

6. Growth Studies

The averages of the age groups are used to draw up a *growth curve* (Fig. 29); the age (age group) is recorded on the abscissa and the length on the ordinate of a coordinate system, and the points obtained for the age groups are connected. A balanced growth curve begins steeply at the zero point, gradually flattens out, and as the fish becomes older, approaches a final value – the average maximal length of the species. However, frequently the observation values are not situated on an 'ideal' growth curve of this kind, but show considerable dispersion, since the inaccuracy of the averages increases as the individual numbers decrease in the higher age groups.

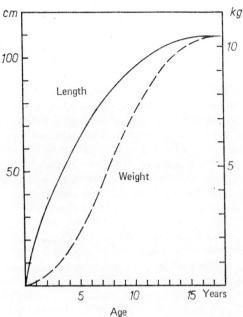

Figure 29. Example of the representation of growth curves (Iceland codfish)

All age groups are not always representatively covered within the range of the study, and in commercial catches the net selectivity often leads to excessively high values for the youngest age groups. Possible errors in the age determinations, especially for old fish, must be added. By forming intermediate classes it is possible to cancel out the irregularities of the curve to some extent. When the measurement series indicate *a differentiation in growth according to sex*, the averages of the age groups are calculated separately for males and females and a growth curve is drawn for each sex.

The age can also be correlated with the *weight* and the average weight of the age groups can be calculated. In this event the graphic picture yields a *sigmoid curve* (Fig. 29): the weight increase grows rapidly initially, then remains constant over a fairly long period; it then gradually decreases again, the weight approaching a final value. If an exact mathematical formulation is desired, it is advisable to use the *growth equations* of Bertalanffy, whose parameters are used to characterize the varying growth of various species and populations. In such calculations reliable averages are required for the higher age groups as well, and it is precisely these which are frequently missing, as mentioned previously.

A frequently used aid in growth studies is the *retrograde calculation* of the length attained at the end of one year on the basis of *measurements of the width of the annular rings* on the scales. It has been shown that the radius of the scale, measured from the central field – where it is first laid down – to the edge, increases in parallel with the length of the fish; a correction for the first growth period is necessary because the scale only appears when the fish has reached a certain size; however, it soon makes up for this delay. If we designate the scale radii at the end of the first, second, etc. year of life as $l_1, \times l_2, \ldots l_n$, and the length of the fish correspondingly as $L_1, L_2, \ldots L_n$, we obtain the equation $l_1 : l_2 : \ldots l_n = L_1 : L_2 \ldots L_n$. Using an ocular micrometer or, preferably, using the projection or photograph of the scale, we measure the distance of the edge and of the individual winter lines from the middle point of the central field, and by means of the formula above calculate each length at the end of the individual year of life. In this manner it is possible to determine the individual longitudinal growth and, from a fairly large number of such retrogressive calculations, the course of growth of a population. This method is suitable above all for fish with large scales (clupeoids, engraulids, salmonids). The l_1 measurement is particularly important for the analysis of catches in which descendants of one species come from various spawning periods – for example, from spring and autumn herring. Retrograde calculations of this type can also be carried out with otoliths, but in that case it is advisable to establish beforehand the correlation between the fish length and the otolith radius for the total length range; by means of a suitable correction, this will enable one to take into account any possible changes in this relationship as the fish size increases.

7. NUTRITIONAL STUDIES

In regard to the growth rate of a fish species, its *type of nutrition* is of interest; it also indicates its position in the food chain. Therefore the stock examination should include some attention to the stomach content of the fish. Even on board ship a brief examination of the food consumed will show whether plankton consumers, benthos consumers or predators are involved, and the most important food animals can be recorded. For more thorough studies the intestinal tract is preserved in a strong formalin solution (1 : 5) and the quantity and composition of the stomach contents are later established; the state of preservation of the partially disgested remains limits the possibilities of determining the species. Identification of the animal groups may be sufficient; very common forms should be identified more closely, if possible, in order to show the relationship of the fish species to the biological community at the site of the catch. An attempt should be made to obtain information on the composition of the food in the form of percentages according to animal groups and species by weighing the individual components, and one should not be confined to general estimations of the frequencies. This is true also for the degree of fullness of the stomach. Numerical data alone does not mean much because there is often a great variation in the size of the food organisms. Since the food often changes as the size of the fish increases, information relating to sizes is always required. It is also very informative to record in what percentage of the fish certain food animals were found.

8. SEX AND MATURITY DETERMINATIONS

In the cartilaginous fish (sharks and rays) the male sex is easily identified because of the claspers (pterygopodia); in the bony fish external sexual dimorphism is very rare, and a different coloration of the sexes at spawning time is not frequent. The sex of fish ready for spawning is easily determined by pressing on the cloaca, causing sperm or eggs to come out. As a rule, however, it is necessary to open the body cavity and to examine the gonads. The generally paired *ovaries* are easily identified as such in maturing females with the naked eye because the eggs shine through the wall of the ovaries. They are of a sac-like or tubular shape and when fully developed fill a considerable portion of the body cavity; in flatfish they stretch far into the tail section, in the form of strands coming to a sharp tip; this can be seen in some species by examining them externally against the light. The *testes* are generally less large, band-shaped or flap-shaped, of triangular form or tubular like the ovaries. The peculiarities of a particular species are found when examining it more closely. Differentiation of the sexes in *juvenile* fish presents considerable difficulties; even the detection of the very small gonads requires attentive work. Their form is generally similar to that of the maturing gonads, but in the young state they are much smaller and more delicate, often of thread-like or band-shaped thinness. At fairly strong magnification it is possible to detect rounded oöcytes with a large, vesicular-shaped nucleus in the ovaries. A *separation of the sexes* is particularly necessary if their behaviour differs, in regard to growth – in which the female sex is often ahead of the male – or in regard to a temporary preference for certain environmental conditions, which may lead to a shift in the normal 1 : 1 sex ratio, especially as a result of spawning migrations. The portion of each sex is given in percent; frequently it is also very informative to set the frequency of one sex – generally of the most frequent one – at 100 and to calculate thus the relative number of the other sex. If the sex ratio in a catch sample is meant to inform us about the stock, or a comparison of several catch samples is desired in this regard, a significance test will be required in order to eliminate the role of chance in the sample selection.

The size and age of the *onset of sexual maturity* should be noted. During the maturation process it is easy to distinguish young from sexually mature individuals. Both for the entire fish group studied and for the individual age groups we set up a series of numerical correlations indicating the manner in which the portions of the young or mature individuals alter as their size and age increases. As a rule it is found that the onset of sexual maturity usually extends over a period of years and over a considerable size span, which is evident in the increasing percentages of sexually mature individuals. The lower and the upper size and age limits are particularly noteworthy. Comparative studies on the onset of sexual maturity in various species and in the populations of one species in different areas of its distribution zone gain in significance when correlated to the average life span and to the temperature and nutritional conditions.

In order better to understand the *maturation process* of the gonads, which is repeated each year after the onset of sexual maturity, we distinguish seven maturity stages which may be briefly defined as follows in accordance with their external characteristics:

Stage I: young; gonads very small, clear as glass, transparent.
Stage II: at rest; gonads small, dully transparent, reddish-grey.
Stage III: onset of maturation; gonads enlarged, non-transparent, blood vessels distinct, ova detectable in the ovarium.
Stage IV: further maturation; gonads considerably enlarged, ovaries reddish to bright orange, tightly filled with polygonally flattened ova.
Stage V: maturation almost complete; gonads of definitive size, ovaries as previously, brittle, ova round, some already glassy, testes white, filled with viscous sperm.
Stage VI: ready for spawning (flowing); if injured, sperm or mature ova as clear as glass flow out.
Stage VII: spent; gonads shortened, flabby, empty, dark red, with traces of sperm or ova.

A transition stage VII/II, designated as a recovery phase after spawning, leads back to resting stage II, from which – after a fairly long pause – the maturation cycle begins again. Maturation stages

III to V in particular are difficult to mark off from each other; it therefore makes little sense to differentiate them even further by designating intermediate stages. Often it is more useful to omit the classification of the maturation process and to differentiate the four principal maturation stages only: immature or resting (I to II), maturing (III to IV), mature (V to VI) and spent (VII). The stages can also be determined exactly when examining preserved material, when it is possible to identify pre-spawning associations as well as the spawning time and the location of the spawning sites; the stage of maturity of the females should carry more weight than that of the males, who mature earlier and remain at the spawning site longer. One should also note whether the eggs mature and are expelled all at once or whether they are ejected in several thrusts following closely on each other, which is the case in most of the species. If ripe (flowing) individuals of both sexes are caught, they can be used for artificial fertilization trials, which will be discussed in greater detail elsewhere. In the fully mature state the gonads, particularly the ovaries, take up a considerable part of the body cavity and make up a very important part of the weight of the fish. By exact weight determinations one can follow the increase in the *percentage weight of the gonads* as maturation increases, thus obtaining an objective measure for this process and for the synthesis capacity of the body, which was previously mentioned when discussing the determination of the length-weight coefficient. Series studies are yielding information on the *duration* of the maturation process, showing that higher water temperatures accelerate the process and lower ones slow it down.

9. Fertility Studies

Here we find a problem which has recently received a fair amount of attention: the *fertility* of a fish, which is of great importance for the reproduction of the stock of one species. Even mere observation shows that the volume of the ovaries, and concomitantly the number of eggs produced, increases with the size of the fish. Apart from the question of the order of magnitude in which the egg numbers of one species range – which necessarily also depends on the *size of eggs* – there arises the further question of the *correlations between the egg number* and the *length* or *weight of a fish*. To investigate this we select females with ovaries which are in advanced stages of ripeness (stages IV and V) and have eggs (the spawn) which are readily separated from the small oöcytes remaining in the resting stage (the reserve stock); under no circumstances should signs of spawning be detectable. The carefully loosened ovaries are best fixed in Gilson solution, but formalin substances may also be used. The well-hardened ovaries are placed in a petri dish, cut open and the eggs are freed from the tissue. They are then isolated from each other and the ovarian tissue is removed as completely as possible. A measured portion of the egg pulp thus obtained is counted out under low magnification; from this portion of the total quantity the number of eggs of the fish is calculated. It has been found simplest and most useful to free the egg mass of excess fluid by using a paper filter and to weigh it, while in a moist state, in a covered glass receptacle. We then remove several (3–4) small *partial samples* from various places in the egg mass, place them individually in a weighing bottle with a cover, and weigh them on an analytical balance. The eggs of each partial sample are counted separately and the results are first converted to units of weights in order to establish their distribution. If the possible sources of error are carefully eliminated (unequal degree of moisture, insufficient mixing) the figures only fluctuate a little and it is possible to calculate the number of eggs with adequate accuracy from the ratio of the sum of the sample weights to the total weight; it is useful to round off the value in accordance with the height of the egg number in each case. The number of females of one species used for studies of this kind should not be too small, i.e., approximately 50 individuals and preferably more, and should embrace the largest possible length range; in this regard one must make sure that the individual size groupings are represented without too much inequality.

One already obtains an idea of the fertility of a species if one calculates the *average number of eggs* for one or several size groupings. If the study material involves a fairly large number, the nature of the *relationship between the egg number and the fish size* should be analysed in greater detail, if possible. The relevant data is obtained by a *scatter diagram* in which the abscissa represents the length or weight, and the ordinate the egg numbers. Since these often show a considerable fluctuation even in individuals of almost the same size, it is advisable to classify the material into some six to ten length or weight groups and to record the averages obtained from the original data for this purpose in a diagram. In this way it is possible to cancel out a considerable part of the dispersion of the individual values, and the points obtained take on an arrangement which allows us to connect them by approximation with a straight line or a curve (Fig. 30). This already indicates whether the egg number and the fish length or weight alter to the same extent, i.e. whether their mutual dependence is of a linear nature or whether the increase of one characteristic occurs at an increasing rate – that is, exponentially.

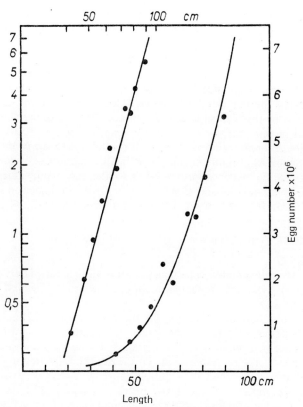

Figure 30. Example of linear and logarithmic representation of the correlation between egg number and length (Baltic – codfish)

One thus has the choice between the linear equation $y = a \cdot x + b$ and the power function $y = a \cdot x^b$, y being the egg number and x the length or the weight. The relationship between the egg number and the weight being approximately linear, and in view of the relationship between the length and the weight, an approximately linear relationship between the egg number and the cube of the length may be expected, corresponding to the formula $F = a \cdot L^3$; a graphic representation of the group values will again demonstrate this, the third power of the length being recorded on the abscissa.

Determination of the fact that the increase in the egg number in relation to the weight or length follows a power function does not exclude the possibility that a simple linear relationship exists within a limited sphere, particularly between the egg number and the length, as a slightly arched curve section can be approximated to the power function by a straight line. It goes without saying that each of these equations is only valid within the observation range, where the simple linear relationships between the egg number on the one hand and the weight, cube of the length and length on the other deserve preference because of the possibility of observing and comparing the magnitudes involved. The use of a power equation with one of three divergent exponents makes a comparison more difficult in this case, as it does in the length to weight relationship. We should therefore always test with the X^2 test to see whether the hypothetical egg numbers calculated in this manner really do offer a significantly better approximation to the empirical values than simple linear relationships. When dealing with biological data one must be careful not to work it too hard mathematically.

10 STOCK ANALYSIS WITH THE AID OF MERISTIC CHARACTERISTICS

The preceding statements have probably shown adequately that in a stock analysis, the single *individual* of a species is observed chiefly a part of a *group*.

It is not actually the specific morphological and anatomical characteristics of the species which are studied, but rather the continuing or cyclic *changes* in certain qualities within the lifetime of the members of a population. This raises the following important question: in the broad distribution range of a species, has *differentiation* of various stocks, strains or 'races' developed? This general zoogeographic problem applies to all faunal areas of the sea; but it is particularly important in connection with commercial fish since the possibility of exchange between adjacent and far-removed populations is of great significance for the choice of regulations aiming at establishing a rational fishing procedure. Consequently fishing biology research has developed study methods which help to establish whether *morphological differences* exist between stocks. The extent to which these are genetically determined or the result of modifying environmental influences is as yet unknown. Differences in the mode of behaviour and spawning habits provide further criteria for racial differentiation, since they ensure the maintenance of closed reproductive associations in the same way as spatial isolation. As in the classification into genus and species, the *measurable and calculable qualities of the variability* play an important part in the subdivision of the species into subspecies, races, local forms, etc., the term given to these categories depending on the importance placed on the differentiation characteristics. The relevant characteristics in fish, particularly in bony fish, are related to some extent to the metamerism of the body: number or vertebrae and rays in the paired and unpaired fins, the lateral scale rows and the scales at the ventral edge, the gill rakes on the gill arches. Insofar as these were formed during the early development stages and have remained unchanged since then, they are preferable to characteristics which alter as growth progresses. This applies particularly to measurable characteristics such as the body proportions, the size of individual body parts (eyes, mouth, operculum, fins) and to the distances from the tip of the nose. In order to convert the measurements into comparable values they are related to the total length and recorded in percentages. Of course the internal organs can also be included in the investigation. The measurements are carried out with the aid of a measuring tape, calipers or a sliding gauge, depending on the size of the fish and the accuracy required. Only fish of similar size may be compared with each other when dealing with characteristics which alter in size. One turns to proportion measurements of this kind when the morphological differences of stocks which are spatially far distant from each other are to be examined to indicate their independence from each other and the numerical characteristics fail us, their variation remaining unchanged.

For all these measurements and calculations the greatest care and accuracy are the first rule. Although measurements are best based on the fresh material, since shrinkage occurs when the specimens are preserved, calculations can also be carried out on preserved fish. Some suggestions are briefly given for this. If the *vertebrae* are to be counted, we boil the fresh fish in one piece or after removal of the fillets and the head depending on the size; one must be careful to leave a piece of the base of the cranium attached to the vertebral column. The boiling should not be too long, so as to leave the vertebral column undamaged. The vertebral column is then cleaned of the remains of flesh by laying it on a sieve and rinsing it under running water or brushing it off carefully in a bowl with water. The vertebrae can then be counted immediately, but it is better to let them dry first. The trunk vertebrae and the tail vertebrae are counted separately; the first vertebra with closed haemal arch is counted as the first tail vertebra. One looks for *coalescence* of two or more vertebrae, which is evidenced by coalescence lines on the vertebrae. The urostyle is counted as one vertebra. Another finding which deserves attention is the occurrence of 'complex' vertebrae at both ends of the vertebral column, particularly at the tail end; these are distinguished by the presence of *accessory spinous processes* (often on one side only), the vertebra being homogeneous although somewhat longer, and are probably attributable to an incomplete separation of the vertebral rudiments during previous developmental stages. The more frequent their occurrence, the lower the number of vertebrae; they are consequently a sign of a *reduction*.

When dealing with small fish it is best to lay bare the vertebral column by monolateral filleting; it is then cleaned with a scalpel, allowed to dry slightly, and the count is made out at low magnification. The same method is used for preserved fish. Selective *bone staining* permits accurate counting even when very small specimens are being examined. After soaking them well they are placed in approximately 70% alcohol to which saturated alizarin solution (in 96% alcohol) is added as well as traces of potash lye if necessary, until the colour changes from yellow to Bordeaux red. The calcified skeletal elements stain dark red and the remainder remain colourless so that the segmentation of the vertebral column and all details stand out clearly, including the first tail vertebra with closed haemal arch and all coalescences and complex formations. Each count is carried out at least *twice* to ensure accuracy.

When counting the *fin rays* one must be careful not to overlook the final ray, which is often very small. When dealing with large specimens the fin is spread out by standing it up on the first ray, the tensor ray, and a preparation needle is used for counting; after every ten rays the skin of the fin is nicked in order to facilitate the re-check count; this is especially advisable for very long fin seams. Here too the *alizarin stain* helps to bring out all the details of preserved specimens. If one is careful not to damage the unpaired fins when the vertebral column is laid bare, it is possible to count and record the vertebrae and fin rays one after the other. The number of rays on the paired fins is usually counted on one side only, as are the lateral *scale* rows and the *gill rakers* on one gill arch, generally the first one. In clupeids the number of *gill scales* between the ventral fins and the anus offers a good criterion for differentiation.

The results of the calculations are collected in tables. In general the variation series show a *normal distribution*, thus following the Gauss error laws, and the usual statistical methods are used for their interpretation. The average condition of a sample in regard to the characteristics tested is identified by the *mean value*, the *standard deviation* or dispersion and the *average error* of the mean. To represent the frequency distribution of a characteristic graphically we select a column diagram or a polygon, more frequently the latter as this method allows us to represent the variation polygons of various samples on the same base for comparison purposes. The mean value is always shown as the vertical (Fig. 31). As mentioned at the outset, the purpose of statistical variability studies of this kind is to determine whether the differences between the mean values of two

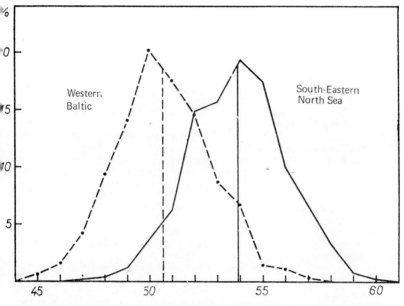

Figure 31. Example of the representation of the variability of a meristic characteristic (number of fin rays in the anal fin of the plaice)

samples are produced by chance when taking the samples (the samples consequently originating from a stock which is *homogeneous* in regard to the characteristics studied), or whether they reflect true inequalities in the samples, which are therefore taken from two different stocks. To establish this we test the *difference of the mean values* by means of the t-test for their *significance*, i.e., for the degree of probability with which the difference in the mean values is statistically significant or not significant. More detailed information on this and on the solution to other previously mentioned statistical problems will be found in general textbooks on probability calculations and statistical methods, to which we refer the reader.

Numerous studies, including some experimental investigations, have shown that in marine fish species with wide distribution areas the *temperature* and *salinity* in particular affect the development of quantitative characteristics in a certain direction. This offers the marine biologist a broad field of research; the field of fishery biology provides tested research methods which, suitably modified, may be used for similar stock information studies in other zoological groups. At the same time this justifies the range of instructions presented here for the examination of a fish stock.

REFERENCES

BEVERTON, R. J. H., and S. J. HOLT (1957). 'On the Dynamics of Exploited Fish Populations.' *Fish Invest.*, Ser. II, **19**, 533.

BÜCKMANN, A. (1929). 'Die Methodik fischereibiologischer Untersuchungen an Meeresfischen.' *Abderhalden, Handb. d. biol. Arbeitsmethoden*. Abt. IX, Teil 6, Verlag Urban & Schwarzenberg, Berlin u. Wien.

DECKER, K. (1955). 'Fische. Exkursionsfauna von Deutschland. III' Wirbeltiere. 1 – 77. VEB Verlag Volk u. Wissen. Berlin.

F.A.O. *Manuals in Fisheries Science.* Food and Agriculture Organization of the United Nations, Rom (in Vorbereitung).

GRAHAM, M. (1956). *Sea Fisheries. Their Investigations in the United Kingdom.* Edward Arnold Ltd., London.

KÄNDLER, R. (1941). 'Untersuchungen über Fortpflanzung, Wachstum und Variabilität der Arten des Sandaals in Ost- und Nordsee, mit besonderer Berücksichtigung der Saisonrassen von *Ammodytes tobianus* L.' *Kieler Meeresforsch.* **5**, 45 – 145.

— (1964). 'Seefische.' P. BROHMER, *Fauna von Deutschland.* 9. Aufl., 444 – 462. Quelle und Meyer Verlag, Heidelberg.

LILLELUND, K. (1961). 'Untersuchungen über die Biologie und Populationsdynamik des Stintes *Osmerus eperlanus eperlanus* (L.) der Elbe.' *Arch. Fischereiwiss.* **12**, 1. Beiheft, 1 – 128.

LAMP, F. (1966). 'Beiträge zur Biologie der Seeskorpione *Myoxocephalus scorpius* (L.) und *Taurulus bubalis* (Euphr.) in der Kieler Förde.' *Kieler Meeresforsch.* **22**, 98 – 120.

MUUS, B. J., und P. DAHLSTRÖM (1965). *Meeresfische in Farben.* (Übersetzung aus dem Dänischen). Bayerischer Landwirtschaftsverlag, München.

ROUNSEFELL, G. A., and W. H. EVERHART (1953). *Fishery Science. Its Methods and Applications.* John Wiley & Sons, New York/London.

SNEDECOR, G. W. (1962). *Statistical Methods.* 5th Ed. The Iowa State University Press, Ames, Iowa.

C. Benthonic Vegetation

H. SCHWENKE

The methodology of a scientific discipline has its specific difficulties in each case, determined by the subject under study. In an examination of the situation in marine botany it must first be mentioned that the benthonic marine vegetation is only accessible to some extent and at certain times. Leaving aside the supralittoral stock, this accessibility is determined by the local tidal range and consequently by the extent of the area which periodically becomes dry. The time limitation is also determined by the tidal rhythm. Moreover, the sublittoral marine vegetation is always covered by water, forcing the observer to use mechanical or optical apparatus which intervenes between himself and the subject under study.

An essentially unavoidable difficulty is presented by the water motion. The work can become much more difficult or impossible in coastal waters in moderately rough seas, moderate surf and strong currents. The vegetation on rocky shores with a constant strong surf is virtually inaccessible.

Another major difficulty is of an institutional nature, i.e., no generally recognized and accepted study methods for marine plant science exist. Directions relating to procedure can only be taken from individual studies, therefore; however, procedural indications of this kind cannot avoid a local imprint, produced by the situation at each specific site. A compilation of the procedures based on this fact consequently indicates the manner in which one may proceed rather than the manner in which one should proceed.

1. TYPOLOGY OF MARINE PLANT STOCK ANALYSIS IN THE BENTHOS

Another inevitable consequence of the lack of methodological cohesion is the fact that the technique of benthonic stock analysis can assume widely differing aspects and aims. We will therefore first attempt to establish a kind of 'typology' of these procedures. At least the three following aspects must be taken into account.

(a) Differentiation between Floristic and Plant Science Study

Most of the well-known specialized monographs are concerned with a floristic examination. Examples are the works of Levring (1937), Jorde and Klavestad (1963), Kylin (1944, 1947, 1949), Sundene (1953), Waern (1952). These studies strive for taxonomical accuracy and floristic completeness. The attempt to achieve the latter often burdens the study with questionable references from the earlier literature. Because of the present difficult position of phycological taxonomy it is generally impossible to attain taxonomical accuracy in the current collections, as well as subsequent clarification of questionable bibliographical references. It must be pointed out that collections made up solely from the shore and dredger samples no longer satisfy present-day floristic requirements. The floristic study of an area must also make use of plant science methods.

The study of the marine benthos in terms of plant science attempts to determine and describe plant synthesis in relation to the particular ecological vegetation conditions and in terms of periodic (seasonal) or aperiodic chronological changes in the vegetation picture. A sketched representation of the undisturbed plant structure at the natural site is a prerequisite for work in marine botany; this presents its own specific problems.

A further prerequisite, which is theoretically self-evident, is a well-defined taxonomic and floristic definition of the plant elements. In practice, however, this cannot always be realized under present-day conditions. Many of the genera which are important in plant physiognomy are taxonomically difficult. We must warn against determinations based on the principle of the greatest probability. It is neater methodologically to identify unclear taxonomical facts as such.

(b) Differentiation between General and Applied Marine Botany

Differentiation between general (basic scientific) and applied marine botany does not play as large a part in the practical stock analysis of the benthonic vegetation as might be expected in theory. We owe many examples of stock investigations and the decisive stimulus for methodological work to applied marine botany (Chapman 1944, Grenager 1964, Hoffmann 1952, Walker 1950, 1954). A disadvantage of such studies is that they aim at the examination of commercially interesting types. This is partially balanced by the fact that these types predominate in terms of vegetational type at least in cold and moderate seas. Furthermore, the techniques used for this purpose largely coincide with production-biology study methods. In any case there is no reason to ignore the interpretations offered by 'applied' studies on the basis of technical methods and plant science. Perhaps it would be more correct to speak of methods of general marine plant science on the one hand and those of applied marine plant science on the other.

(c) Differentiation between Large-Scale and Small-Scale Studies

The differentiation between large-scale and small-scale studies in itself points to particular study methods. The large-scale study techniques must necessarily be carried out from the point of view of plant typology and vegetational type, since evidence relating to the vegetation structure can only be obtained by random sampling or in the form of large-scale profiles. Large-scale studies are necessary in relatively flat marine areas which do not set any depth limits to the marine vegetation for reasons of light ecology, on strongly indented coasts with protruding islands, and in so-called archipelagos. Examples are the western Baltic Sea (Kiel Bay) the coast of Scotland, the rocky Scandinavian coasts. Such studies correspond in many ways to the interests of applied marine botany.

Small-scale studies aim at complete coverage of the vegetation composition within narrowly delimited areas. They are easiest to carry out in periodically dry littoral zones and in shallow coastal waters. The method used is largely the same as that employed for terrestrial plant ecology (quadrat or net studies). However, when applying the terrestrial method the sociological plant terminology should not be taken over uncritically at the same time. Marine botany records above all the plant specimens which together form the particular vegetation structure. Whether one may accurately speak of associations or other sociological units must await interpretative analysis and is not at all self-evident in the field of marine benthonic vegetation. As a rule, small-scale studies will be of the nature of example sections within a larger vegetation structure and will be coupled with large-scale profiles or plant-typology descriptions. Small-scale studies representing an example of the particular vegetation belt in its entirety are possible on rocky coasts with a steep gradient and strong light (such as the Norwegian fjord coasts).

Apart from these more basic aspects there are a number of site-specific factors which can be determined by technical methods under the specific local conditions. These are as described in (d) and (c) below.

(d) Coastal Morphological Aspects

Hard-ground or rocky coasts must be differentiated from soft ground or sandy coasts. Rocky coasts are as a rule strongly overgrown and deeply indented. This indentation results in a great number of small biotopes (such as tide-pools) with a great variation in ecological factors, and such coasts are therefore difficult to study for technical reasons. Sandy coasts are as a rule overgrown in a sparse and sporadic manner and show little indentation. Mixed types are frequent, consisting of fairly large or small rock groups embedded in the sandy shore or rocky bays filled with sand. Examples, Western Baltic, Fehmarn, with large pebbles and individual rock blocks; also the Breton Coast, particularly around Roscoff.

We must also include the local height of the tidal range here, which, together with the angle of inclination of the coastal profile, determines the extent of the littoral zone which periodically becomes dry and thus easily accessible.

(e) Plant Typology Aspects

Finally, in regard to technical methods it is also important to know whether one is dealing, for example, with large seaweed stocks (kelp beds, *Macrocystis pyrifera*, such as those found on the coast of California), or with a North-West Europe large seaweed vegetation (*Fucus, Ascophyllum, Laminaria*) or with a small and calcareous algal vegetation, such as is frequently found in warmer seas.

2. PROCEDURE FOR THE MARINE PLANT STOCK ANALYSIS IN THE BENTHOS

As mentioned previously, all the relevant procedures have a local character. The presentation which follows is based in part on our personal experiences, gained from marine vegetation studies in the Western Baltic. Our classification into shallow-water and deep-water studies is in itself arbitrary, but is based on the fact – decisive for the technique used – that large-scale deep-water work requires a suitable ship. The equipment necessary for littoral and shallow-water work, on the other hand, can be transported by car.

(a) Preliminary Remarks concerning the Floristic Procedure

Floristic evidence material is necessary for any type of study. If possible it is best to collect intact specimens, including the hold-fasts, at the natural site. If possible, living material should be used for the analysis; fixed and preserved specimens present additional problems. It is better to use a moist pack for the transportation of most marine algae, rather than fairly large quantities of water. Damage due to heat should be avoided. If the material is to be kept alive for a fairly long time, temperatures of about 5°C are advisable. Plant material is laid on suitable paper (heavy, well-sized types) in a shallow dish with water, carefully removed and covered with a cloth, enclosed in filter paper and pressed between an adequate number of newspaper sheets; the filter paper and newspaper should be changed after a day. If plant material of this type is to be used for morphological studies, it is advisable to press the plants lightly only and to dry them on paper. Large, hardy specimens should be dried by airing without being laid down, and are then glued down; very small material should be kept in folding paper bags, etc.

The beginner should be reminded once more of the present difficult position of phycological taxonomy. Consequently, determination trials should only be undertaken with great care. No determination literature in the true sense is available. In the North-West European area the closest we can find is Newton (1931, newly reprinted!), Lakowitz (1929, out of print), Kylin (1944, 1947, 1949), and the large area monographs (see References).

(b) Littoral and Shallow-Water Work

As a rule, marine benthonic vegetation consists of a vegetation belt running parallel to the coast, bounded in the supralittoral area by the splash zone and in the sublittoral area by limiting ecological factors, particularly light and substrate. In essence it is therefore possible to cover the whole width of the vegetation belt with its ecologically determined internal structure. For study purposes section profiles or path profiles will be drawn through the vegetation zone. When using the section profile the vegetation elements are shown along an imaginary or actual line; the path profile also includes the surface arrangement within the individual zone sections.

The first step in an examination of a stock is, thus, the organization of the study area, allowing a scaled representation of the vegetation pattern. As in terrestrial botany, measuring lines or surveyor's chains and collecting nets are used. The material of the chains, lines and nets must be resistant to seawater and stable in regard to length. The length of the lines and the size of the nets depends on the extent of the vegetation area. Longer lines, particularly for diving work, are preferably made up of variously coloured 2-metre pieces which have been shackled together. The shackles simultaneously serve as weights. For the haul procedure we use the preprinted cards DIN A4, which have been provided with a DIN A5 half for station data, location, date, etc., and for a survey sketch; the second half carries a predrawn grid. The reverse has been left free for recording a list of species and further procedure data.

In the periodically dry littoral zone this collection work is very simple, but this is not so in the shallow coastal water of seas lacking any considerable tidal range. Here difficulties arise at a water depth of as little as 1 m. Since in our latitudes it is only periodically possible to work in bathing costume, practical protective clothing should

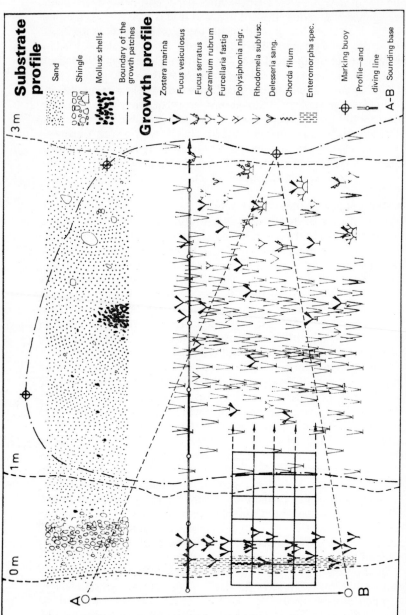

Figure 32. Schematic representation of the littoral method and shallow-water method illustrating the vegetation conditions of the Western Baltic. Path profile based on net studies: separation of growth and substrate profiles.

consist of foam rubber diving suits or simple breast-high waders made of thin material which is pressed against the body by the water pressure and scarcely hampers movement. As a rule, protection against cooling will be required even in the summer. Boats will be required in the depth range between 1 m water depth and the depth at which a (fairly small) research vessel can operate, i.e., generally at approximately 5 m. Rubber boats are not sufficiently stable, but can be practical as a means of transporting equipment in shallow water. A stable wooden boat with an outboard motor is better; the ideal is a launch, if possible with a window on the bottom. If this is not present, one must use a so-called water observation window, i.e., a short tube with a bottom pane, preferably attached to the research boat. In addition a small triangle dredge and a rake or similar object with an attachable shaft for bringing up algae and a few small marking buoys with anchors and adjustable cable lengths are necessary. In shallow-water work with the boat the site determination necessary for mapping purposes presents some difficulties. To take bearings with the hand bearing compass or similar instrument from the boat in relation to landmarks is not entirely simple. We therefore tried out the reverse procedure, in which the bearings of the boat or of marking buoys are taken from the land. It is easy to make up one's own simple bearing discs, consisting of complete circles with sectional divisions provided with an alignment arrangement and mounted on a photographic tripod. On shore we measure a baseline which must be long enough to produce no bearing angle of more than 70°. The angles to the boat are determined from the end points. It is best to establish the connection between the measuring base and the boat by manual radio transmitters. The procedure can then be recorded on land. Fig. 32 shows a graph of this method under the vegetation conditions of the Western Baltic.

Botanical work in the sea without *diving* is no longer conceivable today. A mask, snorkel, and flippers are of very considerable help for marine botanical shallow-water work (see, for example, B. Ernst 1959). For some investigations the *marking* of individual algal thalli is necessary. For this purpose we cut strips of thin coloured plastic sheet, measuring approximately 2 cm in width and 40 cm in length. These strips are easily attached to the larger brown seaweed. The commercially available fastenings for plastic bags are also suitable for this.

(c) Deep-Water Work

In the practice of marine botany deep-water work cannot be clearly delimited. As a rule it requires the use of a fairly large vessel, begins where the simple aids of shallow-water are no longer adequate

– i.e., more than approximately 4-5 m water depth – and extends down to the lower vegetation limit. This may be at 20 m (Western Baltic), to 150 m (Mediterranean).

For a long time the dredge was the only working device. It generally takes the form of a serrated triangle or unserrated rectangular dredge (see Figs. 33 and 38). The form of construction and the structural details are heavily influenced by local conditions and experience. If possible, the dredge is worked over a winch. During this operation one must make sure that sufficient line or cable is

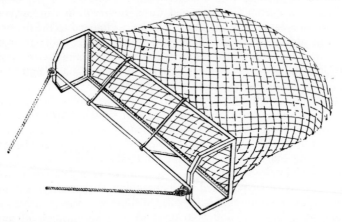

Figure 33. Unserrated rectangular dredge; constructed at the Institute for Marine Science, Kiel University (see also Fig. 38)

paid out so that the instrument actually does reach the sea bottom. On deck a minimum of two fairly large tanks is required to wash and roughly sort the catch; a sufficient number of smaller receptacles for more detailed sorting is also required. We use wooden stacking frames for this purpose, each of which will take six square plastic containers measuring approximately 30 cm in length from edge to edge.

The disadvantages of dredger work are evident. There is no control over the catch, one gains no knowledge of the organization of the vegetation and, further, one cannot be certain that the material collected was grown at the site. Direct observation of the site, which is necessary to cancel out these disadvantages, can now be accomplished by two means: diving and underwater television.

Of the *equipment-diving procedures* used today, only diving with pressurized air apparatus comes into consideration for our work. Helmet diving and the use of oxygen circulation equipment are unsuitable and too dangerous respectively. Diving work from on-board at fairly great water depths requires suitable equipment for the ship (diving ladder, compressor, life-saving equipment), impeccable, complete, and reliable diving gear, strict working discipline under a responsible leader, unconditional observance of all working and security regulations, as well as physical fitness and thorough training of the diver. This makes such undertakings somewhat expensive and cumbersome, but the above pre-requisites are indispensable in regard to the responsibility of the leaders of projects of this kind.

In deep-sea diving, the aims of marine botanical investigation of the stock are similar to those of shallow-water work: study of profiles and test areas, quantitative investigations and the tracing of dynamic botanical processes (seasonal successions, etc.). Neushul and others have developed a special SCUBA diving procedure devised for specific vegetation conditions (giant seaweed and large algal components): the diver swims along a marked base line and transmits his observations through a microphone (which is built into a complete diving mask) to the soundtrack on board the accompanying vessel.

It is also possible to build in small battery-driven, fully transistorized soundtrack equipment in a watertight manner, thus making the diver independent of the cable connection. In the field of marine botany, the *underwater television* is especially useful for providing a survey of the large-scale vegetation organization of an ocean area rapidly, safely and independently of unfavourable water temperatures. Fig. 34 shows the example of Kiel Bay. The pressure-protected television camera, which as a rule is coupled with a remote-control photocamera, is best operated from the slowly moving ship. If underwater visibility conditions do not allow photographic documentation, vegetation typology sketches can be based on observation of the projection screen image. The observation procedure is recorded on a soundtrack similarly to the SCUBA procedure. It is also possible to store the television observations by means of a picture tracing device.

(d) Quantitative Aspects

Until now quantitative stock investigations have been carried out largely with regard to the practical use of marine algae. On European coasts they have been concerned almost exclusively with the commercially interesting large brown seaweeds, such as *Laminaria*, *Ascophyllum* and *Fucus*, and in other areas with the so-called giant seaweed – above all *Macrocystis pyrifera*. As a rule such studies are not especially accurate, as they merely aim at yield estimates for

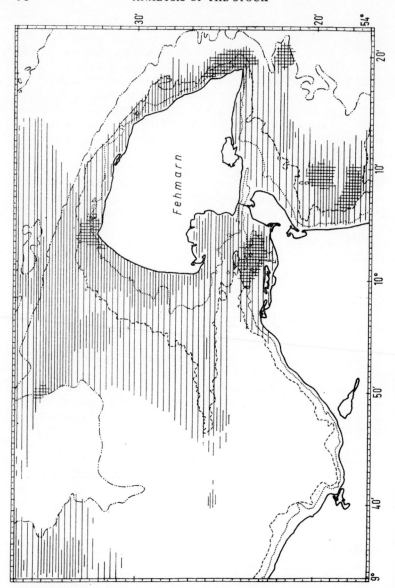

Figure 34. Example of a large-scale benthonic vegetation study with the help of underwater television apparatus; Western Baltic (Fehmarn area). The shaded part represents the ground cover as a percentage of the surface (wide shading: $1-5\%$; narrow shading: $5-30\%$; cross hatching: $30-100\%$. This scale is adjusted to the specific growth conditions of the Western Baltic)

commercial purposes. In the simplest procedure the fresh weight of the algae is related to certain surface units. On the basis of empirical values the fresh weight then leads to the dry weight and, further, to the yield of the technological end product, such as alginate. In the periodically exposed littoral such zonal studies are carried out very easily; in deeper waters their reliability depends greatly on the methods which can be applied. Modern diving procedures and underwater television are of practical interest here. Examples of quantitative studies of this type are the studies of F. T. Walker and co-workers on the coasts of Scotland. In these investigations the seaweed yield was studied in particular in relation to the water depth and to the time of year.

C. Hoffmann (1952) attempted a quantitative stock examination of the occurrence of *Fucus* to a water depth of approximately 6 m on the German Baltic coast between Flensburg and Lübeck. The entire coastal band was followed with a small motor boat, the growth of *Fucus* visible with a simple water observation window being estimated and mapped in three stages, according to yield.

It is easily seen that such studies can never completely cover a study area quantitatively. They are much more in the nature of a random sample, which is sufficient for the practical purpose as a rule. Grenager (1964) has described an interesting mathematical estimation procedure, based on random samples, for the yield predictions in island areas where they are naturally particularly difficult to obtain, such as on the Scandinavian rocky coasts. Remarkably few quantitative studies of the phytobenthos relating purely to problems of plant science or production biology have been carried out until now. In his studies of shallow-water vegetation on the Sorrentine coast (Mediterranean), J. Ernst (1959) conducted a few quantitative investigations (in kg fresh weight per m² for various algal communities; see also further literature references in Ernst). Mediterranean *Caulerpa-Cymodocea* fields were studied by Gessner and Hammer (1960) for production biology purposes. To name a further example, in their littoral studies in the Japanese Sea, Stschapowa and Selitzkaja (1957) carried out quantitative studies divided into species constituents in various horizons over the O-line and for comparison in places of variable surf intensity.

Since the fresh weight determination of the various algae is a problem from the point of view of production biology because of their variable water content, no special indications are perhaps required – but we have nevertheless made a beginning.

(e) Mapping

For the sake of brevity, we include under the term mapping all

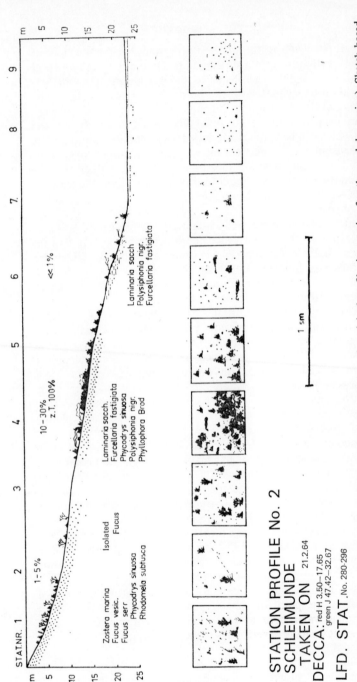

STATION PROFILE No. 2
SCHLEIMUNDE
TAKEN ON 21.2.64
DECCA: red H 3.50—17.65
           green J 47.42—32.67
LFD. STAT .No. 280-296

Figure 35. Example of a growth profile from Kiel Bay (section, the original profile rises again after the mud depression). Sketch based on underwater television – and dredger observations. Lists of species obtained from dredging samples belong to the numbered stations. The percentages relate to the ground cover (see Figure 34). The substrate is shown below the bottom line; in this case, sand (dotted) and mud (broken lines) with small shingle. The vegetation sketches below the profile have been drawn according to underwater television observations; under good optical conditions they can be replaced by underwater photographs

which serves the clear graphic or the cartographic representation of the results obtained from the stock examination of the marine phytobenthos. The classical example is the well-known Nienburg scheme for the vegetation zonation on the French canal coast (Nienburg 1930). The possibility of representing the vegetation organization by a profile of this kind rests on the preliminary assumption that the benthonic vegetation band, determined by an appreciable tidal range, is organized into zones of specific species or species combinations, depending on the emersion tolerance in each particular case. The zonation profile then offers a realistic cross-section, at least through the periodically dry littoral zone. These conditions are especially well realized on the rocky Scandinavian coasts, which generally have a relatively steep slope. Good examples of the use of profile representation are consequently found, for example, in the works of Jorde and Klavestad (1963) and Sundene (1953). Fig. 35 shows a study at Kiel Bay. The vegetation organization on highly indented coasts with variable substrates can only be represented inadequately by profiles of this type.

In such cases, surface sketches or path profiles yield a more accurate picture, but such processes have been used very little in marine botany until now. An example of the representation of fairly small sample surfaces (1 m²) is given by Ernst (1959). Large-scale mapping examples are to be found in Floc'h (1965) for the division of the brown algae growth in an island region facing the west coast of Brittany or in Schwenke (1964) for Kiel Bay.

No agreement has been reached until now in regard to the type of representation. For profiles clear standard pictures of the representative algae are readily chosen, and for surface mapping geometric symbols, but highly conventionalized standard pictures are also used. The latter undoubtedly have the advantage that legends may be dispensed with if they are drawn up instructively. Of special botanical value are presentations which include the principal ecological factors relating to the site, particularly the substrate conditions. Recommendations for the use of geometric symbols in marine vegetation mapping have been published by Davy de Virville (1964). Fig. 36a, b, c shows some of the possibilities of mapping of this kind, which have been realized so far.

REFERENCES

CHAPMAN, V. J. (1944). 'Methods of surveying *Laminaria* beds.' *J. Mar. Biol. Assoc.* 26, 37–60.

DAVY DE VIRVILLE, A. (1964). 'Sur un nouveau procédé de cartographie des algues marines.' *Proc. 4th Intern. Seaweed Symp.*, 175–178.

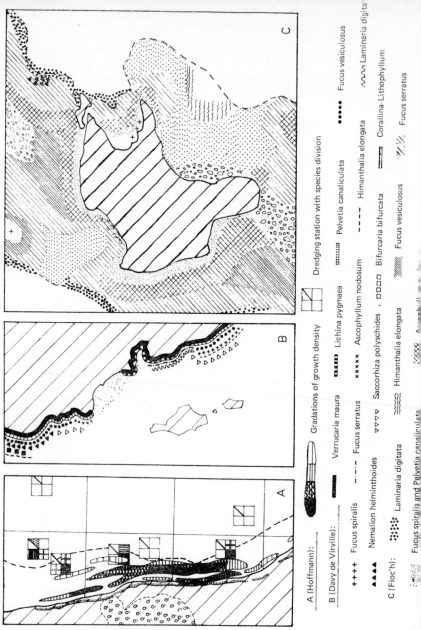

Figure 36. Some examples of vegetation mapping. Example A concerns Kiel Bay, d—
according to the unpublished original maps of C. Hoffmann (see his work 1952). A more det—
explanation can be omitted in view of the nature of the example; the gradations in growth de—
relate to fresh weight quantities, the dredger symbols give the species composition and
quantitative relationships according to a specific key. Example B according to Davy de Vir—
(1964) and example C according to Floc'h (1965); both relate to the French Atlantic c—
(Brittany)

ERNST, J. (1959). 'Studien über die Seichtwasser-Vegetation der Sorrentiner Küste.' *Pubbl. Staz. Zool. Napoli* 30, Suppl., 470–518.

FLOC'H, J.-Y. (1965). 'Répartition des Phéophycées dans l'Archipel de Molène (Finistère). I. Cartes de Molène et de Quéménès.' *Phycologia* 4, 135–140.

FUNK, G. (1927). 'Die Algenvegetation des Golfs von Neapel.' *Pubbl. Staz. Zool. Napoli* 7, Suppl., 1–507.

GESSNER, F., und L. HAMMER (1960). 'Die Primärproduktion in mediterranen *Caulerpa-Cymodocea*-Wiesen.' *Botanica marina* 2, 157–163.

GRENAGER, B. (1964). 'A theoretical approach to seaweed mensuration in insular districts.' *Proc. 4th Intern. Seaweed Symp.*, 191–196.

HOFFMANN, C. (1952). 'Uber das Vorkommen und die Menge industriell verwertbarer Algen an der Ostseeküste Schleswig-Holsteins.' *Kieler Meeresforsch.* 9, 5–14.

JORDE, I., und N. KLAVESTAD (1963). 'The natural history of the Hardangerfjord. 4. The benthonic algal vegetation.' *Sarsia* 9, 1–99.

KYLIN, H. (1944, 1947, 1949). 'Die Rhodophyceen, Phaeophyceen, Chlorophyceen der schwedischen Westküste.' *Lunds Univ. Årsskr.* 40, 1–104; 43, 1–99; 45, 1–79.

LAKOWITZ, K. (1929). *Die Algenflora der gesamten Ostsee.* 1–474, Danzig.

LEVRING, T. (1937): 'Zur Kenntnis der Algenflora der norwegischen Westküste.' *Lunds Univ. Årsskr.*, N. F., Avd. 2, 33, 8, 1–147.

NEUSHUL, M. (1964). 'SCUBA Diving Studies of the Vertical Distribution of Benthic Marine Plants.' *Proc. 5th Mar. Biol. Symp. Göteborg*, 161–176.

NEWTON, L. (1931). *A Handbook of the British Seaweeds*, 1–478, London, Nachdruck 1962.

NIENBURG, W. (1930). 'Die festsitzenden Pflanzen der nordeuropäischen Meere.' *Handb. Seefischerei Nordeuropas* 1, H. 4, 1–54.

SCHWENKE, H. (1964). 'Vegetation und Vegetationsbedingungen in der westlichen Ostsee (Kieler Bucht).' *Kieler Meeresforsch.* 20, 157–168.

STSCHAPOWA, T. F., und N. M. SELIZKAJA (1957). 'Verteilung der Algen im Litoral der Insel Monjeron (Japan. Meer).' *Trudy Inst. Okeanol. Akad. Nauk SSSR* 23, 112–124.

SUNDENE, O. (1953). 'The algal vegetation of Oslofjord.' *Skrifter Norske Vidensk.-Akad.* 2, 1–244.

WAERN, M. (1952). 'Rocky-shore algae in the Öregrund Archipelago.' *Acta Phytogeogr. Suecica* 30, 1–298.

WALKER, F. T. (1950). 'Sublittoral seaweed survey of the Orkney Islands.' *J. Ecol.* 38, 139–165.

—(1954). 'Distribution of Laminariaceae around Scotland.' *J. Cons. intern. Explor. Mer* 20, 160–166.

Bottom-Living Animals:

MACROBENTHOS

E. ZIEGELMEIER

The bottom fauna – the benthos – is involved in the cycle of materials in the sea and plays an important part in food chains as do the plankton and nekton in the pelagic zone (see the reports by Lenz and Kändler).

According to the size of the benthonic organisms we distinguish macro- and microbenthos. All animals which are retained in a sieve with a mesh width of $1 \cdot 0 \times 1 \cdot 0$ mm are classified as macrobenthos, and all smaller ones as microbenthos (see the reports of Gerlach and Uhlig).

All ocean bottoms are more or less heavily populated by invertebrate species, which live either within (infauna) or on the bottom (epifauna). According to Remane (1940) the inhabitants of the phytal are included as a special ecological group; these are animal species which have settled on fixed plants in the shallow, illuminated water or which live in it as motile animals.

Many types of macrobenthos possess skeletons (Echinodermata), exoskeletons (Crustacea) or shells (Mollusca); long after the death of the animals their traces still remain in the form of residual hard structures, which frequently make it possible to identify the species to which they belonged. It was thus possible very early on to recognize many of the larger bottom-living animals, such as those which remain lying in large quantities in the alluvium on the shore after rough weather. Fisheries as well have made it possible to describe many species of invertebrates which arrive on board as part of the secondary catch in the trawling nets. It is therefore understandable that the systematic study of the macrobenthos species has progressed further than that of smaller forms, and that attempts were made early to investigate collection methods for the study of the larger bottom-living animals and the development of suitable sampling equipment. A brief review of the history of benthonic research is presented by Holme (1964).

1. Sampling Methods

The easiest method and the one which does not require equipment of some complexity is the collection of the larger invertebrates in the eulittoral – the true tidal zone – where the animals of the epifauna may be taken from the most varied types of substrate (including a stony and rocky substrate) at low tide and those living within the sediment may be dug out with spade or fork. In most cases the more deeply settling animals may also be revealed. For special purposes, such as studies of the burrow of *Arenicola marina* in the undisturbed sediment, a sampling box has been developed (Fig. 37*) which makes it possible to reveal the burrow (Ziegelmeier 1964).

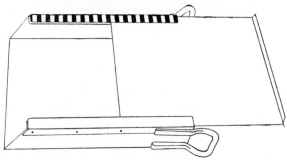

Figure 37. Sampling box with slide half drawn up. Centimetre division on the left runner. The measurements of this instrument made of galvanized sheet iron are 35 to 40 × 28 × 8 cm. The sampling box reveals an *Arenicola* burrow as follows: after finding a burrow the instrument must be positioned in such a manner that the funnel and casting are located roughly in the centre of the digging area. The sampling box with its closed slide is pressed as vertically as possible deep into the muddy ground. The sediment is removed from around the slide by digging. The instrument can be tipped backwards over the edge of the hole and after the water has drained away from the lower part of the area which has been dug out, the slide can be drawn away. Using a very sharp spatula the sediment can now be scraped away in thin layers until the *Arenicola* burrow has been reached

Macrobenthonic organisms adhering to tiles or harbour walls which are not exposed to the air at low tide are collected with instruments similar to hand nets (pile scraper), in which the anterior bow rim of the net bag is provided with a knife-shaped or toothed edge.

More difficult are the sampling methods in the sublittoral zone joining onto the eulittoral zone and in the deeper sea areas.

* We should like to extend our warmest thanks to Mr. Martin Söhl, List/Sylt, for the preparation of the drawings.

Whereas the species of the microbenthos usually inhabit the upper sediment layers (see the report by Gerlach, p. 117), much larger benthonic invertebrates occasionally live very far down in the sediment. This should be considered when developing sampling equipment. We should also take into consideration the purpose for which the collective animal material is to be used, i.e. for faunal studies, as the basis for experiments, for observations on living matter or for breeding. If for the first purpose, all equipment used for the collection of bottom-living animals may be employed, and particularly trawling nets and dredges; for quantitative studies bottom grabs are used, which are built in a great variety of forms and are variously useful.

The number of types of instruments used for sampling the macrobenthos organisms is very large, and within the framework of the present report it would not be possible to present and describe in detail all the equipment developed until now and mentioned in the literature. In this brief review we aim to point out various constructions, the equipment being used in European ocean areas being treated in somewhat greater detail.

Among the most important tools for the collection of samples for purely qualitative purposes are the *nets* and *dredges*. To obtain fairly large quantities of macrobenthonic organisms the most suitable are the trawling nets used by commercial fisheries (see the report by Kändler p. 64 ff). However, because of the wide mesh, only the larger bottom-living animals are found in the secondary catch in these nets. To increase the number of species and individuals in the catch, and also in order to include the smaller forms, the Biological Institute of Heligoland uses the 'Heligoland Trawl', a *trawling net* which they developed (see p. 107); this resembles the otter trawl in construction, with smaller overall dimensions and a narrower mesh width.

At their simplest, dredges are containers which scoop up different sediments at various depths. A bucket with a suitable long line attached to its rim would suffice to bring up sediment with animals in it, the bucket having been dragged briefly over the bottom. This is probably the way the bucket-dredge (Robertson 1968) developed; in this dredge the upper rim in both halves followed a convex curve, somewhat in the manner of a ploughshare, to improve penetration into the substrate, and was provided with a metal rod to prevent the instrument from rolling over the bottom. These constructions are generally not very productive, since the dredge is filled as soon as it has been dragged for a short distance, which makes the further collection of material impossible. The same results are obtained with a *sack dredge*; this dredge consists of a three- or four-cornered frame provided with a bag made of narrow-mesh, strong sacking, en-

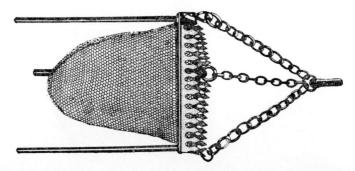

Figure 38.   Toothed dredge. Length of the toothed edges = 55 cm, height of the metal rods = 58 cm. An external bag is provided (mesh width = 60 cm) to protect the internal net (jute or sacking)

closed in a wide-mesh net sack. The edges of the frame are either knife-shaped or have sharp teeth, which is why it is called a *toothed dredge* (Fig. 38). This dredge has been much used in coarse sand for the collection of *Branchiostoma lanceolatum*. More productive are the catches made with a combination of a sack dredge with a bottom egg net, presently used by the Biological Institute of Heligoland (Fig. 39). The sack dredge, acting as a 'forerunner', frightens the lancelets which lie obliquely in the coarse sand out of the ground, and they are then collected by the two devices. The bottom-living animal material in the catch is enriched to some extent by this procedure.

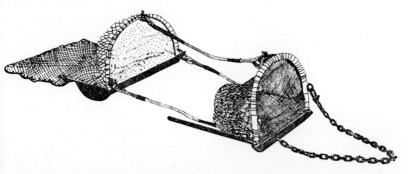

Figure 39.   Combination of a sack dredge (as 'fore-runner') with a bottom egg net. Measurements for the anterior net: height = 35 cm, length of lower knife edge = 45 cm. Mesh widths: internal net = gauze 1 cm, cover = 6 cm. Posterior net: distance from anterior = 85 cm. Width of the bottom egg net opening = 60 cm. Mesh widths: internal net = gauge No. 4 280, cover = 5 cm. Length of net bag: front = 40, back = 90 cm.

Figure 40. Oyster scraper or oyster iron. Mesh widths: anterior net within the rectangular metal frame = 15 cm, gauze of the internal bag = 4 cm, cover of chain links or net, as desired = 4 cm.

Selection and enrichment are attained with a dredge provided with a net bag with fairly wide mesh; the sediment escapes while the dredge is towed, so that only components which are larger than the mesh width are collected. The most commonly used dredge of this type is the '*oyster-scraper*' or the '*oyster iron*' (Fig. 40.) A knife-like downwards slanting edge digs into the sediment while the dredge is dragged over the substrate and 'ploughs' the soil. The lower part of the net bag consists of metal mesh rings, the top part of the net yarn mesh. If possible, this infauna dredge should only be dragged for a brief period since otherwise the animal material will be much damaged by the remainder of the net contents, such as seashells or their fragments (shell gravel), stones, etc.

Less damaging in this respect are the dredges for the collection of animals of the epifauna. The tool which has been found best for this purpose is the '*hydroid dredge*' (Fig. 41); this consists of a net bag with somewhat finer mesh which is held open rectangularly by means of an iron frame. Chains are attached to the net at the top and bottom along both long sides. This dredge can be towed behind the slowly travelling ship for 15 minutes without causing any great damage to the hydroids, bryozoans, ascidians, asteroids, snails and other epifaunal organisms contained in the net bag.

As with any equipment which is dragged over the substrate, when launching (lowering) the hydroid dredge one must be careful to

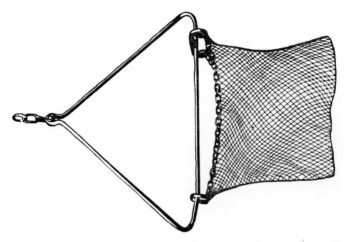

Figure 41. Hydroid dredge. Width of net aperture = 80 cm, height = 40 cm, length of external bag = 60 cm. Mesh widths: internal net = 1 cm, external net = 4 cm

pay out a sufficient amount of cable – approximately three times as much as the water depth. Further types of epifaunal dredges are described by Holme (1964).

Whereas the previously mentioned tools can generally only be used for studies of the faunal stock, *bottom grabs* have been built for quantitative studies; operating on the principle of a digger, these dig out a particular area from the ocean floor. The bottom grab method goes back to the classical studies of the Danish researcher, C. G. Joh. Petersen (1910 to 1918). His original intention was to establish quantitative studies of the macrobenthos – through assessing productivity – what kind and how many bottom-living animals at various places and in various sediments are available to support the food fish.

The later works of Petersen led to the establishment of *bottom-living animal associations*. These were named according to species which – on the basis of their stock density – are characteristic to certain ocean substrates and are absent or infrequent in neighbouring areas.

As a result of the incentive provided by Petersen, numerous investigations following his lead were carried out subsequently, and a great number of 'bottom communities' were named (see Remane 1940, Thorson 1957). Many publications are available, some of which also discuss the usefulness of the studies for future research.

The communities or associations were often established on the basis of only one or a few consecutive sample collections. Statistical studies of ocean areas with a great variety of bottom conditions were obtained. These studies are roughly comparable to instantaneous photographic snapshots, which only retain momentary spatial conditions. In a study of the bottom-living fauna this signifies quantitative detection of the number of species and individuals at the time, and their size, weight and volume.

These quantitative faunal and taxonomic studies provide virtually no information on the dynamics existing within the communities, their development and disappearance, and on the biotic and abiotic factors which determine changes. To trace the changing behaviour of an association in regard to space and time, we must pattern our studies on the quick-motion method as it exists, for example, in motion pictures: i.e., individual pictures are joined together and analysed. The dynamic course of a spatially firmly

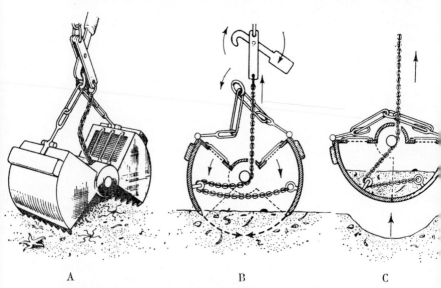

A   B   C

Figure 42.   Petersen bottom grab (from Holme 1964). Mode of action in three stages. A = the open instrument closely above the ocean floor. B + C = after the digger has touched bottom, the catch which holds the bottom grab open while it is being lowered is released. When the digger is being raised, the two digger pans are closed by the pulling action of the chain. If we consider the tensile point of attachment and track of the chain which first runs over the immovable digger pan axis, subsequently over a movable roller, and is finally attached to the other digger pan on the inside, we can readily see that when this instrument is drawn up out of loose sediment the bottom grab is already somewhat raised, which causes the digging depth to be reduced and the abundance of macrobenthos in the digging area to be smaller

defined ecosystem must be studied through continuous qualitative and quantitative investigations carried out over a period which is as long as possible.

For quantitative studies Petersen constructed a *bottom grab* whose mode of operation is shown in Fig. 42. When Petersen began his studies the digging surface (the area of the ocean floor surface to be dug out) was 0·1m²; this was later changed to 0·2 m². Unfortunately the digging depth and the grab content are not sufficient with this instrument, particularly on sandy bottoms. Only up to 5 cm is dug out from solid, partly muddy sand. This is due to the closing mechanism, which is not very good. The rising of the grab from the bottom is connected with its closing in such a manner that while the grab is closing it is already rising, which causes the digging depth to be reduced.

Much better results are obtained with the Van Veen *bottom grab* (Fig. 43), in which the closing mechanism and the digging effect are coupled in such a manner that when the instrument is lifted up the penetration into the ground is increased. With the Van Veen grab almost double the digging depths were sometimes obtained in very solid sediment. When using this bottom grab three principal points

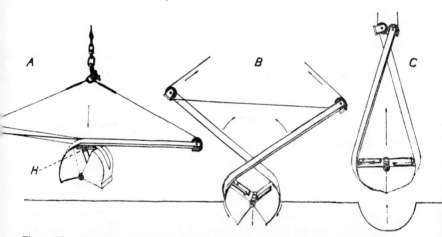

Figure 43. Van Veen bottom grab (0·1 m²) which, in contrast to the Petersen grab, digs deeper even in loose, sandy sediments because of its better closing mechanism. A = open digger immediately above the ocean floor. When the instrument touches bottom, the lock lever (H), which keeps the digger pans open, tips downwards. B + C = when being hauled up, the digger is closed in a scissor-like manner by the cable running over two rollers attached to the top end of the pans. During this process the two digger pans dig in a pincer-like movement which is deeper than that of the Petersen digger. The best length of the pans is 100 cm. (measured from the upper edge of the digger pans to the axis of the rollers)

must be taken into consideration: (1) The two side pieces of the grab pans must correspond in length with the size and the weight to ensure stability while the grab is being lowered and correctly positioned on the bottom. (2) The cables at the ends of the side pieces should not be attached but should be guided over rollers. (3) Perfect functioning of the Van Veen bottom grab depends not least on the experience and the knowledge of the ship's captain and those members of the crew who operate the equipment. When these requirements are met it is possible to obtain excellent samples even in poor weather – in some cases at a wind force of 5 to 6 – and also at depths exceeding 1000 m.

To reduce the transmission of the ship's movements to the equipment while sampling in rough seas, the cable of the instrument is guided over a pulley which is attached to a cable at the end of the beam, with an intervening 'expander'.

The two bottom grabs described above (for further types see Holme) are used to collect most samples of more or less disturbed sediment. Undisturbed samples can be obtained with *hydraulic cores*, but because their digging area is too small, they are not suitable for quantitative studies of macrobenthos; better results are obtained with the Knudsen bottom scoop with a digging area of $0 \cdot 1$ m² to 30 cm digging depth (cf Hagmeier 1930, Holme 1964).

The best tool for this purpose is the Reineck *sampling box* (Fig. 45), which makes it possible to obtain undisturbed sediment samples of 14×28 cm area to 50 cm digging depth. Because of the size and the weight of the sampling box, however, it is hardly possible for smaller vessels to use the instrument.

2. Further Treatment of the Samples on Board

The best method for separating the sediment from the macro-benthonic organisms and abiotic constituents of the substrate is *sieving*. For our studies we use one set of three sieves with mesh widths of 5, 3, and 1 mm.

After being hauled up, the closed Van Veen bottom grab is set on a table-like dredge stand (Fig. 44) above a round aperture whose diameter is equal to the length of the lower edges of the dredge. The sieve set is placed below the round opening. To avoid losses when the digger is opened (in rough seas, for example), a short tube made of sail-type material set below the opening reaches into the top sieve of the set, into which the total dredge contents fall. After the dredge table stand and the bottom grab have been rinsed, the sieve set is placed in a fairly large round wooden bucket (balje) and continuously shaken up and down in order to wash out the sediment. It is not

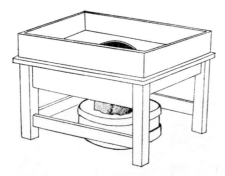

Figure 44. Dredge stand. Total height = 60 cm, edge length of the square stage plate = 80 cm. Diameter of the circular aperture = 30 cm

advisable to use the strong water jet of the deck washing pump for the rinsing procedure as this damages animals or causes the smaller forms to be pressed through the sieve mesh.

The remaining components in the individual sieves make up the sieve content, which must be separated into the living macrobenthos (living constituents = biomass) and the strainer residue. The best and most unexceptionable method, although the most time-consuming one, is the *selection* of the living animals. In earlier studies this was even carried out on board during expeditions to assess productivity. The disadvantage of this, however, is that a number of smaller species and individuals are overlooked. A second drawback is that when the sieves are kept on deck for a fairly long period, organisms such as the polychaetes slip down through the mesh, thus reducing the biomass in the quantitative samples. Thus, the polychaete component in a bottom grab sample depends on the amount of time which a sieve has spent on deck or in the balje before selection.

When the study of productivity was resumed within the framework of the studies carried out at the Biological Institute of Heligoland, the following method was used from 1949 on: the total content of the two or three sieves (in areas where the sand is clean two sieves are sufficient) of one set is placed in glasses with the aid of a small spatula immediately after collection of the sample. Particularly suitable for this are the commercial glass-topped bottles available in 1, 1·5, and 2 litre sizes. The samples are then fixed for later work in the laboratory with 4% to 5% seawater formalin which is as neutral as possible. Alcohol fixation is not advisable because (a) considerable shrinkage occurs in some cases; (b) the bristles of the polychaetes generally become too brittle and break off easily, which hinders later identification. With this sampling technique it is possible to

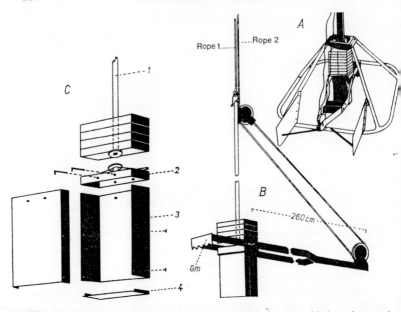

Figure 45. Reineck box grab. A = instrument with the pyramid-shaped control seat, which has a base measuring 300 × 300 cm. Its purpose is to maintain the grab box in a vertical position during sampling. B = instrument ready for use. Cable 1 for lowering the grab, cable 2 for closing the grab knife (Gm) and for raising the instrument. C = individual parts of the box grab: 1 = weighing rod with weights (to 750 kg), 2 = box shoe, 3 = open box, 4 = base for closing the box while the grab knife is opening

collect a fairly large number of samples during the bonitation voyages with relatively limited labour, which was never possible with the earlier method – involving the immediate sorting of the catch on board – particularly in unfavourable weather.

To allow substrate re-testing of the bottom conditions, a small, unchanged *sediment sample* should be taken from each catch before rinsing on the digger stand; when using the van Veen grab this is easily done by means of the two flaps found on the upper side of the two digger pans, which can be firmly closed with wing nuts. These sediment samples are also fixed in 4% – 5% seawater formalin and are also useable for studies of the microbenthos.

3. INTERPRETATION OF THE SAMPLES IN THE LABORATORY

For selection of the bottom-living animals from the fixed sieve contents, flat tanks or fairly large photographic trays are especially suitable. It is advisable to wash some of the formalin off before beginning the sorting procedure.

During this first selection the use of a hand lens is often sufficient. It usually reveals all the forms which secrete sticky mucus, such as smaller living polychaetes which adhere to small stones, to mollusc shells or to small pieces of shell gravel, which are very often overlooked when sorted on board.

Consequently, when using the methods described here (instruments which function better for the collection of quantitative samples and improved techniques for the sample analysis) the results of present studies can only be compared very conditionally with those of earlier investigations.

A further sorting of the sieve contents into portions in petri dishes at somewhat stronger magnification under a binocular preparation microscope and with suitable light ensures that no macrobenthonic organisms have remained in the sieve residue.

The types of livestock can now be determined and counted, and the sizes of the mollusca, echinodermata, and crustacea can be determined, as well as the *gross weight* and *gross volume* for the total stock and for the individual species and groups; the water-displacement method of the formalin-fixed animals is used for the determinations. All shellfish and snails are weighed and measured volumetrically with their shells, hermit crabs without the gastropod shell, and the tubicolous polychaetes and coelenterates without the tube. Only the volume of the sieve residue is measured. The weight and volume are determined after any adhering residual water is removed by means of filter paper.

There is absolutely no uniformity in the terminology used by bottom-fauna investigators with regard to the quantities of living organisms obtained with the help of the described quantitative apparatus. 'Standing crop', 'biomass', 'livestock' are the terms used for the quantity of living material per surface unit in a particular area at a given time. Much more important, however, is the need to compile – as far as possible – all the data when interpreting the samples; this includes the species picture, weight, volume, numbers for the crop density and, if at all possible, a great deal of size data in regard to the individual macrobenthos organisms.

In addition to the gross weight determination described above, where only slight differences are revealed between formalin-fixed and living material after comparative weighing, many investigators of the benthos determine the dry weight. If possible, the drying should take place without shells and other skeletal elements at the commonly used temperature of 105°C.

A further method is the determination of the organic substance of the individual macrobenthos invertebrate specimens, particularly when it is attempted to determine the value of the species as food animals, as for example, for bottom-living fish (cf. Thorson 1957).

However, these studies (of potential productivity) only form the basis for an understanding of the development and losses within one year in a particular area. Real determination of the production depends upon observations carried out in rotation at the shortest possible time intervals, and co-ordinated with studies of the micro-benthos and plankton.

REFERENCES

FRIEDRICH, H. (1965). *Meeresbiologie.* 'Eine Einführung in die Probleme und Ergebnisse.' 1–436. Gebr. Borntraeger, Berlin-Nikolasee.

HAGMEIER, A. (1925). 'Vorläufiger Bericht über die vorbereitenden Untersuchungen der Bodenfauna der Deutschen Bucht mit dem Petersen-Bodengreifer.' *Ber. dtsch. Meeresforsch.,* N. F., **1**.

— (1930). 'Die Züchtung verschiedener wirbelloser Meerestiere.' ABDERHALDEN, *Handb. d. biol. Arbeitsmeth.,* Abt. IX, Teil 5, Heft 4, Lieferung 326, 465–598.

— (1951). 'Die Nahrung der Meerestiere.' *Handb. d. Seefischerei Nordeuropas* **1**, Heft 5 b, 87–242.

HOLME, N. A. (1964). 'Methods of Sampling the Benthos.' *Adv. mar. Biol.* **2**, 171–260.

PETERSEN, C. G. J. (1913). 'Valuation of the sea. II. The animal communities of the sea-bottom and their importance for marine zoogeography.' *Rep. Danish biol. Sta.* **21**.

REINECK, H. E. (1958). 'Kastengreifer und Lotröhre "Schnepfe", Geräte zur Entnahme ungestörter, orientierter Meeresgrundproben.' *Senck. leth.* **39**, 45–48.

— (1963). 'Sedimentgefüge im Bereich der südlichen Nordsee.' *Abh. Senckenb. naturf. Ges.* **505**, 1–138.

REMANE, A. (1940). 'Einführung in die zoologische Ökologie der Nord- und Ostsee.' *Tierw. d. Nord- u. Ostsee* **1**, Teil I a.

RIEDL, R. (1963). 'Probleme und Methoden der Erforschung des litoralen Benthos. Verh. Dt. Zool. Ges. 1962 in Wien.' *Zool. Anz.* **26**, Supplementband. Leipzig.

THORSON, G. (1957). 'Bottom Communities (Sublittoral or Shallow Shelf).' *Mem. geol. Soc. Amer.* 67, **1**, 461–534.

ZIEGELMEIER, E. (1964). 'Über die Wohnbau-Form von Arenicola marina L. (Freilegen von Gängen in situ mit Hilfe eines Stechkastens.)' *Helg. Wiss. Meeresunters.* **11**/3–4, 157–160.

MEIOBENTHOS

S. A. GERLACH

The attention of the marine biologist has only recently turned to the minute fauna on the ocean floor, although individual systematic studies on certain animal groups were carried out as early as the nineteenth century. The work of Remane (1933) and his students clearly showed that the ocean floor is inhabited both in the tidal area and in the deeper zones by a microfauna with a great number of species (Fig. 46), which consists of representatives of the Protozoa (Foraminifera, Ciliata), Hydrozoa (*Protohydra*, *Psammohydra*, *Halammohydra*, etc.), Bryozoa (*Monobryozoon*), Turbellaria, Gnathostomulida, Gastrotricha, Nematoda, Kinorhyncha, Polychaeta (particularly 'Archiannelida', *Hesionides*, *Psammodrilus*, *Stygocapitella*), Oligochaeta (Enchytraeidae, Tubificidae), Tardigrada, Copepoda (above all Harpacticoidea), Ostracoda, Amphipoda (*Bogidiella*), Isopada (*Microcerberus*, *Ingolfiella*), Tanaidacea, Cumacea, Gastropoda (*Microhedyle*, *Pseudovermis*), Holothuria (*Leptosynapta*) and Ascidia (*Psammostyela*). This compilation is not complete and is increased every year by the addition of new finds.

With regard to the forms which inhabit the interstitial system between the individual sand grains, the English term 'interstitial fauna' or the French term 'faune interstitielle' is satisfactory; the German term is 'Lückenfauna' or 'interstitial fauna'. However, a world of micro-organisms also populates the algae, rock fissures and inhabits the superficial layers of muddy ocean bottoms where an interstitial system in its true sense does not exist. That is why the designation 'microfauna' or (more specifically) 'meiofauna' or 'meiobenthos' should be used.

The concept 'microfauna' is generally used to contrast with the concept 'macrofauna' and is based on considerations of method. To obtain the macrofauna the marine substrate sediment is rinsed through a wire sieve with a mesh width of 0·5–2·00 mm (see the section by Ziegelmeier). The animals retained by this sieve are described as macrofauna, and those which pass through the mesh are known as microfauna.

The English marine biologist M. R. Mare (1942) is responsible for

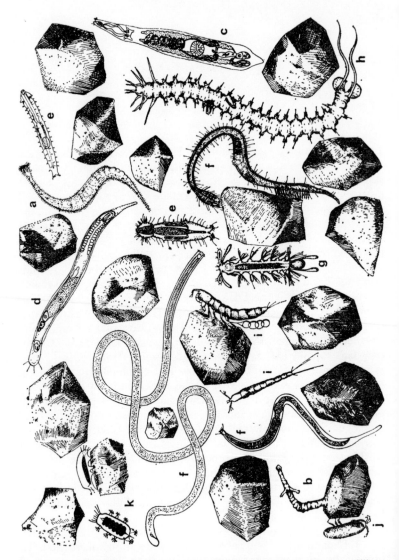

Figure 46. Characteristic representatives of interstitial marine fauna. *a*, Ciliates (*Trachelorhaphis*); *b*, Hydrozoa (*Psammohydra*); *c*, Turbellaria (*Kalyptorhynchia*); *d*, Gnathostomulida; *e*, Gastrotricha; *f*, Nematoda; *g*, Archiannelida (*Neril-lidium*); *h*, Polychaeta (*Hesionides*); *i*, Copepoda, Harpacticoidea; *j*, Ostracoda; *k*, Tardigrada (*Batillipes*). (Drawing by Hagen Westphal)

the term 'meiobenthos', used to describe all the organisms which – although passing through a sieve of the mesh width described – are retained by a finer sieve of approximately 45 mesh to 1 cm or with a mesh width of 60 μm. Among these are the representatives of the zoological groups described, with the exception of small ciliates. Such ciliates also pass through the fine sieve, as do single-celled algae and bacteria, and are described by Mare as microbenthos in the narrower sense.

This definition has not been universally accepted and at present 'microfauna', 'meiofauna' or 'meiobenthos' and 'interstitial fauna' are used as identical terms.

On floors of pure sand where specimens of the macrofauna are often rare, representatives of the microfauna make up a considerable part of the animal population. On muddy bottoms with their abundant macrofauna population one generally finds approximately a hundred times more individuals of the microfauna than animals of the macrofauna. Samples are available from various ocean areas where the individual number of microfauna per square metre approaches one million, and from muddy shallow areas it is reported that the number of nematodes alone exceeds 400 per cubic centimetre of sediment. However, the live weight attained by individuals of the microfauna even at a great population density is small – ranging between less than 1 and 20 g per square metre; it represents approximately 1 to 10% of the live weight of the macrofauna. The role played by the individuals of the microfauna in the circulation of matter on the ocean floor is largely unknown. However, we can be certain that it is of relatively greater significance than is shown by a simple weight comparison with the macrofauna, since the metabolism is more active in relation to the smaller body size and the propagation intensity is greater than among individuals of the macrofauna.

1. SAMPLING PROCEDURE

As with the macrofauna, the methods used to obtain the micro-fauna vary greatly and depend on circumstances. Frequently it is possible to use the same methods as those employed for the collection of the macrofauna; this is particularly true for the use of bottom grabs. However, all sampling methods must take into consideration the fact that in muddy regions of the ocean floor the individuals of the microfauna live very largely in the surface centimetre of the sediment; only a few forms, particularly the nematodes which are very resistant to lack of oxygen, live at a depth of more than 5 cm below the surface of the substrate. This limitation to the sediment surface does not hold true for bottoms of pure sand, particularly in regions of coastal ground water on a sandy shore, where a characteristic interstitial fauna inhabits the water mass of the system several metres below the surface of the shore.

(a) Sampling in the Tidal Zone

Because of its direct accessibility at low tide, sampling in the tidal zone presents the least number of difficulties; in particular it is easily possible to clear away thin sediment layers one by one and to examine them for their fauna. For quantitative studies we use a thin-walled measuring tube of known width (Fig. 47), which is used to cut out a sediment cylinder. Complicated slicing instruments have been developed by various investigators (Fig. 48). In sandy shore areas, where even larger animals are present below the surface, a coring tube can be hammered into the sand for quantitative studies. However, the strong friction on the grains of sand limits this procedure to a depth of approximately 30 cm, and consequently it is preferable to dig fairly large holes on the shore and to remove sand samples quantitatively from the wall of the hole.

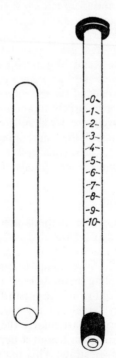

Figure 47. Coring tube for quantitative sampling of meiofauna

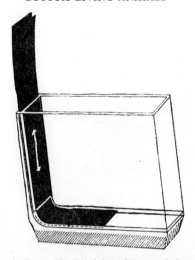

Figure 48. Thiel meioslicer. The Plexiglass ('Perspex') box, which is provided with cutting edges and is open at the bottom, is stuck into the sediment. The bottom aperture is closed by slipping a thin copper plate into a groove

For qualitative studies in coastal groundwater it has been found useful to dig a groove reaching the groundwater level. The groundwater then flows from the surrounding sand layers into the groove and brings with it part of the groundwater fauna, which can be fished out from the groundwater with a hand net made of plankton netting (Fig. 49). Further methods have been described by Husmann (1956) and Delamare Deboutteville (1960).

(b) Sampling from Deeper Regions

For the study of hard bottoms and coral reefs the free diving method is almost the only possible one; in shallower zones a mask, snorkel and fins are required, and in deeper zones breathing apparatus. Smaller stones with their growths can be transferred carefully by hand into a jar or a bag made of strong plastic material. It is also possible to tear off or cut off bunches of algae and to carefully break off branches of coral, bringing them up together with their microfauna in jars or bags. As the animals of the microfauna are generally capable of maintaining themselves on the substrate when water movement begins no losses take place with this sampling method.

Muddy and sandy sediments can also be collected by free diving; in particular it is possible to carefully remove and to collect the surface layer up to certain depths by means of a plankton hand net. Of course, the dredge – and digging methods, which can be used

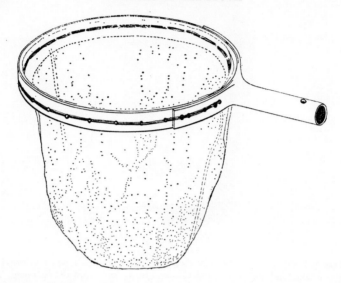

Figure 49. Hand net with bag made of plankton netting. Diameter approximately 25 cm

without diving from the ship, are well suited to provide a picture of the microfauna.

If only a qualitative sample is required, an ordinary triangle dredge often meets the purpose if the net bag is made of strong, coarse sacking. A heavy iron bucket with a strong handle can also be used successfully. It is important for the sampling equipment not to dig too deeply into the sediment, since only the surface layer is inhabited.

Special sampling equipment takes this characteristic of the microfauna into consideration. In sledge dredges (Fig. 50) runners prevent the instrument from sinking into the sediment. The edge is so positioned that up to a determined depth it conveys the sediment into the net bag made of plankton gauze, or the net bag glides freely over the surface of the sediment and a row of small grapnels in front of the

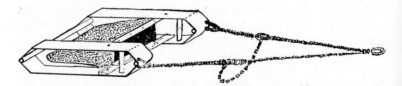

Figure 50. Ockelmann sledge dredge

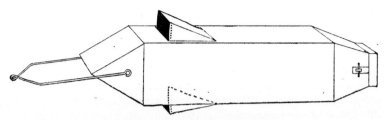

Figure 51. Karling bottom plane

net bag opening cause the bottom deposit to be stirred up and brought into suspension.

Bottom planes (Fig. 51) are instruments of stable construction, successfully used on pure sand bottoms especially.

Dredges are not suitable for quantitative sampling. For this purpose we either use bottom grabs or equipment developed especially for the study of the microfauna. A bottom grab must be constructed in such a manner that the sediment surface remains as undisturbed as possible when the sample is brought up; it must have flaps large enough to allow cutting out of sample cores from the sediment by means of slicing instruments (Figs. 47 and 48), exactly as in the tidal zone. These requirements are met by the van Veen grab (see p. 111, Fig. 43). The Reineck box grab (Fig. 45) can also be used, as can various types of coring tubes. A variety of smaller impact tubes have been developed especially for the quantitative study of the microfauna; in some cases several tubes have been combined in a single instrument. However, it must be remembered that, especially when working in greater ocean depths, only large and heavy instruments can be taken into consideration; these instruments will be used first of all for examination of the microfauna, and can at the same time bring up macrofauna.

2. Sorting of the Material

The organisms of the microfauna are so small that as a rule they cannot be identified with the naked eye, but only with a lens; a binocular dissecting microscope is best for this purpose. The most effective way of examining the sediment under the microscope is in petri dishes marked with lines on the bottom. The dish can be pushed by hand under the lens in such a manner that each part of the bottom of the dish is scrutinized in turn.

When working under a dissecting microscope it is best to use a capillary pipette provided with a small rubber cap. Unfortunately many representatives of the microfauna have adhesive organs which they use not only in order to adhere to grains of sand or to the bottom of the petri dish, but also to attach themselves to the internal wall of the pipette. It then requires much patience to remove the individual animals from the petri dish and to place them either in a solid watchglass, a culture dish or on a slide; sometimes it is better to work with a brush. For the selection of fixed animals one may use either capillary pipettes or a bamboo splinter, an eyelash or a thin insect needle (entomological pin) bent into a hook-shape at the end, in a suitable holder.

The animals of the microfauna are easiest to distinguish from sediment particles when they are alive and thus standing out because of their movements. It is therefore better to examine the samples soon after collection. In samples from cold seas the fauna can be kept alive for a long time if the samples are kept in the refrigerator.

If the sample cannot be sorted while alive, it must be preserved. A weak formalin solution (4%) is more suitable for this than alcohol, because many forms – such as nematodes – shrink under the influence of the alcohol and are subsequently scarcely identifiable. Staining of the sample with Bengal pink has been found useful. This stain colours the cytoplasm a bright red, so that when the sample is being sorted the animals of the microfauna are easily found. Staining with Bengal pink also allows, for example, to distinguish between living Foraminifera and dead shells which no longer contain any cytoplasm and consequently do not stain. The details are described in the report by Uhlig.

Only those representatives of the microfauna which have a strong body wall can be fixed sufficiently with formalin or other fixative to allow for further systematic study. Delicate forms without skeletal material, such as ciliates, Turbellaria, Gastrotricha, etc. usually do not tolerate this treatment. It is therefore important to realize that if a complete view of all the representatives of the microfauna contained in one sample is desired, the sediment must be examined under the dissecting microscope without further manipulations and while the animals are still alive.

There is no procedure which will concentrate the microfauna in the sediment in order to make the study more convenient without damage or loss of one of the animal groups as a result of the procedure. These inadequacies in the method are accepted, however, in order to avoid the expenditure of time required to sort an untreated sediment sample. In the main, concentration procedures can be classified as follows.

(a) Sieving

We have already mentioned that the 'meiofauna' are those representatives of the fauna which will go through a sieve of 0·5–2 mm mesh width but are retained by a plankton net of approximately 50 μm mesh width.

In particular, when the sample sediment is very fine-grained the sieve procedure can be very useful because the sediment particles will flow through the mesh into a suitable hand plankton net (Fig. 49) during rinsing of the sample; consequently, when the sample is later sorted under the preparation microscope, these particles do not disturb. The sieving itself can be carried out very carefully and attentively either by rinsing the hand plankton net with its sediment in the sea or in a container with seawater until the water runs clear; but if delicate specimens are not of importance, one may proceed less carefully by turning a jet of tapwater directly onto the sediment contained in the hand plankton net. If this method is used for only a few seconds and the residue is immediately transferred into a petri dish with seawater, robust specimens such as nematodes and crabs are not damaged, nor are some of the turbellarians and gastrotriches.

When sieving through plankton netting it must be noted that under certain circumstances the mesh width will be larger than the largest diameter of the small, elongated forms, such as nematodes. Although the body curvature of these animals frequently prevents them from slipping through the mesh, one must nevertheless anticipate a certain loss of small forms and juvenile specimens. According to McIntyre (1964), 38% of nematodes slip through a mesh width of 75 μm; however, since these are the small specimens, their weight is only equivalent to 18% of the total weight of the nematodes.

Thus, when the sieve method is used for quantitative studies it is advisable to establish by random checks what part of the microfauna slips through the mesh.

(b) Rinsing

At least among specimens of the microfauna which do not possess solid shells, the specific gravity is generally smaller than that of the coarser sediment particles; when stirred up, the animals remain in suspension longer than do sand grains. It has therefore been found useful to bring into suspension a sediment sample kept in a container, such as a bucket, by carefully stirring it by hand, or – somewhat more energetically – with a spoon. After a few seconds the sand grains settle on the bottom of the container and the supernatant water, which contains the microfauna and other lighter particles in suspension, can be poured through a hand plankton net (Fig. 49); thus, the rinsing procedure is followed by the sieving procedure.

If rinsing is carefully carried out and the sediment is stirred up by only moderate movement, even sensitive animals can be preserved;

however, when rinsing sand samples it must be realized that many animals of the microfauna possess adhesive organs and rapidly attach themselves to neighbouring sand grains when a disturbance occurs. Such animals are not obtained by rinsing or will only occur in a small number. It is therefore advisable to carry out rinsing in two stages: at first carefully, as described, and subsequently more vigorously, either after fixation of the sample with weak formalin or after narcotization of the remaining microfauna with approximately 6% (isotonic) magnesium chloride solution (Mg Cl_2. 6 H_2O).

(c) Environmental deterioration

It has already been mentioned that the representatives of the microfauna on the ocean floor live almost exclusively in the top centimetre of the sediment, as it is only this layer which contains sufficient oxygen in the pore water because of the exchange with the supernatant water. Most of the representatives of the microfauna actively seek out this surface layer, which is favourable to their existence, and we can make use of this behaviour to increase the yield of our sediment samples.

If a sediment sample is placed in a glass container and covered with seawater, the animals of the microfauna migrate into the surface layer of the sediment and can be carefully removed from there by pipette, together with this layer.

The level to which the glass container is to be filled with sediment varies, as does the optimal time interval after which the surface layer is to be siphoned off. Samples consisting of pure sand can be filled into suitable containers up to a level of 10–20 cm; one day later some of the fauna has become concentrated, but even many days later a further concentration of animals takes place at the surface, namely to the extent to which the oxygen conditions in the pore water deteriorate. Samples of muddy sediment and those with a strong admixture of decomposable organic material must be worked up more rapidly since putrefaction processes occur rapidly in these cases.

Samples of algal vegetation can also be worked up successfully by the 'deterioration' method. It has been found useful to keep algae samples in dishes under seawater; by using increasing temperatures together with continued respiration while preventing assimilation and ensuring inadequate water exchange it is possible to cause representatives of the vagile microfauna to migrate out of the interstitial system of the algae and to collect at the surface, especially on the side which is nearer the light. Particularly when working up algae by this method it is important for the collection of the expelled animals to keep pace with the deterioration of the oxygen conditions (Riedl 1953). Sensitive forms appear after as little as one hour and must be siphoned off if they are to remain alive. A certain slowing up of the deterioration and concomitantly a slower appearance of the fauna, spread over a longer period, can be obtained by cooling or by the addition of fresh seawater. It has also often been found useful to place the samples in the dark, since light-sensitive animals then leave the interstitial spaces more readily and can be siphoned off more easily. One must also make sure that the samples are kept free from vibration.

The environmental deterioration method does not always yield a quantitative result. On the one hand it is only usable for the motile forms, which can migrate actively out of the zones of unfavourable living conditions, and on the other it does not deal with the forms which (like some ground water inhabitants) resist oxygen-poor conditions or which are too slow in their migration and die off in the sediment or in the algae. In particular, sensitive species are not capable of penetrating fairly large masses of mud.

(d) Further Methods

The marine substrates to be studied are as different as the animal groups of which the microfauna is composed. Consequently, a large number of further methods has been described and is continually being increased, dealing with procedures for separating the representatives of the microfauna from the substrate. Some of these methods have been described in the report by Uhlig, and we should merely like to mention here that they are suitable not only for protozoa, but also for various other representatives of the microfauna. Thus, many animals as well as ciliates can be driven out of the sediment by the specific introduction of fresh seawater. We must also mention the bait method: if we place a killed *Tubifex* or a piece of meat in a dish where, for example, predatory turbellaria are concealed in the sand, they will shortly collect around the food and can be caught easily. Methods of driving the animals out of the sediment by the use of alternating current are being developed; but more detailed studies are still required to establish the extent to which the many methods of substrate biology can be transferred to the conditions on the ocean floor.

(e) Combination of the Methods

It is almost always advisable to use not one method alone, but a combination of various procedures:

Sand: 1. careful rinsing
               2. environmental deterioration
               3. rinsing by increasingly thorough methods

Mud: A. part of sample: climate deterioration
      B. sieving and rinsing
Algae: A. part of sample: climate deterioration
      B. part of sample: rinsing by increasingly thorough
      methods

REFERENCES

Ax, P. (1966). 'Die Bedeutung der interstitiellen Sandfauna für allgemeine. Probleme der Systematik, Ökologie und Biologie.' *Veröff. Inst. Meeresforsch. Bremerhaven, Sonderbd.* **2**, 15–65.

DELAMARE DEBOUTTEVILLE, C. (1960). *Biologie des eaux souterraines littorales et continentales.* Paris.

HUSMANN, S. (1956). 'Untersuchungen über die Grundwasserfauna zwischen Harz und Weser.' *Arch. Hydrobiol.* **52**, 1–184.

MARE, M. F. (1942). 'A study of a marine benthic community with special reference to micro-organisms.' *J. Mar. Biol. Assoc.* **25**, 517–554.

MCINTYRE, A. D. (1964). 'Meiobenthos of sub-littoral muds.' *J. Mar. Biol. Assoc.* **44**, 665–674.

REMANE, A. (1933). 'Verteilung und Organisation der benthonischen Mikrofauna der Kieler Bucht.' *Wiss. Meeresunters., Abt. Kiel, N. F.* **21**, 161–221.

— (1952). 'Die Besiedlung des Sandbodens im Meere und die Bedeutung der Lebensformtypen für die Ökologie.' *Verh. Dt. Zool. Ges. Wilhelmshaven 1951*, 327–359.

RIEDL, R. (1953). 'Quantitativ ökologische Methoden mariner Turbellarienforschung.' *Österr. Zool. Z.* **4**, 108–145.

SWEDMARK, B. (1964). 'The interstitial fauna of marine sand.' *Biol. Rev.* **39**, 1–42.

THIEL, H. (1966). 'Quantitative Untersuchungen über die Meiofauna des Tiefseebodens.' *Veröff. Inst. Meeresforsch. Bremerhaven, Sonderbd.* **2**, 131–147.

WIESER, W. (1960). 'Populationsdichte und Vertikalverteilung der Meiofauna mariner Böden.' *Int. Revue ges. Hydrobiol.* **45**, 487–492.

PROTOZOA

G. UHLIG

The first serious studies of marine protozoa go back to the turn of the eighteenth century. In those years the interest of well-known biologists was turned principally to the micro-organisms of the plankton, a microcosmic marine life developed into an astonishing variety of forms (Radiolaria, Foraminifera, Tintinnidae). A comparable stimulus for the marine protozoologist occurred only with the discovery of the ocean floor as the habitat of a multiform microfauna. Numerous individual reports have shown that the protozoa of marine bottoms represents both qualitatively and quantitatively an essential element of the meiofauna. However, we are not in a position as yet even to approximate their significance in the cycle of matter or as an organism group in the marine ecosystem.

There has been no lack of attempts to subject the meiobenthonic protozoa to as comprehensive a study as possible. The subtility of numerous forms, their small cell size and the variety of their modes of life require special management and a well-considered procedure. Insofar as qualitative studies are concerned, these difficulties can definitely be overcome, although they require a corresponding investment of time and technical aids. The greater the progress made by stock analysis, which has largely been oriented qualitatively until now, the greater the need for an understanding of the quantitative dynamics of benthonic protozoa.

A complete listing of the methods developed until now would go beyond the boundaries of this book. We intend to point out a few important and proven working procedures, and in the process will emphasise some methods which open up practicable ways for quantitative investigation of the benthonic protozoa.

The ocean floor is populated by representatives of all classes of protozoa: flagellates, rhizopods, sporozoa, and ciliates. To deal with the sporozoa, which exist parasitically, requires special techniques which will not be considered here.

There is as yet little information on the *flagellates* as representatives of the benthonic microfauna (Dragesco 1965). It can be demonstrated that numerous dinoflagellates often dwell in the interstitial spaces of various sand bottoms. Since they will even pass through a fine sieve with a mesh width of 60 μm, they are very difficult to capture by means of the methods commonly used up till now.

Among the *rhizopods*, the Foraminifera form an important constituent of the microfauna. The largest population density was obtained in the region of high-water mark (Phleger and Walton 1950, Richter 1964)., where Richter found that up to 4,600 individuals were distributed over 10 cm³. Which environmental factors influence the distribution of the species has not yet been clearly established. Obviously special importance must be ascribed to the temperature, salinity, water movement and nature of the sediment (Rottgardt 1952, Richter 1964, Lutze 1965). Little is known at present with regard to the distribution of Amoebea, Testacea, and Heliozoa. Because of their small size, their sluggish locomotion and the lack of degrees of differentiation, these forms are often overlooked and are also technically very difficult to obtain.

Numerically, probably the best represented group of benthonic protozoa is that of the *ciliates*. For example, a calculation by Borror (1965) yielded approximately 2,000 ciliates in 36 ml interstitial water from the inter-tidal region. The ciliates have a world-wide distribution. Motile and sessile forms occupy very different substrates. Numerous sessile forms (Peritricha, Folliculinidae, Suctoria) live as epiphytes or epizoa on marine algae or on various members of the macrofauna. Exclusively vagile forms are found in the sand interstitial system or on muddy bottoms. Not infrequently various ciliates also appear as symbionts, commensals or parasites in annelides, arthropods, echinoderms and other representatives of the benthonic macrofauna.

1. SAMPLING PROCEDURE

It is not always easy to make a suitable selection among the numerous methods and instruments which have been developed for the collection of bottom samples. To the degree to which qualitative aspects predominate in a stock study, the investigation of the benthonic protozoa can proceed essentially in the same manner as previously described for the meiofauna (see the report by Gerlach p. 117). If interest is directed predominantly towards the sessile forms, the sampling methods employed for the macrofauna can be useful for obtaining the various substrates. Often the indirect approach, such as the plate method, also yields satisfactory results: several plates are placed in a firm holder which is lowered to the ocean floor after being marked with a buoy (cf. Fig. 66). If the settling sequence is taken into consideration, the rate of coverage sometimes yields information on the quantitative distribution of the sessile animal in the particular area.

Insofar as the local conditions allow it, the free diving method with simple equipment (mask, snorkel, and fins) or with pressurized air equipment should always be taken into consideration.

The use of grab equipment for quantitative investigations requires special care. Some representatives of the microfauna, especially flagellates, numerous small ciliates, as well as acoelomate turbularians, are definite 'swimming' forms. They possess few or no adhesive organs or organelles and can be transported partly actively and partly passively during slight percolation of finely grained or coarsely grained sediments. Consequently grab equipment which allows percolation through the sediment sample while it is being raised, is only poorly suited for this purpose. Apart from this, one must make sure that the equipment is well sealed, for if the interstitcial water drains out some of the motile animals may be lost. The Reineck box grab (cf. Fig. 44), which has already been mentioned in another section, has been found very good in this respect.

For ecological studies, especially those involving Foraminifera, coring tubes or gravity weights are often used. These instruments too make it possible to take an almost undisturbed sample, but if the coring surface has a fairly large diameter (more than 30 cm^2) they become very cumbersome and heavy, so that routine studies with them are scarcely possible. For a fairly low population density Lutze (1965) consequently recommends the use of a modified van Veen jaw-grab (1000 cm^3). The upper part of the grab is provided with flaps which, after removal of the supernatant water by suction, allow direct withdrawal of the surface sediment. According to his data, the wet volume of the removed sediment surface roughly corresponds in cubic centimetres to the size of the surface used in square centimetres.

The use of suitable grab equipment from shipboard can only provide a random sample type of picture of the distribution of the microfauna. Successive sampling within a narrowly confined area is left more or less to chance. Especially in coastal areas the frequent changes result in the development of a variety of sediment structures, which may also overlap. Interpretation of random samples from such areas is accompanied by the danger of incorrect generalizations. This is a broad area of activity within which sports diving interests may be guided into scientific channels. In deeper zones sampling can also take place by means of a television camera (Riedl 1963).

When diving with pressurized air equipment the cutting wedge shown in Figure 52 has been found useful. The longer lateral face of the cutting wedge which is made of V2A steel, can be opened by

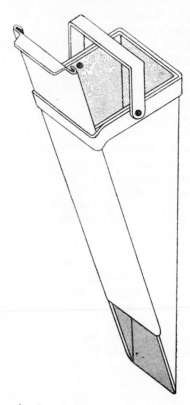

Figure 52. Cutting wedge for obtaining undisturbed sediment samples. The device is particularly useful when used by free divers

means of a slide. The open wedge is driven into the sediment either vertically or horizontally, then closed by means of the slide and withdrawn from the sediment by grasping the handle. After a cover has been placed on it it can be pulled up on board and kept in a suitable container until further examination. The wedge yields almost undisturbed sediment samples (sand, mud) which, when studied, also furnished information on the vertical or horizontal distribution of the microfauna.

2. Sorting of the Material

The methods described in the preceding report (see Gerlach) for the sorting and working up of the meiobenthos material are applicable to a limited extent only to the benthonic protozoa. A certain concentration of the forms is indeed possible, but the methods of sieving, rinsing and environmental deterioration always capture only a part of the single-celled microfauna. In the future too it is scarcely likely that the isolation of the total microfauna will be realized by means of a single, elegant method.

(a) Sessile Forms

The purely sessile representatitives of the benthonic protozoa require a special method in accordance with their nature. The previously mentioned settlement plate method is probably the one best suited to furnish approximately quantitative results. Other substrate samples must be inspected carefully under the dissecting microscope, and, if possible, with an inverted or transmitted light microscope (Amoeba, Heliozoa). In regard to inhabitants of calcareous shells (Mollusc Shells; Sea Barnacles, Serpulids, etc.), the animals (such as Folliculinidae) may be isolated after preliminary fixation by dissolving their shells in diluted hydrochloric acid. Otherwise the animals must be carefully separated from the substrate by means of a fine glass or steel needle.

In the wider sense the Foraminifera too can be grouped with the sessile microfauna. However, they are definitely capable of carrying out small movements – a characteristic which is particularly useful in the tidal zone (Richter 1964). If one wishes to obtain the sand-inhabiting Foraminifera, the sieve method can be used successfully. It should be noted, however, that empty, i.e., dead shells are also sorted out during this procedure so that a false picture of the actual stock may be obtained. The slime should be washed from the sediment samples which have been fixed in 50% alcohol by using fine sieves (0·07 or 0·1 mm) and the residue should be immersed for some 10 minutes in a 0·1% aqueous solution of Bengal pink (Rose bengal, a higher homologue of eosin). The excess dye must then be washed out again carefully (approximately 10-15 minutes). Since only the animals which contain cytoplasm take on the dye, i.e., those still living before fixation, they can be readily distinguished from the empty shells (Walton 1952).

Recently Lutze (1964) advised an abbreviated procedure: 1 g Bengal pink is dissolved in 1 litre methylated spirits and this stain-fixative mixture is poured over the sediment sample immediately after its collection (2 parts sample volume to 1 part alcohol). Fixation and staining are carried out in wide-neck plastic bottles which are well shaken for 1-2 minutes. After obtaining the wet volume, the slime is washed out of the samples with lukewarm water over a 0·1 mm sieve and the residue is dried at 80°C in the drying chamber.

Any suspended material which may be present (organic detritus, chitinous, or plant remains, etc.) can be poured off previously and dried separately. After drying, the Foraminifera shells can easily be concentrated by sprinkling the sediment sample into a container filled with carbon tetrachloride (CCl_4 flotation). Whereas the sediment sinks, the shells remain at the surface and can be filtered off.

(b) Motile Forms

Since a large number of the smallest representatives of the meiofauna pass through the usual sieves together with the finest particles, filtering and washing procedures are not very suitable, at least for quantitative studies of the motile and flexible protozoa. Even careful management of the washing method cannot prevent damage or destruction of sensitive and fragile forms; the larger types of ciliates are particularly subject to this. There is no doubt that the *environmental deterioration* method (see p. 127) represents an elegant concentration procedure; however, the problem of separating the concentrated animals from the sediment has not yet been solved with this.

For a *qualitative* survey it is sufficient – following climate deterioration – to remove the upper sediment layer from a fairly large sample, to rinse it carefully and to decant the supernatant seawater into a petri dish.

For the *quantitative* separation of the microfauna in general, and especially for the motile protozoa, a method which has already proved itself in many cases was recently developed (Uhlig 1964, 1966). In this new procedure the sediment sample to be studied is placed in a Trovidor tube which is closed on one side by a tightly stretched nylon gauze (Fig. 53). The mesh width of the gauze must be adjusted to the average grain size of the sediment. The filter tube is filled up with crushed seawater ice. A cotton wool layer is placed between the sediment and the ice to prevent excessively sudden cooling of the top sediment layer. The filter tube is suspended in a culture dish filled with some filtered seawater in such a manner that the nylon gauze is just barely in contact with the water surface. If the water level rises because of the addition of the melted water, the filter tube, which is suspended from a crank, is raised or the culture tray, which is set on a sinking stage, is lowered. The latter arrangement is preferable on board ship.

Under the influence of the melting water which slowly percolates through the sediment sample, the motile microfauna abandon the sediment actively or passively and fall undamaged, without any significant contamination, into the prepared culture dish. The essential factors for driving out the microfauna are the initial rapid

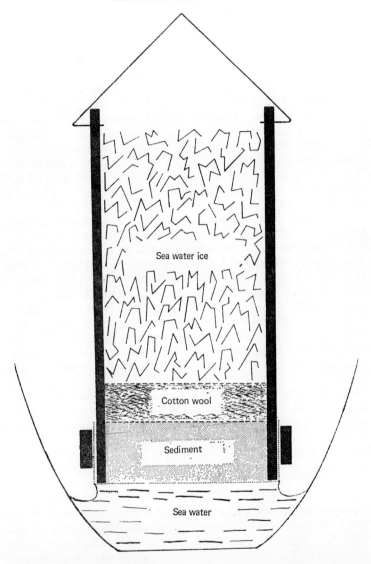

Figure 53. Extraction procedure for driving out the microfauna from fine-grained to coarse-grained sediments by means of seawater ice

increase in salinity, followed by the gradual decrease in salinity of the melting water, and the weak vertical current of the melting water in the sediment.

The extraction method can be considerably modified by suitable alterations to adjust it for specific problems. Thus, the extraction can also be carried out with concentrated seawater (50-60 ‰). Instead of the cotton wool layer we insert a Seitz filter disc (K5), which ensures a percolation rate adjusted to the melting water. However, the yield is generally somewhat slower than when sea-water ice is used.

At present studies are in progress to attempt to combine the extraction method with the membrane filter method. If this combination is found practicable – which the initial positive results suggest – it will be possible to analyse quantitatively several sediment samples in a relatively short time. The specimens, which have been fixed on membrane filters, stained and embedded, can be kept as illustrative material.

The extraction method may exceed in effectiveness all the methods used previously for the isolation of the motile benthonic protozoa. It has been shown that especially resistant groups of the remaining meiofauna, such as the nematodes, are only partially collected. However, it is easily possible to isolate all the animals which were not extracted, i.e., the larger animals as well, after the extraction with the help of the previously described sieving or rinsing methods.

The extraction method is at its most effective only in fine-grained to coarsely-grained sediments; but can also be applied to mixed sediments. The substrate samples for testing must have a capillary structure. The method does not work on pure mud or ooze samples. Since the microfauna populates only the uppermost, loosely-deposited layer in such sediments, this layer can be removed and distributed on a sterile, fine-sand sample in the filter tube. In this manner satisfactory qualitative results can be obtained even from mud samples.

3. PROCESSING OF THE MATERIAL

After careful separation of the benthonic protozoa from the substrate has been carried out, the protozoologist begins his proper and often time-consuming work. Only a few protozoa can be identified immediately on the basis of their conspicuous and typical differentiation characteristics. Studies carried out in recent years have shown that until now only a small part of the benthonic protozoa has been dealt with systematically and taxonomically. According to Corliss (1961), until 1960 a total of some 6,000 ciliate species was described. On the basis of an estimate by Dragesco (1960) the sand interstitial system contains far more than 600 different ciliate species, of which 400 types are known to science at present. In the other protozoa groups the ratio is probably even more unfavourable.

The majority of the larger representatives of the meiobenthos (Hydrozoa, Turbellaria, Gastrotricha, Nematoda, Polychaeta, Copepoda, etc.) can be identified in the living state, and generally also in the fixed state, without the use of expensive preparation techniques. Within narrower limits this is also true for shelled or testaceous protozoa. The great majority of meiobenthonic protozoa, however, can only be definitely identified when – after careful preliminary sorting of the material – live observations and subsequent preparation (staining) of the animals complement each other meaningfully. It must be emphasized that ultimate elucidation of the systematic classification of a form can generally only be achieved through culture experiments (see the report by Grell, p. 248). This indicates that *it is useless to attempt to ascertain the stock of motile unshelled protozoa from fixed sediment samples.*

The relevant literature offers a broad selection of various experimental procedures for processing the protozoa, so that further discussion may be omitted here. However, it seems advisable to complement these by pointing out a few modern techniques in particular.

A chamber developed by American protozoologists has been found extremely useful for the live observation of micro-organisms. This intelligent design, produced for the last few years under the name of 'Roto-compressor' by the Biological Institute of Philadelphia (2018 N. Broad Street), makes it possible to retain even the smallest, very motile animals without damage and even to lay the animals in a particular way, as desired. In the closed chamber system the animals can be kept for several hours, microscopically examined (high-power immersion objectives included), drawn and photographed.

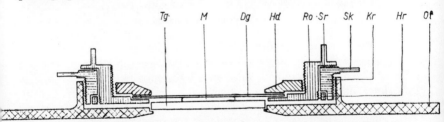

Figure 54. 'Roto-compressor' chamber for immobilizing motile micro-organisms. This device allows live observations for a number of hours. Dg = coverglass, Hd = support for the coverglass, Hr = support for the rotor, Kr = compressor ring, M = medium (object), Ot = slide, Ro = rotor, Sk = peg in the compressor ring, Sr = peg in the rotor, Tg = supporting glass. (With kind permission of the Biological Institute of Philadelphia)

Figure 54 shows the construction of the chamber in cross section. A round disc made of supporting glass (*Tg*) is glued into the metal stage (*Ot*), which is provided with a screw thread. The screw thread raises the compressor ring (*Kr*), in which the rotor (*Ro*) glides like a turntable on a film of grease. The coverglass (*Dg*) is held in the rotor by a support ring (*Hd*). A further support at the base of the rotor (*Hr*) connects the rotor with the compressor ring.

The object is placed on the supporting glass with a small drop (*M*) of seawater. The compressor ring (*Kr* + *Ro*) is slowly turned into the screw thread of the stage until the object under examination between the supporting glass and the cover glass is lying straight. If the rotor is held fast at the pegs, which have been inserted (*Sr*), and the compressor ring at these pegs (*Sk*) is turned further, the object can be pressed vertically without producing a shearing effect. By skilful manipulation with the rotor and the compressor ring it is possible to place the object in the desired aspect. After the chamber is opened the animals can be removed and further treated.

The fewer the differentiating characteristics shown by a particular species, the more demanding and the more difficult its identification. The exact species diagnosis must be based on as many criteria as possible (life cycle, ecology, modification capacity, cytology, cytochemistry, sex, etc.), which can generally only be established through culture. Unfortunately this method can only rarely be put into effect in general practice. It is therefore important to explore all possibilities by means of modern microscopic procedures (phase contrast microscopy, fluorescence microscopy, interference microscopy) and suitable staining techniques which may serve to clarify, bring out and contrast the particular differentiation characteristics.

The introduction of the *silver impregnation technique* has been found exceedingly useful for the study of ciliate systematics. The silvering of certain 'argentophile' structures (basal bodies of the cilia, cortical fibrillary systems, etc.), is based on an accumulation of silver ions which are precipitated as an insoluble deposit by reducing substances and which cause the structures to turn black or brown.

The method originally developed by Klein as a dry procedure was later supplemented by a wet procedure by Chatton-Lwoff. Unfortunately these metal stains do not lead to equivalent results in all ciliates. The frequently modified Chatton-Lwoff method fails in approximately 30% of the known ciliate species (numerous holotrichous and heterotrichous ciliates). Most recently Dragesco (1962) succeeded in transferring a modified form of the Bodian-protargol – or silver proteinate technique, known from neurohistology, to protozoa. With this procedure it is possible to bring out clearly the so-called silver-lines system even in the previously unsuccessful cases. Since the impregnated structures are not always identical when the two latter methods are used, the use of both procedures is always indicated in questionable cases.

The numerous new reports brought out in recent years show that in the future it will not be possible to omit the practices of silver impregnation. The Chatton-Lwoff technique has only rarely been referred to in standard texts in the German literature, and the protargol technique even less so. It would therefore seem useful to describe briefly the staining process for both procedures here.

(a) The Chatton-Lwoff impregnation method
1. Fix the specimens for 2 minutes in Champy, drain off the fixative.
2. Wash twice in Da-Fano solution (marine), rinse in distilled water.
3. Place specimens on a grease-free slide, drain off water.
4. Add 1-2 drops warm (40°C) 10 % gelatine solution, followed by 2 drops seawater; place this on the animals, mix and drain off rapidly, before the gelatine sets (gelatine layer of approx. 200 μm thickness).
5. Cool slide until the gelatine has set and place in refrigerator for 5 minutes in cold 3 % silver nitrate solution (keep in the dark!).
6. Rinse slide with cold distilled water and place in shallow basin with distilled water; water level 2-3 cm.
7. Irradiate the slide 10-20 minutes with the UV lamp (longer in sunlight) until the objects are sufficiently dark (control!).
8. Rinse with cold distilled water, alcohol series, embed.
   For points 5-7 always keep below 10°C!

(b) The protargol technique (cf. Tuffrau 1967). The procedure described here represents a modification of the method recommended by Dragesco and other investigators. It is based on as yet unpublished studies by the author which definitely indicate that the mode of action of protargol impregnation is dependent on the pH.

(A) FIXATION OF THE SPECIMENS
1. Smear a thin layer of albumin glycerine on a grease-free slide, allow to dry and draw the slide through a flame three to four times (no browning!).
2. Fix the animals in Bouin, Champy or saturated sublimate solution (different fixatives are also possible), then rinse with distilled water.
3. Place the animals on the prepared slide and drain off as much water as possible (the animals should not be dry!).
4. Place one drop diluted (approx. 50 %) clear egg white on the animals, mix somewhat and drain off as much egg white as possible.

5. Allow a small drop of 1:3 formol (40%)-alcohol (96%) to fall directly on the egg white from approximately $\frac{1}{2}$ cm height, fix for 1 minute, then dip in a container with formol-alcohol (1:3) and post-fix for one hour.

6. Transfer the slide through 70% alcohol into water.

(B) BLEACHING

1. 5 minutes in 0·5% potassium permanganate solution, followed by rinsing in distilled water.

2. 5 minutes in 0·5% oxalic acid, followed by rinsing in distilled water.

(C) IMPREGNATION

Protargol: The most useful has been found to be 'Proteinate d' Argent' (Establ. Roques, 80 Rue Ardoin, Saint-Quen – Seine).

Preparation of the solution. Place distilled water in a shallow basin and sprinkle on the protargol powder (do not stir!). Allow to stand until completely dissolved. Impregnation is carried out with a 1% protargol solution, which is always prepared fresh if possible. The pH of the solution should be between 8·3 and 8·6. Old or used solutions have a pH of less than 8·0, and are no longer usable.

1. Several copper sheets of slide size are placed in a stand (slide stand) alternating with the slides. Approximate weight ratio of the protargol solution to the copper = 12:1 (with the copper the size of the surface is of decisive importance!).

2. The stand is placed in a suitable container and protargol solution is poured in until the slides are entirely immersed. A gentle turbulence is produced in the container by means of a magnetic stirrer.

3. Preferably with the help of a pH meter or with special indicator paper, impregnate until a pH of 8·0 to 7·8 is attained (approximately 15 to 30 minutes at 20°C).

4. Rinse the slide well in distilled water.

5. *Develop* for 5 minutes in 1% hydroquinone + 5% sodium sulphite or any photographic fine-grain developer desired.

6. *Gold toning* briefly dip the slide in 1% gold chloride and rinse rapidly, then place in 2% oxalic acid for 5 minutes.

7. *Fix* for 10 minutes in 5% sodium thiosulphate, then rinse thoroughly in distilled water.

8. Follow, if necessary, by nuclear staining (Feulgen, etc.); otherwise through alcohol series to embedding.

The bleaching described (B1, 2) is particularly advisable for pigmented forms or those of low transparency, but can otherwise be omitted. The gold toning (C6) often increases the contrast of the metal colouring, but is not absolutely necessary.

REFERENCES

ARNOLD, Z. M. (1954). Culture methods in the study of living foraminifera. *J. Paleontol.* 28, 404–416.

CORLISS, J. O. (1961). 'The Ciliated Protozoa.' *Characterization, Classification and Guide to the Literature.* Pergamon Press, London-New York. 310 S.

— (1961). 'Fixing and staining of Protozoa.' LACY, D. S., and S. O. PALAY, *The microtomist's Vademecum* (Bolles Lee). 12 Aufl. London.

CUSHMAN, J. A. (1948). *Foraminifera. Their classification and economic use.* Harvard Univ. Press, Cambridge, Mass., U.S.A. 605 S.

DRAGESCO, J. (1960). 'Les Ciliés mesopsammiques littoraux (Systématique, morphologie, écologie).' *Trav. Sta. biol. Roscoff* (N. S.) 12, 1–356.

— (1962). 'L'Orientation actuelle de la Systématique des Ciliés et la Technique d'Imprégnation au Protéinate d'Argent.' *Bull. Microscopie Appl.* 11, 49–58.

— (1965). 'Etude cytologique de quelques Flagellés mésopsammiques.' *Cah. biol. mar.* 6, 83–115.

FJELD, P. (1955). 'On some marine psammobiotic ciliates from Drøbak (Norway).' *Nytt Mag. Zool.* 3, 5–65.

KAHL, A. (1933). 'Ciliata libera et ectocommensalia.' GRIMPE, G., und E. WAGLER, *Die Tierwelt der Nord- und Ostsee, Lief.* 23 (*Teil II c₃*), 29–146. Leipzig.

LUTZE, G. F. (1964). 'Zum Färben rezenter Foraminiferen.' *Meyniana* 14, 43–47.

— (1965). 'Zur Foraminiferen-Fauna der Ostsee.' *Meyniana* 15, 75–142.

PHLEGER, F. L., and W. R. WALTON (1950): 'Ecology of marsh and bay Foraminifera. Barnstable, Mass.' *Amer. J. Sci.* 248, 274–294.

RICHTER, G. (1964). 'Zur Ökologie der Foraminiferen I.' *Natur u. Mus.* 94, 343–353.

RIEDL, R. (1963). 'Probleme und Methoden der Erfoschung des litoralen Benthos.' *Verhandl. d. Dt. Zool. Ges.* Wien 1962, *Zool. Anz.* 26, Supp.-Bd., 505–567.

ROTTGARDT, D. (1952). 'Mikropaläontologisch wichtige Bestandteile rezenter brackischer Sedimente an der Küste Schleswig-Holsteins.' *Meyniana* 1, 169–228.

TUFFRAU, M. (1967). 'Perfectionnements et pratique de la technique d'imprégnation au protargol des Infusoires Ciliés.' *Protistologica* 3, 91–98.

UHLIG, G. (1964). 'Eine einfache Methode zur Extraktion der vagilen mesopsammalen Mikrofauna.' *Helgol. Wiss. Meeresunters.* 11, 178–185.

— (1966): 'Untersuchungen zur Extraktion der vagilen Mikrofauna aus marinen Sedimenten.' *Verh. Dt. Zool. Ges. Jena* 1965, *Zool. Anz.* 29, Suppl.-Bd.

WALTON, W. R. (1952). 'Technics for recognition of living Foraminifera.' *Contr. Cushman Found. Foraminifera Res.* 3, 56–60.

E. Fungi

W. HÖHNK

Many fungi live in the water. We have known this for some 200 years with regard to freshwater, and have recently discovered it with regard to the ocean. The presence of organic substances and of the minimum materials, organically bound nitrogen and phosphorous, allows us to anticipate the presence of fungi in all marine areas, from the coast to the depths of the ocean. Sample collections have confirmed this assumption.

Marine mycology has forged forwards from the coast in width and depth. In our brief discussion we shall follow in the same direction. We begin in the coastal area and treat the algae as long-known fungus carriers, then the shore plants as terrestrial factors imperilled by fungi, and finally wood as a commercial material which is continuously losing in value. The sampling methods described can also be applied to other substrates and values. The collection of oceanic substrates from the bottom, its fauna and samples from free water follow each other in the discussion, which touches on both qualitative and quantitative aspects.

Marine fungi can only be detected with the help of optical aids. Our discussion consequently ranges from the collection methods under natural conditions to the identification of the fungi in impure cultures. Living material is necessary for our studies, as we wish to understand the life cycle of the organisms obtained. Only in special cases is preserved material satisfactory.

1. METHODS OF STOCK ANALYSIS AT THE COAST

On rocky or stony coasts it is above all the *algae* which are attacked by fungi. They are collected at ebb tide in the tidal belt or from on board ship, or when diving with face mask and snorkel or aqualung; or they are found floating on the sea. Small forms and parts of larger ones are collected and red as well as brown or green algae are included.

Delicate or small pieces should be picked and handled carefully by means of forceps; all substrate should be placed immediately in one or two-litre receptacles containing water from the original site or its neighbourhood. Discoloured or spotty substrate portions often indicate a poor yield. Only a few algal fronds should be placed in each container and should be kept in diffuse light and at a water temperature similar to that of the original site until the subsequent microscopic inspection. We transport 200 ml specimen tubes in a water bath covered with damp cloths and empty them into well-aerated aquaria with a water temperature of approximately 17°C. Results were equally good when 3 to 5 small algal fronds were placed in 2-litre glass jars with large mouths covered with wide-mesh gauze; these jars were then lowered from the *landing stage* to the bottom (water depth approximately 2 m) and kept there.

Microscopic examination of the cut portions placed in a petri dish or on a slide by means of forceps must begin immediately because of those fungi which are sensitive to changes in environment. The more tolerant fungi will later predominate on the remaining pieces. If thicker substrates are involved, sections can also be prepared. Fungi were also detected in the containers of *Ascophyllum nodosum*, for example.

Apart from obligatory or facultative parasites, the samples also conceal saprophytes. To concentrate these we add bait such as *Pinus* pollen, pieces of chitin, small parts of boiled algae or boiled cellophane, parts of plant seeds, etc. They often show a fungus growth after a few days.

Approximately 50 types of Phycomycetes have already been found in and on algae, as have members of the uniflagellates and of the biflagellates, and almost as many Ascomycetes, Fungi imperfecti and several Myxophycetes.

At many places along flat coasts the *Phanerogams* hang over the edge of the foreshore or the high water line into the saline water.

A phenomenon originally ascribed to the fungi had a catastrophic effect in this field: the stocks of *Zostera*, which were widely decimated on both sides of the north Atlantic, became diseased and disappeared; this brought about both the loss of a material used for food and other purposes and the conversion of the extensive seagrass fields with their rich organic ground and rich fauna into highly dangerous quicksand.

This situation stimulated the study of the fungi living on underwater plants. The following have already been identified: *Phragmites communis, Zostera marina, Z. nana, Z. hornemanniana, Ruppia maritima, R. sortellata, Zannichellia palustris, Z. Spp., Diplanthera wrightii, Posidonia oceanica, P. caulini, Halophila ovalis, H. baillonis, Thalassia testudinum, Spartina alterniflora* and *Juncus maritimus.* The list can easily be extended, which should be done with the following procedures in mind.

The cut central stem or leaves, which are less sensitive than the frequently delicate algae, can be transported in plastic bags with water from the original site or which are moist inside. Careful handling with regard to sterility should be kept in mind.

In the laboratory the leaves are selected and classified after spotting and staining, then cut into small pieces; they are then cleaned by scraping off the epiphytes and solid particles with sterile instruments (knife, scalpel, or lancet), and rinsed in sterile water.

Apart from microscopic examination *in toto* and in sections, which reveals the mycelium and the fruit bodies in the tissue, small pieces of the attacked tissue are pressed into or placed in sterile nutrient media (agar base or fluid). The nutritional requirements of fungal organisms vary greatly and therefore require different solutions and nutrient bases.

A profitable substrate for fungal samples is *wood*. On the coast it is used for groynes, for landing stages, and lighthouses, and is also available from wrecks which have been brought up.

Frequently the outer layer – a few millimetres thick – of the parts of the squared timber, round logs or planks covered by water is penetrated by fungal mycelium. Using a knife, we cut off splinters or small blocks or we remove a core measuring a few centimetres from the substrate by means of an Auger drill and transport it in sterile bottles or plastic bags which have been moistened on the inside. These pieces are then placed in petri dishes (10 cm diameter) or shallow covered glass pots and kept underwater in sterile membrane-filtered water from the original site. If kept for a fairly long period the water must be renewed at intervals of a few days, or the material must be placed in a container provided with running filtered water.

Even when inspected for the first time with the binocular microscope at high magnification or with the microscope at low magnification a spotty growth of hyphae or a velvety thick stock of conidia may be apparent; spherical or almost spherical fruit bodies may be lying open in the substrate, or the drainage neck of the fungal fruit bodies – which were formed in the wood below the substrate – may be found penetrating through the substrate cover. If the wood surface does not show a fungal growth immediately, it will appear later within several weeks. The possibility of contamination by aerospores should be avoided or kept to a minimum during inspection.

Considerable attention is being directed to the damage done to the commercial material, wood. The problem is not limited to the detection and identification of the organisms, but it is also attempted to establish effective prevention measures on the basis of the physiological behaviour of these fungi towards external factors at the site of origin and towards impregnation materials and protective coatings.

For these purposes test blocks, sections of plank or columns (with a lateral length of a few centimetres) of various types of wood, treated by different preservation methods and of various ages are suspended individually or next to each other in a frame at selected places in the water. Such places are, for example, harbour walls, the buoys at the channel boundaries with their anchor chains, light towers, and light ships. The wood is collected after a number of days, weeks or months, according to the growth at the site or according to a previously chosen time interval. They are transported in sterile glass tubes, glass containers or moist plastic bags, and subsequently treated as described above.

The settling rate, the increasing influence of the mycelium, the association of the fungi and their changes as a result of various external factors then become evident.

The fungi growing in the wood usually do not settle at all or only settle after a time on the bait added to the impure cultures, i.e., pieces of hemp thread, straw scraps, hemp seeds, pieces of chitin and pollen. Occasionally ascospores of pycnospores floating at the water surface also show the presence of wood fungi.

In some cases these ascomycetes and the fungi imperfecti of the wood can be determined without a pure culture. If, however, the pure culture is required for further studies, a solid nutrient base is used.

2. QUALITATIVE DETECTION OF OCEANIC FUNGI

The hosts and substrates for fungi which we have considered will only be encountered by us in the expanse of the sea as random floating material and are therefore not suitable for establishing the presence of an oceanic mycoflora. For this purpose we use the fungal saprophytes from the *ocean floor* and from the *free water*. A research ship provided with a laboratory for microbiological work is required for this purpose.

We preferred to begin with *bottom samples* for the first studies because the bottom is richer in organic substance than the water and its cellular structure offers favourable settling possibilities for the fungi, which appear there in concentrated form.

Several models of bottom grabs are used for lifting the bottom samples. We have preferred the van Veen grab (Fig. 55; cf. also Fig. 43). Its two halves are provided with an angle iron as a lever for each half jaw bucket and move around a common axis which is set in the upper edge of the jaw bucket. Before use the angle irons are pressed apart laterally, which open the buckets. The instrument is maintained in this position by a catch bar. It slides down into the

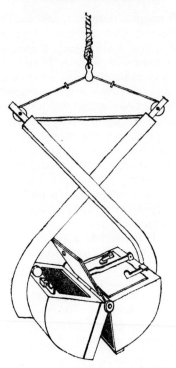

Figure 55.   Van Veen bottom grab

water along a wire cable from which it is suspended. When the bottom has been reached, the grab sinks into it. As soon as it has come to rest, the catch bar disengages. When being raised the levers approach each other until the box is closed, and the grab remains in this position until opened on board on the worktable or on deck.

If the wind and currents are favourable, i.e. weak, the instrument can be used at depths of several thousand metres; for example, it was used in the Indik in February 1965 at a depth of 3,450 m. To avoid loss of time caused by faulty grabbing as a result of insufficient cable length, the use of a pinger is highly advisable at greater depths. The pinger sends out continuous electrical impulses underwater and consequently its route can be followed exactly on the shipboard recorder. If the pinger indicates that the bottom is about to be reached, the grab suspended a few metres below the pinger will also close.

If the bottom is muddy, the grab will be full; if it is sandy or clayey, little sediment will have been taken. In that case the space above the sediment will be filled with water from the depths. Both are useful. The water is scooped out with a sterile beaker and poured into sterile bottles. After passing through a membrane filter it is used as water from the original site for later work with the samples. We remove the bottom samples through the windows of the grab bucket with spoons (standing ready in alcohol) or directly with the sampling beakers when the grab is open; usually four samples per original site are taken.

Sampling beakers made of plastic material and with a 45 or 90 ml capacity (Fig. 56, *b* and *c*) have been found excellent for this purpose. In order to slow down – if not to prevent – the loss of moisture, it is advisable to seal off the projecting cover with 'Vaseline'. In occasional test series the samples in the plastic beakers have continued to yield a fungal harvest after four years, and this does not represent the upper time limit as yet.

Apart from the station work there is often sufficient time to make preparations from selected sites or from all the original sites. Using a spatula, we remove a limited amount of sediment, approximately 5–15 ml from each sample beaker of the station, throw it into a sterile 100 ml Erlenmeyer flask, add approximately 15–20 ml sterile seawater (which has been brought along) and add a knife-tip of sterile *Pinus* pollen. The flask is then closed with a cellulose stopper and a small rubber cap (Fig. 56 *a*).

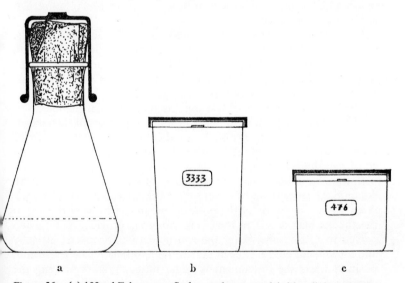

a   b   c

Figure 56. (*a*) 100 ml Erlenmeyer flask as culture vessel (with cellulose stopper and rubber cap). (*b*) and (*c*) beakers made of plastic (with 90 and 45 ml capacity respectively)

After 12–14 days the flask is due for examination. Attacked pollen grains are removed from the flasks by means of a platinum-wire loop, placed in a drop on a slide, and the fungal growth on these baits is examined under the microscope at medium magnification.

What previously was surprise is now expectation. The fungi are found, generally of low growth; later taller growth appears. It is advisable to take microscope photographs immediately, as experience has shown that when the baits are first attacked, forms are visible which it is later useless to look for. This fact must continue to be tolerated at present because on an expedition the time is regulated by a work programme and is not sufficient to carry out isolation and cultivation trials. The photographs also help to clarify whether a double or even a complex succession of dominant forms occurs in one flask over a protracted period.

Near Greenland and Iceland the bottom grab has brought up a relatively rich *secondary catch*, together with the substrate, from depths of up to 500 m. Fragments with animal growth, part of corals and also sea cucumbers, sponges, starfish, etc. were brought up from stony sediments. All yielded fungi.

A good subject for work is the sea cucumber. After approximately one hour of captivity in a closed preserving bottle without water it ejects its intestine. We cut this off and place its contents dropwise on pouring plates, on a solid nutrient base for smears and in a nutrient fluid for concentration.

From the sponges (*Geodia gigas*) we took small, young individuals, held them with the left hand above an agar dish, cut them through and allowed drops of the body fluid to fall on agar plates or into reagent bottles with nutrient fluid; we also cut out small portions, scraps or slices and pressed them into agar bases.

The intestinal contents of fish were used in a similar manner.

The viable preparations are kept in the refrigerators of the laboratory at $\pm$ 4 or 8°C. They contain mixed cultures. The preparation of pure cultures is generally the work of the Institute. The yield usually consists of higher fungi, particularly yeasts.

Bottom sampling from the *deep seas* of the often stormy north Atlantic was carried out by means of the Pratje impact corer (Fig. 57). A nozzle is screwed on at the bottom and an end piece at the top of the tube, which is 2·5 m long. The nozzle is provided with a flap curved to fit the tube. After the instrument has dropped vertically and a sediment core has been taken, the flap falls into a horizontal position while the instrument is being raised, thus preventing the slipping and loss of the bottom core. The closing mechanism at the top has a double-walled construction which consists of a box which rises when the instrument is lowered and is pressed downwards by the water pressure into a bed with sealing rings when the instrument is raised. The water streams through from bottom to top while the tube is falling; but the reverse stream, from top to bottom, is blocked

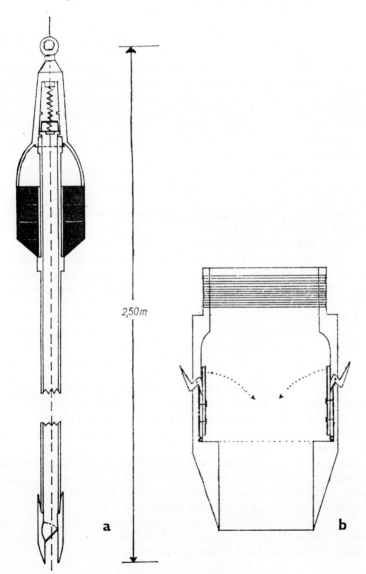

2,50m

a

b

Figure 57. Impact corer: (*a*) Pratje model; (*b*) the nozzle with new semicircular lock

when the tube is rising, and the core is firmly enclosed. The weights increase the falling rate and regulate its control; they are placed at the top, so that they will not slow down the penetration into the soil.

The impact tube is usually dropped at the rate of 1 to 1·7/sec. If the bottom is muddy or soft, the sediment core is long; if it is earthy and solid, the core is shorter; if the bottom is sandy, there will be no sample or only a small one.

The manufacturer includes a set of three brass tubes. They are slipped into the steel tube and are arrested by support rings at the top and bottom. If samples from only certain parts of the core are desired, instead of the round tubes their longitudinal halves are used; when they arrive on board after the core has been taken, the top half is removed and the substrate is collected from the desired spots with a sterile spoon or spatula.

Instead of brass tubes plastic tubes are also used. They facilitate the work considerably. The length of the core is immediately visible. The topmost sediment layer, which is always important for these investigations, can be removed undisturbed, as can samples of stratification sites and of certain depths in the core.

As a result of technical problems with the locks, it has happened on some occasions that while the instrument was being raised, the core slipped out above the water surface, before arriving on deck; as a result only its upper portion or none of the core was usable. In the tightly organized timetable of an expedition this means the loss of a station.

Consequently the manufacturer* has now provided the headpiece with a stronger spring, and has also constructed a semi-circular lock for the nozzle. This yields better results; the impact tube also works in sand. The tube diameter is somewhat larger.

For further processing the sediment removed is pressed into sample beakers in the laboratory or on board, using a stamp with a diameter which is somewhat smaller than the inner diameter of the brass or plastic tube; the beakers are immediately covered and remain closed until required for use.

We have found the impact tube to work reliably; only among the Faroe ridges were faulty attempts unavoidable. The rocks there are bare, and one is fortunate to find a trough or recess with a few decimetres of eroded soil among the rocks.

However, the essential part of the ocean is the salty water. It contains fewer nutrient substances than the substrate, and their distribution is not regular. The nutrient substances occur more abundantly in

---

* Firma Hydrowerkstätten, Kiel-Hassee, Uhlenkrog 38. This company will also supply bottom grabs and the Weyland bottles described later

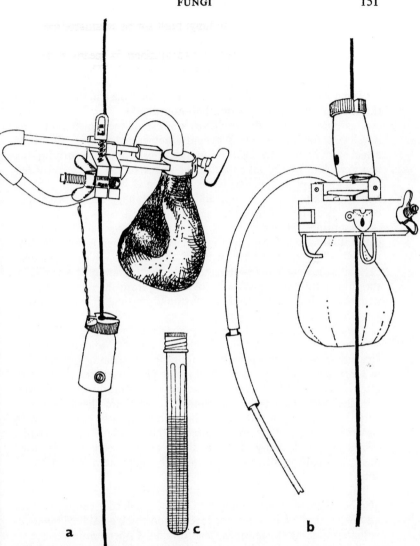

Figure 58. Cobet-Weyland water bottle: (*a*) closed and empty; (*b*) open and full; (*c*) screw-top tube with capillaries

turbid water near the bottom and in plankton patches than in clear water. However, because turbidity and plankton thicken, pull apart, divide and dissolve in the same manner as clouds in order to repeat a similar pattern elsewhere, the fungi must also be obtainable from unclouded water.

Water samples are collected by hydrographers by means of the Nansen or tipping bottle (comp. Fig. 4 and see p. 27). These bottles remain open while being lowered. It is therefore not certain whether the fungi come from the depths shown by the cable length or whether they have been pulled up from shallower depths.

This uncertainty is dealt with by the use of the Cobet bottle, as developed further by Dr Weyland (Bremerhaven) (Fig. 58). It is attached to a wire cable by its mounting, which simultaneously holds a neoprene ball of 150 ml capacity. The neck of the ball is provided with a short glass tube, which is lengthened by a rubber tube to approximately 30 cm. Its free end has a capillary whose projecting end has been heat-sealed. During use this capillary is placed across the mounting and held by loops. When the messenger slides down the cable it breaks the capillary and releases another messenger suspended below the mounting, which then goes through the same process with the next lower bottle. After the destruction of the capillary, the rubber tube stretches and moves away from the mounting. The neoprene ball then fills relatively fast, and particularly so because it has been pressed in before use.

The balls with rubber tubes are prepared before the trip. They are sterilized for 30 minutes in autoclaves at 120°C. and the internal space is freed of organisms by evaporating in them a few drops of water which were previously sprayed in. They are taken on the trip after being sealed with cellulose and sheets of tinfoil. The capillaries are filled with water through a cannula; several are placed in a row, with their opening at the bottom, in a screw-top tube half filled with water. In this way they are sterilized in the autoclave with the balls. We always took many of these along, as they can only be used once. The balls can be used repeatedly, being re-sterilized on board ship.

This Weyland bottle can be used in a series on the cable and can also be included in the tipping-bottle series; indeed it can be used for any procedure which does not involve too thick a cable. This saves valuable time at the stations. The original Cobet bottle cannot be used in a series. Our instrument worked reliably in all cases, its operation is easy and its durability is good.

The water which has been brought up is immediately emptied into several 100 ml Erlenmeyer flasks to which bait (usually *Pinus* pollen) is added to concentrate the fungi. Since clear seawater is poor in fungi and faulty preparations are therefore not infrequent, it is advisable to divide the total contents of the neoprene ball into as many flasks as there are 20 ml in the contents. The baiting must be considerably longer than 14 days – until an epidemic-like frequency of the fungi has been attained in each drop sample.

The usual large nets are used for *plankton hauls*. It must be remembered, however, to remove the required plankton before the washing with which planktologists rinse the plankton from the net walls takes place. If this is not done, pollution originating from shipboard can result. Only a few plankters should be placed in a single Erlenmeyer flask, or gas development will be disturbed.

Apart from the use of the kinds of bait which will attract saprophytes, it is also advisable to undertake the microscopic examination of the zooplankton, as they are occasionally attacked by parasites.

3. Quantitative Determination of the Density of the Fungi in the Water and Sediment

All the methods mentioned have been used for the qualitative identification of the fungi in or on the substrate. In addition, they provide gross indications of the fungal density.

Gaertner has now described a method whereby the number of fungal infection units in 1 litre of water or substrate at the station can be calculated. His estimates are based on comparisons of dilutions of the water from the station or suspensions of portions of the substrate which have been determined in terms of quantity. For the water samples he gives the arrangement of the culture bottles shown in Figure 59.

Stages	1	2	3	4	5	6	7	8	9
Sterile Sea water	–	–	–	12·5 ml	19·75 ml	21·875 ml	23·44 ml	24·22 ml	24·61 ml
Station Water	100	50	25	12·5	6·25	3·125	1·56	0·78	0·39
Number of Bottles	10	10	10	10	10	10	10	10	10

Figure 59. Explanation in text

From left to right there are nine columns, each containing 10 bottles. In the first column each bottle contains 100 ml, in the second bottle 50 ml, in the third bottle 25 ml of water from the station. Starting with the fourth column, this water is mixed with sterile seawater in such a manner that the portions of station water fall by one half from step to step (up to 0·39 ml in the ninth) and the quantity of seawater increases in compensation to the same extent, so that the bottles always contain 25 ml water.

The mixtures are prepared in the laboratory, i.e. the sterile sea-water is previously placed in the bottles in the required quantities. At sea the portions of station water are added by pipette immediately after being drawn up and a knife-tip quantity of *Pinus* pollen is added as bait to each bottle. The prepared bottles are examined under the microscope after a minimum of 14 days and are kept in a cool place during this period.

For the analysis the number of bottles containing fungi at each dilution step is important. In regard to the fungal density we have the following equation:

$$1 \text{ ml seawater contains } \frac{n}{a \cdot b} \text{ fungi}$$

$a =$ the quantity of seawater introduced into each repeat, the seawater being from the last dilution step containing fungi;

$b =$ the number of repeats of the dilution step;

$n =$ the number of repeats (containing fungi) of the last dilution step with fungi.

Thus, 1 litre seawater contains: $= \dfrac{n \cdot 1000}{a \cdot b}$ fungi.

The numbers of organisms for sediment sections were also determined. Instead of seawater in decreasing quantities, sections of sediment were placed in bottles with 25 ml sterile seawater. The sediment sections had been removed from the sample beakers with forceps

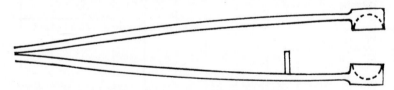

Figure 60. Gaertner quantitative forceps

which had a cup on the tip of each arm* (Fig. 60). The two cups together form a cavity with a known volume. The volumes of a forceps set consist of:

$$0 \cdot 420; \ 0 \cdot 198; \ 0 \cdot 0996; \ 0 \cdot 0490 \text{ and } 0 \cdot 0244 \text{ ml.}$$

The decreasing series of volumes descends stepwise, each volume being half of the preceding capacity.

* Manufacturer: Firma W. Ludolph, Bremerhaven, Am Alten Hafen 75.

Within the steps given, the fungi were found to number from 2570 to a few in seawater and from > 20400 to 238 per litre in sediment.

As containers for the mixtures Erlenmeyer flasks of 300 ml capacity were used for step 1, Meplate bottles of 100 ml capacity for step 2, and Meplate bottles of 50 ml capacity for all the other steps. They are kept in an especially prepared box in which all the bottles required for one station are arranged in a series, as shown in Figure 59; they remain standing in the same place from the beginning until inspection. Only the covers are screwed on and off during filling.

Pinus pollen, which has frequently been found useful – particularly for several families of the phycomycetes, was used as substrate. The numbers relating to fungal density therefore concern only the so-called pollen fungi.

Valuable illustrations and discussions of the role of fungi in the conversion of matter in the sea can be expected from these quantitative determinations.

REFERENCES

(Further literature is mentioned in the works below)

GAERTNER, A. (1966). 'Vorkommen, Physiologie und Verteilung "Mariner Niederer Pilze" (Aquatic Phycomycetes).' Veröff. Inst. Meeresforsch., Bremerhaven. Sonderband II (6. Meeresbiol. Symposion), 221–235.

HÖHNK, W. (1952–1956). 'Studien zur Brack- und Seewassermykologie I–VI.' Veröff. Inst. Meeresforsch., Bremerhaven, Bde. 1–4.

— (1959). 'Ein Beitrag zur ozeanischen Mykologie. In: Die Expeditionen von F. F. S. "Anton Dohrn" und V. F. S. "Gauss" im Intern. Geophys. Jahr 1957–58.' Dt. Hydrograph. Z., Erg.-Heft, Reihe B, Nr. 3, 81–87. Hamburg.

— (1966). Weitere Daten zur Verbreitung der marinen Pilze. Veröff. Inst. Meeresforsch., Bremerhaven, Sonderband II, 209–220.

JOHNSON, T. W. jr., and F. K. SPARROW, jr. (1961). *Fungi in Oceans and Aestuaries.* Weinheim.

MEYERS, S. P., P. A. ORPURT, I. SIMMS, and L. L. BORAL (1965). 'Thalassiomycetes VII. Observations on fungal infestation of Turtle Grass, Thalassia testudinum König.' *Bulletin of Marine Science* 15, 548–564.

SPARROW, F. K. (1960). *Aquatic Phycomycetes.* 2. Aufl. Ann. Arbor.

F. Bacteria

W. GUNKEL and G. RHEINHEIMER

Although it must be assumed that bacteria play an outstanding part in the circulation of matter in the ocean, very little is known about them as yet. This is particularly unfortunate because it is only with great difficulty that bacteria can be excluded from other investigations, especially as they may influence the results. This is due, not least, to the relatively complicated and demanding methodology of microbiological studies, such as the use of sterile working techniques, which are as yet unknown to many marine biologists. Only the principal working methods can be described in the space available here; therefore the reader is referred to the basic literature on this subject, particularly to the work of Zobell, which is still the best introduction to this field, and to that of Oppenheimer, Wood and Brisou (see also the list of references in Section IV C, p. 289 ff). For the general bacteriological working instructions and nutrient media we mention the *Manual of Microbiological Methods*, and as an introduction to general microbiology we recommend Pelczar-Reid and Rippel-Baldes.

1. GENERAL (see also Section IV C, 1)

If at all possible, the samples for bacteriological examination should be worked up immediately after collection. If kept for even only a few hours, the results will be greatly falsified and altered. Consequently, when samples are taken from free water, they must be worked up on board ship even during a day trip. This requires adjustment of the working procedures to the conditions of a rolling ship. If at all possible, the bacteriological study should be carried out in a separate room, which must be kept meticulously clean. Objects which produce or catch dust should be removed. It is desirable to cover the timbers with plastic sheets. UV lamps should be installed and turned on some time before the work begins. Frequent wiping of the floor, table, and walls with a disinfectant is necessary. Places which are difficult to reach, such as screw joints and the backs of cables, should be treated with a spray disinfectant (e.g.: *Luzol Spray*). All apparatus must be secured against falling (gratings for petri dishes, flasks, etc.). A gimbal table with an attached cardanic plate is desirable for the installation of the pouring slates. A gas connection will only very rarely be available. The use of benzene or alcohol burners is dangerous and should therefore be avoided. The 'Labogaz' burner with one-way butane cartridges was found particularly useful for this work. A few strong and manageable wooden crates or carrying frames should be available for all the apparatus and flasks which are required on board ship. Instruments which are used for temperature control (autoclaves, water baths) must have robust temperature regulators which are not sensitive to the rolling motion.

The work on board ship often makes great physical demands on the individual. When carrying out bacteriological investigations the work at sea is not terminated when sampling is completed, but continues with activities which require great cleanliness and precision. All possible preparation should be done on land. For example, it is advisable to write out the procedure sheet on land and to tick it off on board ship. Nutrient media should be prepared on land and poured into the flasks in the required quantities. Labels required for numbering the test flasks or petri dishes should be made out beforehand. For the repeated steps, such as the arrangement of the dishes, exactly the same sequence should always be followed in order to avoid confusion.

In any bacteriological work it must be remembered that all the equipment and materials are in themselves contaminated with bacteria. Consequently, all the apparatus, solutions, and nutrient media which come into contact with the test material must first be sterilized. This is usually carried out in the following manner.

1. Nutrient solutions, agar nutrient media, sample collection bottles with rubber tops, small tubes with bakelite caps are sterilized with saturated water vapour in the autoclave for 15 minutes at 120°C. This is equivalent to 1 *atm*. Containers of more than 1 litre capacity require longer for sterilization (for example, filled 4-litre containers require 30 minutes).
2. Glassware, such as pipettes, small tubes with cotton wool or cellulose stoppers, petri dishes, etc., must be sterilized for 2–4 hours at 160–170°C. It is advisable to use suitable metal boxes for petri dishes and pipettes or to wrap them in parchment paper.
3. Synthetic materials only rarely tolerate heat sterilization. Treatment with 80% alcohol is indicated for these.
4. By passing the mouths of small tubes and bottles through a flame any adhering organisms are killed, which makes it possible to transfer or decant without contamination.

5. Sensitive nutrient solutions can be freed of organisms by filtering through bacteria-proof filters.

Continuous checking of the sterility of the nutrient media dilution water, etc. is indispensable when working on board ship. The possibility of aerial contamination is tested by exposing petri dishes with nutrient media to the air for 30 minutes without cover.

2. SAMPLING FOR BACTERIOLOGICAL INVESTIGATIONS

When collecting samples for bacteriological studies, contamination by extraneous bacteria must be prevented with certainty. Special water bottles which tolerate sterilization are therefore required. Since most metals are bactericidal, they cannot be used for these bottles. When taking sediment samples, the contaminated surfaces of the samples, which have come into contact with a sampling instrument (grab, impact corer) must be removed with a sterile, hot spatula.

(a) Sample Removal from the Water

In the simplest instance samples can be taken by means of sterilized bottles or, for example, a bucket freed of organisms by cleansing with alcohol. However, these can only be used as emergency measures. The same is true for the use of a sterile Meyer water bottle. A whole series of different types of water bottles and modifications are described in the literature. Our report must be confined to three of the most commonly used types (see also report II E, p. 142).

The most widely used is the Zobell water bottle (Fig. 61). This bottle consists of a metal angle (preferably brass, possibly cadmium plated to prevent corrosion), to which a sterilized, thick-walled bottle (0·33 litre beer bottles are very suitable) is attached. The bottle is closed with a top which consists of a pierced rubber stopper, a glass angle, a vacuum tube and a distally scaled small tube. This small tube runs through a lever mechanism which is activated at the desired depth by a messenger. The powerfully sprung lever breaks the glass tube and the vacuum tube (which is under tension) removes the rest of the small tube, which functions as an inlet aperture, to some distance from the water bottle. As a result of hydrostatic pressure and because of the partial vacuum within the bottle (the bottles are closed after sterilization while still hot) the water penetrates into the bottle, which is full after a maximum of 30 seconds. The rebound lever actuates a device which releases the next messenger. It is thus possible to attach many water bottles below each other, and these take almost simultaneous samples. It is also

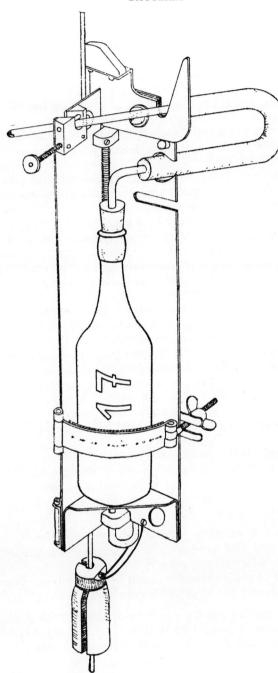

Figure 61. Slight modification of the Zobell bacteriological water bottle. Explanation in text

possible to insert the bacteriological water bottles into a hydrographic series, which is especially necessary in depth series in order to save time. The distance between the individual water bottles must be at least 1 metre in order to allow each messenger sufficient strength to break the tubes. When the bottle is pulled up the hydrostatic pressure decreases; therefore there is no risk of water penetrating from lesser depths. If larger quantities of water are required, the support clamp can be exchanged. It is possible to use thick-walled bottles of up to 2 litre capacity without difficulty. At depths exceeding 100 m the glass bottles must be replaced by rubber or neoprene containers – so called horn bulbs.

The Niskin bottle (1962) works on an entirely different principle (Fig. 62). It consists of two metal wings connected by a strong spring. A sterile, closed bag made of plastic foil is attached to the wings by slipping the wings into pockets on the bag. The cable runs through the support. At the desired depth a messenger activates the cutting device. At the same time a lock is released, the spring causes the wings to spread, and water is drawn into the bag through the tube. When the opening is in a certain position the closing device is activated and the tube is firmly closed.

A reversing thermometer (not shown in the illustration) is connected with the closing device and checks whether the bottle has successfully worked at the desired depth. With this bottle it is possible to remove up to 4 litres of water from any desired depth.

Sorokin (1964 a, b) has built bottles which can also be used in the deep sea. They consist of one or several glass containers filled with sterile salt solution. At the desired depth a messenger opens a sealed tube and a suction device (rubber bulbs or a piston) removes the salt solution; the seawater then flows in, taking up the space.

(b) Sample Removal from Sediments

The samples can be removed by means of bottom grabs, impact corers, piston corers and bucket grabs (survey of instruments in Barnes 1959 and Hopkins 1964; see also Figs. 42, 43, 45, 55, 57). Two points are important:

1. The natural stratification of the sediment must be retained. Especially the first centimetre is endangered because of its consistency, which is often not very solid.
2. Since most of the instruments cannot be sterilized, any external layers which may be contaminated must be removed.

Ideal for removing the top decimetre is the Reineck box grab mentioned elsewhere (see report II D, p. 112 ff; Fig. 45). A one-way glass or plastic medical syringe is used to remove the sample from

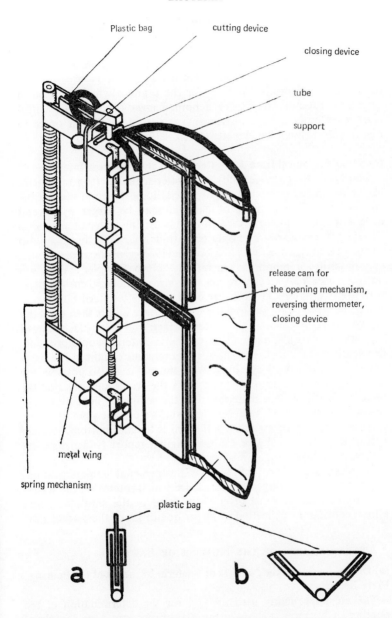

Plastic bag

cutting device

closing device

tube

support

release cam for
the opening mechanism,
reversing thermometer,
closing device

metal wing

spring mechanism

plastic bag

a

b

Figure 62. Simplified representation of the Niskin bacteriological water bottle:
a, b = diagram of the mode of action; a = closed before release; b = plastic
bag filled with water. For further explanations see text

the grab. The front edge is cut off from the outer section, which makes it possible to remove coarse-grained material in accurate quantities. The syringe is sterilized in 80% alcohol. Not less than 2 ml sediment is removed. The sediment is placed in a container with 100 ml sterile seawater. Vigorous agitation is necessary to distribute it uniformly. At least for the samples which are rich in organic substance, the use of a homogenizer (such as the Ultra-Turrax or other mixing instrument) is very advisable in order to disperse the bacterial aggregates (Gunkel 1964).

(c) Sample Removal from Air-water and Water-sediment Interfaces

The air-water (surface film) and water-sediment (top millimetre of the sediment) interfaces are of special importance in the marine environment. The number of bacteria may be larger by several orders of magnitude here than in the underlying water or sediment. This is not necessarily true only of stations near land, but may also be the case in the open ocean. Sieburth (1965) gives examples of this and describes a modified 'Garret Fly-Screen Sampler'. This is a narrow-mesh V2A steel wire net of approximately 30 cm diameter with approximately 36 mesh per centimetre; after sterilization it is placed perpendicularly below the water surface and is then carefully drawn vertically through the water surface (the net surface toward the water surface). A net made of synthetic material would probably be even better for this. The surface film remains caught in the mesh and can be poured off by tilting. With this net it is also possible, for example, to collect thin oil films from the water surface. For the removal of the surface material from the sediment Sieburth uses a so-called 'ooze sucker', a device which sucks a certain quantity of the surface detritus into a rubber ball when it touches bottom. For the construction of this device see the original report (which includes further references).

In Miami, Florida (U.S.A.) Niskin (personal communication) developed a drum-shaped plastic device which rotates on the water; this device runs at some distance in front of the ship's bow and allows continuous collection of larger quantities of the surface film.

3. DETERMINATION OF THE NUMBER OF BACTERIA

(a) Determination of the Number of Bacteria by means of the Pouring Plate Procedure

The most frequently used method for the determination of the number of bacteria is the pouring plate procedure. One sample of the test material is mixed with a seawater agar nutrient medium. The bacteria are isolated, placed and allowed to grow into colonies which can be easily counted with the help of a magnifying glass. The individual steps involved in the procedure are summarized in Fig. 63. After the collection of the sample (1), the bottles are vigorously agitated, the mouths of the bottles are flamed and a dilution series (2) is prepared; 1 ml of the sample is added to 9 ml sterilized seawater and the test tube is agitated vigorously. The number of bacteria per millimetre has now been diluted to $1/10 (= 10^{-1})$. This step is repeated with each of nine new sterile pipettes, until the anticipated number of bacteria per millimetre is less than 300. We attempt to have no more than 300 colonies per plate, as a reciprocal influence will otherwise develop. For removal to the plates (3), one ml from several dilution steps is placed in a number of parallel plates (petri dishes) and the nutrient medium is added (4). The sample and the nutrient medium are mixed by tilting the plates. At least for work at sea it is advisable to use pre-sterilized plastic petri dishes which can be destroyed after one use. If this work is carried out at sea, a rolling table (cardanically suspended table surface) with gratings is very helpful as it prevents slipping of the plates and of the agar in the plates while the nutrient medium is solidifying.

The nutrient medium most frequently used in marine biology for bacterial count determinations, developed by Oppenheimer and Zobell, is numbered 2216 E. It consists of 5 g peptone, 1 g yeast extract, 15 g agar agar, 10 mg iron phosphate, 750 ml aged seawater and 250 ml distilled water. It has a pH of $7 \cdot 6$. Sterilization requires 20 minutes at 120°C.

Since peptone, yeast extract and agar are not chemically defined but vary in their composition in accordance with their manufacturer, if possible, it is best to use the products of a large company who market these in a standardized form. For most investigations, particularly in the United States, the products manufactured by Difco and Co., Detroit, are used.

'Aged seawater' must be used for the nutrient media. This water is obtained by filtering seawater through a paper filter and keeping it in the dark in spherical glass flasks for several months. Bactericidal and organic substances which are naturally present in seawater are broken down during this time. Seawater prepared from salts is not well suited for bacterial cultures. 10 ml of the nutrient medium is poured into each test tube under sterile conditions. If necessary the medium is liquefied at 100°C and adjusted to a temperature of $42 \cdot 5$°C, measured as accurately as possible. The temperature adjustment is necessary because most marine bacteria are very sensitive to temperature and if the quantity of nutrient medium is larger or the pouring temperature is higher, cooling will be slower and some

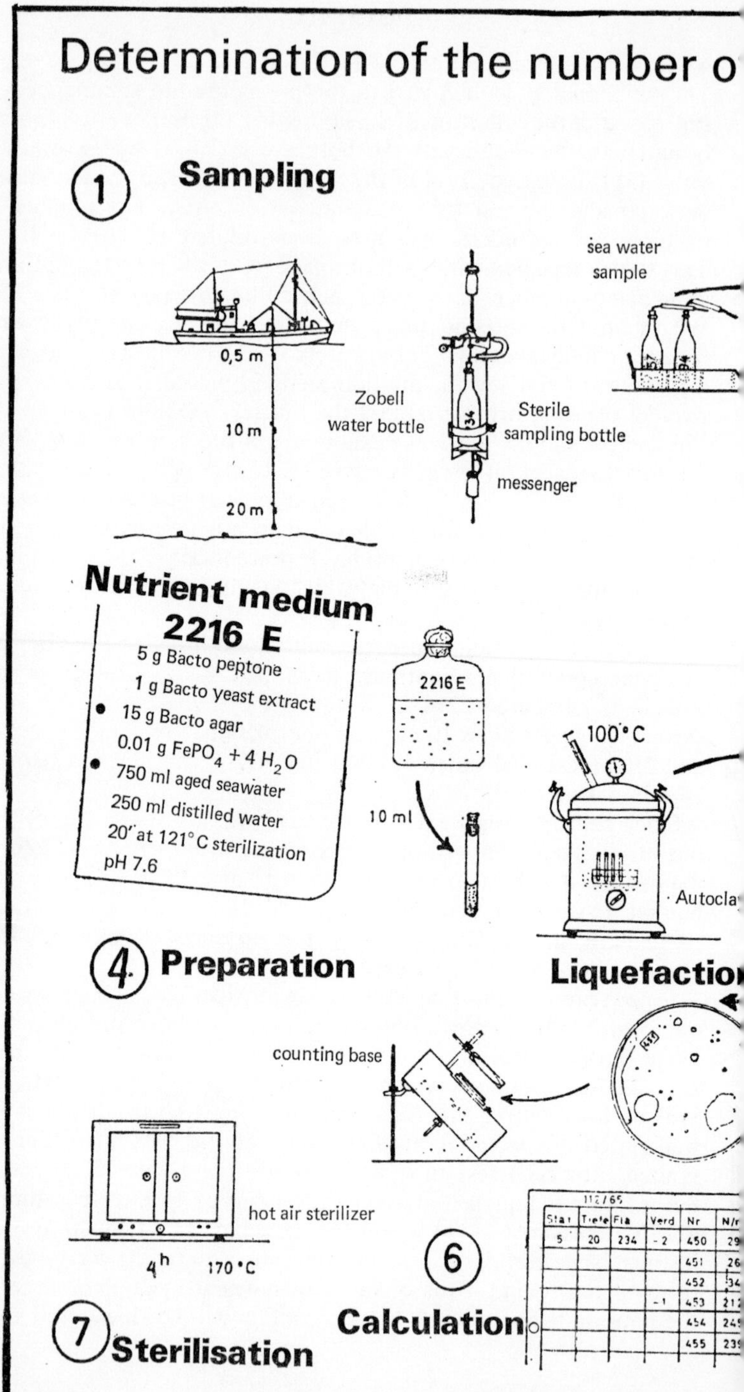

Determination of the number o

(1) Sampling

0,5 m

10 m

20 m

Zobell water bottle

Sterile sampling bottle

messenger

sea water sample

Nutrient medium 2216 E

5 g Bacto peptone
1 g Bacto yeast extract
- 15 g Bacto agar
0.01 g $FePO_4 + 4 H_2O$
750 ml aged seawater
250 ml distilled water
20' at 121°C sterilization
pH 7.6

2216 E

10 ml

100°C

Autocla

(4) Preparation

Liquefactio

counting base

hot air sterilizer

4^h 170°C

(6)

(7) Sterilisation

Calculation

11½/65					
Star	Tiefe	Fia	Verd	Nr	N/r
5	20	234	- 2	450	29
				451	26
				452	34
			- 1	453	212
				454	24
				455	239

Figure 63. Diagram of the pouring plate procedure for determining

bacteria in sea water samples

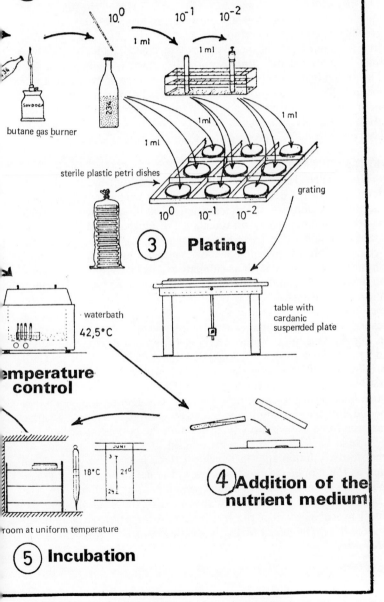

② Preparation of the dilution series

10^0 10^{-1} 10^{-2}

1 ml   1 ml

butane gas burner

1 ml   1 ml   1 ml

sterile plastic petri dishes

1 ml

10^0 10^{-1} 10^{-2}

grating

③ Plating

waterbath
42,5°C

table with
cardanic
suspended plate

temperature
control

18°C   21ᵈ

room at uniform temperature

④ Addition of the nutrient medium

⑤ Incubation

the number of bacteria in seawater samples. Explanation in text

of the bacteria will consequently be killed. The plates are incubated for 3 weeks at 18°C or at a maximum of 20°C. Especially during the winter months, when the water temperature is lower, a lower temperature would seem more suitable; however, the required incubation time would consequently be much longer. Moreover, this would make it difficult to compare trials from different seasons. The majority of the colonies normally appear within the first 7 days; longer incubation is necessary, however, in order to allow the slow-growing bacteria to develop into macroscopically visible colonies. For the count (6) the petri dishes are placed under the indirect light of a ring-shaped fluorescent tube which is built into a box. The individual colonies show up against a dark background. A reading glass is used for the magnification. The results are recorded as the number of bacteria per millilitre. The cleansing and sterilization of the glassware used, such as test tubes, pipettes, etc. (2–4 hours at 160–170°C) prepares for the next cycle (7).

(b) The MPN Method

The MPN (most-probable-number) method was originally developed for the determination of coliform bacteria. It is generally suitable not only for the determination of the number of aerobic bacteria – but depending on the choice of nutrient medium – is also usable for anaerobic bacteria for example, and in particular for determination of the number of bacteria belonging to certain physiological groups, such as sulphate reducers, cellulose degraders, oil degraders, chitin degraders, H_2S producers. The method requires nutrient solutions without agar. Its disadvantage in comparison with the pouring plate method is that it requires more time and that the results are more scattered. The principle of the method is illustrated in Fig. 64. One ml seawater is removed from test flask A and added to 9 ml sterile seawater. By repeating this step a dilution series (B) is produced, dilution taking place at a 1:10 ratio each time. Small 20 ml bottles contain fluid nutrient medium (C). Three (or five) bottles are inoculated with 1 ml each of a dilution step. During the incubation period the bacteria multiply by many powers of 10. A positive reaction (such as turbidity) develops in all the bottles into which one or several bacteria have been introduced. For the interpretation we choose the dilution step in which all three bottles are positive, as well as the number of positive ones in the next two dilution series. With the help of the numerical combination obtained (in the example shown: 3, 2, 0) and with regard to the dilution step, the 'most probable' number can be obtained from tables, such as those printed in the 'Standard Methods' or the '*Manual of Microbiological Methods*'.

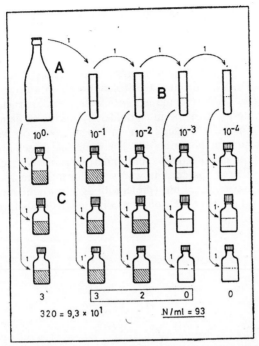

Figure 64. Diagram of the MPN method for determination of the number of bacteria: A = water sample; B = dilution series; C = test bottles

(c) Surface Smearing

A further possible method for determining the number of bacteria is surface inoculation. This has the advantage that all the bacteria in one petri dish have exactly the same growth conditions. 0·1 ml of the test liquid is placed on each plate, which has previously been covered with agar nutrient medium; the liquid is then uniformly distributed over the surface with a sterilized Drigalski spatula (immersed in alcohol and flamed). Smearing must be continued until the film of liquid is no longer visible. The following problems may occur: swarming colonies spread over the entire surface; if the plate is somewhat too moist, the colonies slip off; if it is too dry, growth is poor. In addition – and this is probably the main reason why it is scarcely used for marine biological studies – the method is only usable when the number of bacteria in the test liquid is at least one thousand per millilitre, and this is only rarely the case.

(d) Membrane Filter Culture Methods

Membrane filters have been much used in microbiology in the

last decade. They are spongy, porous discs of some 150 μm thickness, made of cellulose esters of different pore widths, which make it possible to undertake the concentration of micro-organisms. Their use is especially advisable when the bacterial densities in the natural environment are low. The membrane filters are sterilized by boiling for 20 minutes in abacterial distilled water or autoclaving for 5 minutes at 120°C (in a container with water) and are then placed on the glass interplate of the filter holder by means of flamed forceps (Fig. 65). The filter holder is set on top of a suction bottle and a moderate vacuum is produced. The water is drawn through the membrane filter. Particles and micro-organisms are retained. With the filtered material upwards, the membrane filter is then laid on a nutrient medium (preferably agar nutrient medium) and incubated. The nutrient solution diffuses from below into the

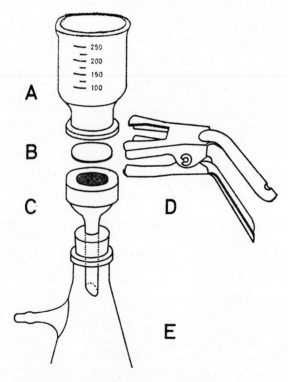

Figure 65. Millipore filter holder with bacteria-proof filter: A = graduated upper part; B = bacteria-proof filter; C = lower section with glass interplate; D = holding device; E = suction bottle

membrane filter, and macroscopically visible colonies develop. Black filters are used in order to detect small, colourless colonies. To make small colonies visible on white filters it is advisable to stain the colonies subsequently with methylene blue solution. Greater detail is available from the manufacturers (for example, the information provided by the Membranfiltergesellschaft, Göttingen).

(e) Determination of the Bacterial Number by Microscopic Count

Direct counting of the bacteria contained in samples is possible if they are present in high density and definitely identifiable as such. This is normally the case in pure cultures and in concentration cultures. Petroff-Hauser counting chambers which have a chamber depth of 20 μm, can be used. Normal blood count chambers have a depth of 100 μm and are unsuitable. The use of phase contrast procedures makes it possible to count the bacteria without staining them. Membrane filters allow the concentration of bacteria from samples whose organism density is too low. The bacteria can then be examined and counted on the filter – which has been made transparent (see further below) – after the largely concentrated material has been transferred into a counting chamber, or on a slide (Jannasch and Jones 1959).

On the other hand, it is not possible to obtain an unexceptionable determination of the number of bacteria in freshly collected seawater by microscopic examination. The small size of the bacteria, confusion with non-living particles of the same size, interference produced by the always abundantly present detritus, and the impossibility of distinguishing between living and dead bacteria have prevented this method – which at first appears very obvious to the layman – from being widely used in marine microbiology. To this should be added the significantly greater amount of time required, as well as a strong subjective factor in judging what is and what is not a bacterium. We nevertheless mention this method since it is fairly often used in fresh water and may yield information on bacterial forms and organizations, such as the appearance of chains, and growth on particles. In accordance with the number of bacteria anticipated, given quantities of seawater are filtered through a bacteria-proof membrane filter; particle-free filtered formalin is added to the water prior to the filtration, so that the end concentration of formalin in the water sample is not less than 0·5%. The filter is then dried on filter paper and placed on filter paper discs saturated with distilled water. Most of the salts are thus removed. Staining is carried out by placing on filter paper discs saturated with methylene blue solution, for example. It is then briefly dried with filter paper and bleached on filter paper saturated with distilled water, so that only the organisms retain the stain. Subsequently the filter is dried for 15 minutes at 45°C. A piece measuring approximately 15 × 15 mm is cut out, placed on the slide and made transparent. With membrane filters made of cellulose nitrate this can be achieved in various ways:

1. By correcting the refractive index, as for example by the addition of immersion oil or of a mixture of 1:1 cedar-wood oil to oil of cloves, embedding in Entellan or the use of embedding media recommended by the optical industry for phase contrast microscopy. The advantage of some substances is that they produce a permanent preparation at the same time (under certain circumstances, edge with coverglass varnish).
2. By dissolving the filter, as for example with a 9:1 mixture of chloroform ethyl alcohol.

Specific information can be obtained from the data published by the Membrane Filter Association.

Extra-thin coverglasses are used for covering. Salt removal, staining and bleaching can also be carried out directly in the filter holder by adding the required liquids. Several squares must be counted, until a minimum of 400 cells per filter has been attained.

In this connection we must mention the possibility of detecting bacteria by fluorescence microscopy, using acridine orange. This method requires a great deal of experience. Its description here would take up too much space and we must therefore refer the reader to the pertinent literature (Strugger, S., *Fluorescence microscopy and microbiology*, Hanover 1949).

4. GROWTH METHODS

A further possibility of obtaining information on bacteriological populations in seawater and sediment is through the study of the growth, i.e., the adhesion of micro-organisms to surfaces such as slides or other materials introduced into the environment. It is not always simple or possible to interpret the results; senseless errors of interpretation have been made ('Krassilnikovia'). The small size of the organisms, detritus, etc., can also cast doubt on the value of this method.

(a) Growth on Slides

Rossi and Cholodny were the first to place slides in the ground, remove them after a certain time and count the microscopic population. Zobell and Kriss used this method more extensively. Fig. 66 shows a simple supporting device for slides, which makes it possible

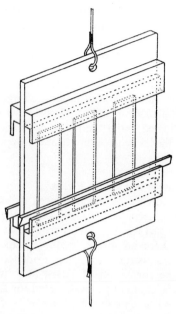

Figure 66. Holder for slides used with the growth method (based on Zobell)

to lower the scrupulously clean slides to the desired depth by means of a cable. After several hours to a few days the slides are removed and examined under the microscope. Forms are found, such as stalked bacteria, which cannot be detected by the plate method. Apart from difficulties relating to microscopic interpretation, it must be mentioned that in this method the glass surface also represents an 'unnatural' object. In addition, only some of the bacteria present are capable of adhering to the slide or to something similar. Nevertheless valuable information is gained, especially by a comparison of the values obtained from different stations or at different times by this method with the values obtained by other methods or by chemical determinations. Wood (1965) suggested the use of coverglass as growth surfaces, these being grown on seawater nutrient agar after removal from the environment. With the help of the microscope it is then possible to discover which bacteria are capable of reproducing.

(b) Growth in Square Capillaries

Perfilieff suggested the use of capillaries with a square cross-section, which are introduced into the sediment and left in it for a fairly long time. The conditions in the capillaries should largely correspond to those in the sediment; in particular, an equilibrium with the sediment should develop. The capillaries can be studied by direct microscopic examination after removal. The picture is not mpaired by the irregular form of the sand grains and the spatial distribution is not disturbed. With the help of other thinner capillaries the overgrown square capillaries can be perfused with a great variety of nutrient solutions; a type of 'microchemostat' is thus created. The changes can be followed under the microscope. For the preparation of the square capillaries see Reumuth and Sopper (1963) (*Mikrokosmos*, *52*, pp. 56–63).

(c) Growth on Cellulose and Chitin Pieces

Cellophane or chitin pieces can be placed in sediments or in the water column and be examined directly under the microscope after careful cleansing. Apart from a general growth, the micro-organisms of special interest here are probably those which are capable of breaking down the substrates, which is evidenced by growth into the substrate or the formation of holes, and by some degree of dissolving of parts. Cellophane is well suited for this purpose, but it must be remembered that we are speaking only of true native cellulose.

Pieces of crab shell or the dorsal shell of *Sepia* are suitable as chitin after they have been decalcified and treated with alcohol, and possibly with saline as well. Apart from its use for the study of bacterial settling, this method also provides a relatively simple way of obtaining cellulose – or chitin – degrading bacteria, since the substrate acts as 'bait'.

5. DETERMINATION OF THE NUMBER OF BACTERIA IN CERTAIN PHYSIOLOGICAL GROUPS AND 'EXPONENTIAL DETERMINATION'

Apart from the number of bacteria, the quota of various physiological groups is of particular interest; i.e. how many of the bacteria present are capable of carrying out a particular function (such as cellulose breakdown). A conceivable way of establishing this by re-testing colonies which were obtained by the pouring plate method (with the Lederberg stamp technique, for example) is not possible in most cases. The specialized bacteria are not detected by this method. A special nutrient medium is required for each group, which only allows the growth of that group; for example, for studies of oil-degrading bacteria, solely inorganic salts are used, apart from oil as the only carbohydrate and energy source. The MPN method has been found excellent for the determination of the number of different groups. Lack of space does not allow us to describe the individual nutrient media in detail and the reader is referred to the pertinent literature. A great number of such nutrient media is listed by Schlegel (1965), and for the marine environment in particular by Oppenheimer (1963).

Another method is that of exponential determination, i.e., the determination of certain reactions shown by the bacteria in a water sample under given conditions, such as the liberation of $NH_4 +$ from proteins, the reduction of sulphate, the degradation of cellulose, etc. (further details in Rheinheimer, 'Microbiological Investigations in the Elbe between Schnackenburg and Cuxhaven'. *Arch. Hydrobiol.*, Suppl. XXIX II, 1965, pp. 181–251).

6. CLASSIFICATION OF THE BACTERIA

As a rule, the determination of bacteria is not possible on the basis of their morphological characteristics alone; in the great majority of cases their metabolic performance must also be taken into account. In addition to a thorough microscopic examination, numerous physiological and biochemical tests must consequently be carried out. The first requirement is well-developing pure cultures of the bacteria to be identified (see Section IV C, p. 275). Their growth on various nutrient media is observed, and in particular the form and pigmentation of the colonies. This is followed by a microscopic examination of the form, size and motility of the cells. With the help of suitable dyes we ascertain the absorption and adhesive capacity of certain stains (gram staining, acid fastness); we also determine the presence of flagella and any capsule and spore formation. The following biochemical tests are carried out, among others; oxygen reaction, gelatine liquefaction, nitrate reduction, H_2S formation from cysteine and sulphate, indol formation, $NH_4 +$ formation from protein or protein units, acid and gas formation from sugars, alcohols and glycosides, starch hydrolysis, degradation of cellulose, agar, chitin, hydrocarbons, catalase and oxydase formation and antibiotic sensitivity. It is also advisable to determine the temperature ranges, salt content and pH, and particularly their optima. All the results are collected in tables for interpretation. Each bacterial strain tested thus receives a 'profile' which is the more reliable the more data it contains. Computers have recently been used to evaluate the data. With the help of this characterization the identification of the bacteria can then be attempted. However, this requires much experience and will frequently only be possible for a specialist. The procedure for carrying out the individual tests is described in Chapter VII: 'Routine Tests for the Identification of

Bacteria' in the *Manual of Microbiological Methods*, where further references are also given. *Bergey's Manual of Determinative Bacteriology* by R. S. Breed, E. G. D. Murray, and N. R. Smith (Baillière, Tindall and Cox Ltd. London, 1957) and '*Identification of Bacteria and Actinomycetes*' by N. A. Krassilnikov (V.E.B. Gustav Fischer Verlag, Jena, 1959) are used for the classification. The procedures required for the determination of the characteristics of the bacteria are listed in particular in Skerman's '*A Guide to the Identification of the Genera of Bacteria with Methods and Digests of Generic Characteristics*', (The Williams and Wilkins Co., Baltimore 2, Maryland, 1959), and in the *Manual of Microbiological Methods*.

REFERENCES

(For basic references see also Section IV C, p. 289 Rheinheimer-Gunkel).

BARNES, H. (1959). *Oceanography and Marine Biology. A book of techniques.* Ruskin House. George Allen & Unwin Ltd., London.

BRISOU, J. (1955). *La microbiologie du milieu marin.* Ed. Med. Flammarion, Paris.

GUNKEL, W. (1964). 'Die Verwendung des Ultra-Turrax zur Aufteilung von Bakterienaggregaten in marinen Proben.' *Helgol. Wiss. Meeresunters.* **11**, 287–295.

HOPKINS, T. L. (1964). 'A survey of marine bottom samplers.' *Progress in Oceanography*, Vol. 2 (M. Sears ed.). Macmillan Co., New York.

JANNASCH, H. W., and G. E. JONES (1959). 'Bacterial populations in seawater as determined by different methods of enumeration.' *Limnol. and Oceanogr.* **4**, 128–139.

KRISS, A. E. (1961). *Meeresmikrobiologie.* VEB Gustav Fischer Verlag, Jena.

KÜSTER, E. (1913). *Anleitung zur Kultur der Mikroorganismen.* Verlag B. G. Teubner, Leipzig und Berlin.

NISKIN, Sh. J. (1962). 'A water sampler for microbiological studies.' *Deep-Sea Res.* **9**, 501–503.

PELCZAR, M. J., and R. D. REID (1965). *Microbiology.* McGraw Hill Book Comp., New York.

PELTIER, G. L., C. E. GEORGI, and F. L. LINDGREN (1955). *Laboratory Manual for General Bacteriology.* John Wiley & Sons Inc., New York, Chapman & Hall Ltd., London.

RIPPEL-BALDES, A. (1955). *Grundriß der Mikrobiologie.* Springer Verlag, Berlin-Göttingen-Heidelberg.

SCHOLES, R. B., and J. M. SHEWAN (1964). 'The present status of some aspects of marine microbiology.' *Advances in Marine Biology.* Vol. 2 (F. S. Russell ed.), 133–164. Academic Press, London and New York.

SEELEY, H. W., and P. J. WANDEMARK (1962). *Microbes in Action. A Manual of Microbiology.* W. H. Freeman and Comp., San Francisco-London.

SHEWAN, J. M., G. HOBBS, and W. HODGKISS (1960). 'A Determinative Scheme for the Identification of Certain Genera of Gram-negative Bacteria with Special Reference to the Pseudomonadaceae.' *J. app. Bact.* **23**, 379–391.

SIEBURTH, J. McN. (1965). 'Bacteriological Samplers for Air-Water and Water-Sediment Interfaces.' *Ocean Science and Ocean Engineering*, 1064–1068.

SOROKIN, J. I. (1964). 'A Quantitative Study of the Microflora in the Central Pacific Ocean.' *J. Cons. int. Explor. Mer* **29**, 25–40.

— (1964). 'On the Primary Production and Bacterial Activities in the Black Sea.' *J. Cons. int. Explor. Mer* **29**, 41–60.

Section III

UNDERWATER OBSERVATION AND PHOTOGRAPHY

H. SCHWENKE

The underwater world has sometimes been described as a continent in itself, whose exploration is equal to the adventure of space travel. This dramatic description expresses above all the fact that to man – and this naturally includes the researcher – this area of the earth is not easily accessible. Man's understanding of the world below the water surface depends to a decisive extent on the technical possibilities which allow him to penetrate into it indirectly by optical means or directly by mechanical equipment.

The suitable methods available to the marine biologist at present are only peculiar to him and to his work to a very small degree; rather, they are shared by others: technology, sports, hobbies and science cannot be separated in this area.

With some generalization we may probably say that all areas and all depths of the sea are now accessible to optical observation methods and those involving technical apparatus – but with one decisive limitation for the marine biologist; in contrast to the interested amateur, he can only rarely freely select the observation site, observation object and observation conditions on the basis of aesthetic or similar grounds. The pressing scientific problems which he faces makes it necessary for him to come to terms with his project under the given – and sometimes poor – circumstances. It is therefore clear why we must mention here the principal difficulty (as it is the one which essentially cannot be eliminated), involved in all direct or indirect optical underwater observations, namely, water turbidity and colouring – whether occasioned by sediment eddies, plankton, or other particles or dissolved substances. We now turn to a brief description of the principal methods. The division into observation and photography is largely for didactic purposes, since in practice the two procedures are generally combined.

1. UNDERWATER OBSERVATION

(a) Underwater Observation with Simple Aids

Underwater observation with simple aids plays an important part in marine botanical shallow-water work. The principal obstacles for the optical penetration into shallow water depths are the possible movement of the water surface and reflection phenomena (mirroring). Both disturbance factors can best be eliminated by placing a surface-parallel glass plate in an anti-reflection shaft and setting it on or immediately below the water surface. This simple and long-known principle is realized in the great variety of 'water observation windows', in modern free-diving masks, swimming mattresses with inbuilt panes and, finally, in the great variety of boats with inbuilt bottom windows. A special variation of the simple optical methods is shallow-water air observation (and air photography) from aeroplanes. If the water surface is calm, a visibility depth of 6 m can be attained in the western Baltic. This procedure is of particular interest for plant science and for applied studies in marine botany and has been much used (see Chapman 1944, Neushul 1965, Floc'h 1965, Schwenke 1966).

(b) Underwater Vehicles and Stations

Underwater vehicles and stations combine simple observation through windows with methods which are generally technically complicated and expensive; the latter are designed to transport one or several observers (protected from pressure) into great water depths, illuminate these artificially and make the observation instrument as mobile as possible. They include pressure capsules which are either lowered from a ship, such as the diving sphere 'bathysphere' of W. Beebe and the 'benthoscope' of Barton, or are suspended from freely swimming, controllable buoyancy bodies (filled with light petrol like the bathyscaphe 'Trieste' of Piccard, principle of the 'underwater balloon'). Included also are all types of undersea boats, and especially the various kinds of special small U-boats for research purposes. The development of these special vehicles began in about 1960 in America with the 'Aluminaut'; the end is not yet in sight. These are not purely observation vehicles, but also serve as instrument transporters, samplers and deep-sea working equipment. Finally we should mention the manned high-sea buoys which either float at the surface (such as Cousteau's 'Bouée laboratoire') or the permanent stations anchored to the sea bottom, such as the American 'Sealabs'. There is no question that these developments, most of which are extraordinarily complicated and expensive, represent nothing more than technical exercises at present. Gradually they will become of decisive importance for the whole field of marine research. For the marine biologist stations of the 'Sealab II' type are of particular interest because – anchored at depths of 50 to 60 m in the trials carried out until now – they can serve as permanent stations for free divers. They thus both allow a thorough study of a particular area and benefit a larger number of researchers (assuming physical fitness); and apart from the station itself, this is obtained without an expenditure which greatly exceeds that of the diving commonly undertaken today.

(c) Diving

Diving with or without simple aids is the best known, oldest and basically the easiest way for man to explore the sea bottom for short periods and in shallow depths. Simple aids of long standing are diving goggles, lines and weights (used by pearl-, sponge-, and mother-of-pearl divers). However, it is also known that among Greek sponge divers, for example, over exertion often leads to permanent physical damage. It is therefore quite evident that the modern, widely disseminated form of free diving for scientific purposes with mask, snorkel, flippers and, possibly, a heat-preserving suit should only be undertaken by physically healthy and adequately trained people, and that the guide lines obtained from experience in regard to the diving depths, diving time, and diving behaviour should be observed. In particular, it should generally be known that lengthening of the snorkel (preferably a straight model without valve) beyond 50 to 60 cm is a threat to life! A large number of reports in marine biology have already pointed to the usefulness of this simple method, particularly for ecological and underwater community studies. For marine botanical practice I refer to the reports of J. Ernst (1959) and J. Kornas and others (for example, 1960).

The main purposes of diving with technical equipment are to increase the diving depth and to lengthen the diving time. The bell and diving helmet, which were previously the only existing gear, have been used sporadically by marine researchers in order to inspect the sea floor. In modern practice equipment of this type is generally ruled out.

The decisive turning point and probably the main step forwards in this field has been the introduction of the free diving procedure. The pioneer works of Hans Hass, Cousteau, Drach, etc. are too well known to require special description here. Apart from particular methods which are uninteresting or unsuitable for general practice,

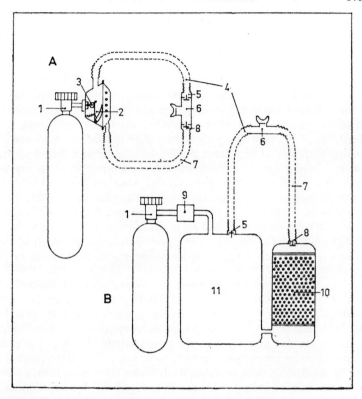

Figure 67. Diagrammatic sketches of diving apparatus: A = pressurized air apparatus; B = oxygen circulation apparatus; 1 = bottle valve; 2 = regulator; 3 = valve; 4 = inhalation tube; 5 = inhalation valve; 6 = mouthpiece; 7 = exhalation tube; 8 = exhalation valve; 9 = reduction valve; 10 = calcium filling; 11 = respiration bag

we distinguish two types of free diving at present: diving with pressurized air equipment and diving with oxygen circulation equipment (see Fig. 67). The latter is used especially for military requirements, since the presence of the diver is not betrayed by rising air. For sports and scientific purposes, however, only diving with pressurized air equipment comes into question, as it is much less dangerous and also makes it possible to attain greater diving depths.

Nevertheless, the following must be stated immediately, in opposition to the enthusiastic and romanticized propaganda encountered on this subject: although pressurized air diving is readily described as 'entirely safe', this is not really true of this method either. Not only must we consider whether individual instruments or parts of instruments can now be said to be foolproof; diving, in its totality, is a technically complicated undertaking from which the possibility of faults and errors, as well as faulty management, cannot be entirely excluded. It is indeed possible for anyone to buy diving equipment and to operate it as desired. However, we are speaking only of diving as it is conducted within the framework of a scientific institution; apart from questions of a personal and technical nature, this involves others concerning responsibility, discipline, safety and, not least, questions relating to insurance. Any individual interested in diving with special apparatus should definitely join a diving sports group. Diving with instruments cannot be learned theoretically by following a set of written instructions. Absolute prerequisites are physical fitness (medical examination for diving fitness!) and adequate training in swimming, including lifesaving and first aid. Only then does the basic theoretical and practical training with the equipment follow, including diving lifesaving procedures. The above yield the following principles for instrument diving within the framework of an institution:

Only trained divers may be used. No one, including a trained diver, is to be forced to dive against his will. Both general and special work and safety instructions relating to the particular context must be evolved and be made known adequately. A responsible diving leader must ensure that they are carried out and strictly observed. For each diving attempt an adequate number of divers in accordance with the safety regulations must be available. If diving takes place from a ship, the ship must be equipped for this purpose (diving ladder, lifesaving equipment). The diving equipment must be complete and in good condition. It consists of protective clothing suitable for the local temperature conditions (at present these are generally wet suits), the pressurized air diving equipment and further accessories: mask, snorkel, flippers, lead belt with one-hand rapid removal device, a good diving knife and a reliable depth indicator. Depending on the place and type of the dive, further equipment might be: underwater compass, watch, decompression meter, safety lines, buoys, underwater lamps, photographic and sound track apparatus, as well as sampling and recording gear (preferably sheets made of synthetic material, to be used with a soft pencil). The diving equipment itself consists of one or several pressurized air bottles containing 5 to 16 litres at 200 to 250 atm. The most important component of the equipment is the so-called automatic lung, a pressure-regulated valve with compensation chamber which reduces the bottle pressure to the correct respiratory pressure at any given time. The equipment is carried on the back with the aid of a carrying frame, the breathing tube being conveyed from there to the mouthpiece. For greater depths (more than 15 m) a double-tube system is required. The pressurized air bottles – generally two 7-litre bottles are used – are provided with a main valve and a reserve. Empty bottles are prepared for re-use at filling stations or with a special compressor device. The instruments and equipment should be checked regularly. This is of particular importance when heavy general use is being made of instruments and equipment in an Institute. In such cases it is best to employ a diving leader as chief officer.

In regard to the *physiology of diving* we can only mention the most important points here. The human body is burdened by mainly two factors underwater; the hydrostatic pressure and, on occasion, low temperatures. There is little danger that at the usual or even the greatest (instrument) diving depths the pressure will affect the human body as a whole; the blood and vascular system as well as the body cavities filled with gas or fluids are virtually incompressible. However, three physiological phenomena must be remembered when diving:

Equalize the pressure in order to protect the eardrum by swallowing movements (train!). People who have a permanent eardrum perforation should not dive!

Furthermore, beyond a diving depth of approximately 40 m there is the danger of the 'euphoria of the deep'. A euphoric floating feeling leads to careless actions which endanger life. Consequently one should never dive alone at such depths, and beginners in particular should be carefully watched!

Finally, the diver must realize that if he remains at fairly great depths for a protracted period, nitrogen will be dissolved in the blood. This is unavoidable and not dangerous in itself if one makes sure that when coming up the nitrogen does not bubble out because of an excessively rapid pressure decrease, which produces Caisson disease (gas embolism). The decompression times shown in tables must be observed rigorously in each case, therefore. Table 7 shows a few examples in order to demonstrate the relationships between diving depths, diving duration and decompression time.

The phenomenon described brings us to the physiological question of the most suitable respiratory gas to be used for instrument diving. This can be stated briefly and concisely: for normal diving use there is no reason to use anything other than pressurized air. Special respiratory gas mixtures, such as those which have been used in modern underwater stations (Cousteau) or by the Swiss H. Keller for depth diving trials are not relevant to the average diving practice.

When diving in cold seas it is most important to avoid partial or total hypothermia, which first paralyses the ability to concentrate and make decisions, and can ultimately result in a life-endangering situation! If suitably outfitted, trained divers can now work even in arctic waters under the ice.

TABLE 7

Decompression times in relation to the diving depth and diving duration

Diving depth m	Diving duration min	Time spent at the following steps min			Total decompression time min
15	120	—	—	2	2
20	60	—	—	3	3
	90	—	2	10	12
30	25	—	—	—	0
	60	—	16	16	32
	75	—	27	21	48
40	30	—	10	15	25
	60	13	28	29	70
50	10	—	—	7	7
	20	—	—	28	28
	40	16	27	30	73
60	10	—	—	16	16
	20	—	20	35	55
	30	15	26	42	83

(d) Underwater Television

Underwater television first met with little approval in the field of marine biological work. This may have been due partly to the fact that such apparatus is relatively expensive, and it must also be remembered that as a result of the small number produced, the many special requirements of the buyer and the expensive developmental work, the production costs are naturally very high.

The application possibilities are indeed manifold, including for example large scale vegetation pictures, biotope and behaviour studies of marine animals, as well as instrument function tests, perhaps in fisheries biology. Underwater television instruments can be used while suspended freely from slow-moving (or drifting) ships or from anchor stations; they can also be drawn along the sea bed on runners or operated on stationary ground frames. Some ground frames also have remote-controlled horizontal and vertical camera motion devices. Finally, most apparatus is arranged in such a manner that at least the television camera equipment (very light underwater) can be carried by the divers.

In view of the small number of models produced and the rapid technical progress, it makes little sense to describe special types of apparatus. The author relies especially on various instruments produced by the IBAK company, H. Hunger, Kiel. Other well-known manufacturers are Hydro-Products, San Diego (California) and Pye Ltd., Cambridge (England).

Instead, let us turn to the general construction principles of such instruments. Almost all of them have in common a suitable, i.e. manageable, commercial television photography camera which is as disturbance-proof as possible and is built into a cylindrical pressure casing. In the IBAK instruments the case has a conical plexiglass ('Perspex') window at the front; the posterior bottom plate carries the cable connections and by means of a flange and O-ring joint, makes up the seal of the pressure casing. In addition to the photographic camera, the casing contains the particular control elements required for the structure of the instrument involved, the remote control driving system for the optical foci and a humidity detector (humidity warning device). The instruments are usually combined with an automatic underwater photography device: a similar cylindrical pressure casing contains a small-picture camera (such as the 'Robot' system) together with a remote control device for the lens, distance and hoisting mechanism, as well as a humidity warning device.

Finally, a lighting device is included among the underwater parts; at its simplest this consists of two underwater floodlights (such as iodine vapour lamps of 500 W each) or more expensive arrangements may be used; very efficient underwater flash instruments are now available (electron flashes to 360 W sec).

Special underwater cables are used for transmission of the signal, control, and energy. The type and construction of these cables depends on the construction principle of the particular instrument. So-called shallow water instruments (as, for example, the IBAK Instrument Ingatlas 8a) work by video-frequency picture transmission, the control device of the photography camera belonging to the section above the water. In this case multicore special cables are used for the whole transmission system (such as 48 strands for the IBAK instrument 8a). When used in connection with a Vidikon photograph tube, this is the simplest and cheapest way. However, such cables are very cumbersome and – above all – very heavy when extended at considerable lengths above the water; moreover, they require expensive underwater socket connections. Their usefulness reaches its limit after as little as 400 m, for physical reasons. Because of frequency-determined damping, video-frequency picture transmission over cable lengths of more than 400 m is generally not possible. For greater depths – to 2000 m with special intermediate amplifiers for all depths – high-frequency picture transmitters must therefore be used. As a result, important control elements of the electronic apparatus must be built into the underwater casing. The transmission system requires only a relatively thin (such as a 17 mm) double coaxial cable which transmits, in addition to the picture signals, the operating current and the remote control signal (through impulse-controlled pacing switches). The above-water part of these instruments is very easily manageable as it consists merely of a wiring deck with a monitor set on top; the underwater part as well is not larger than that of video-frequency instruments (but the purchasing costs are approximately double).

Only a few words remain to be said in regard to the electronic and technical wiring principles of the photographic camera, insofar as they are required for a basic understanding of the equipment and for evaluation of its capacities.

The first underwater television cameras developed in England contained picture recording tubes from the Image Orthikon System. Since this method was readily subject to disturbance and the wiring was technically complicated, the use of Vidikon tubes (with antimony-trisulphide-mini-resistron) by the IBAK Company (Kiel) for their apparatus was an advance; these tubes, which had just been developed, had a simpler wiring system and were very robust in the form in which they are generally used in industrial and commercial television monitoring installations (such as the Grundig 'Tele-Eye') In addition to their well-tried Vidikon instruments this company now also builds Orthikon systems again, particularly because of the considerably higher light-sensitivity of this procedure. With a camera of this type it is possible, for example, to work in the Baltic at approximately 20 m depth without floodlights, if the visibility and light conditions are good. This is an important advantage for many biological (zoological and fisheries biology) uses. For photographic pictures a high-capacity flash instrument is used in this case.

For the method of using this equipment on board ship we may state the following: special limnological work (such as that of Ohle 1960, Schröder 1961) has shown that equipment of this kind can be used even on small vessels under very makeshift conditions. The fact that the underwater portion is very light at least in the water and consequently easily manageable, and that the power consumption (at about $1 \cdot 5$ Kw, including floodlight) is not too large is an advantage; a small, transportable, petrol-driven motor outfit can therefore be used to supply the power. The choice of the correct cable length is important for the equipment, since a case with 100 m special cable weighs 3-4 hundredweight. This can lead to irritation during improvised applications (transport!).

For use on well-equipped research vessels, the following principles hold true: the observation, operation, and control equipment should be kept in a darkened room. The use of sound-tape equipment is useful for recording the stepwise procedure. Apart from the amount required for the remaining work on board ship, the current generator of the vessel should provide an adequate amount of energy which is frequency-stable and disturbance-free. Because of the fine control required if the distances on the bottom are small, the underwater equipment should be driven over an electrical winch. The cable and wire must be prevented from twisting (possibly by the use of lock clamps); the permissible flexion radius is to be watched. An intercom system should be set up between the observation room, the winch operator and the bridge. If the work is being carried out while the vessel is travelling or drifting, the navigating procedure should be guided by an assistant on the bridge. This guidance includes: the mention of time signals (every two minutes, for example), the observation and reading of the echosound or echograph and the reading and recording of the position, which is nowadays generally obtained with the help of a Decca radio navigational instrument if the study area lies within the range of a Decca chain. Figure 68 shows a block wiring diagram of the IBAK instrument Ingatlas 8a as well as an overall picture for use on board.

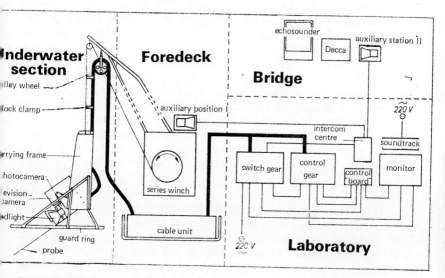

Figure 68. Block wiring diagram of an underwater television unit (Ingatlas 8a) and overall picture for use on board ship

Underwater television instruments can also be provided with (simple) sampling or grab devices. This makes specific sampling possible. In general the appropriate underwater camera will be used for a permanent record of the underwater observations. Screen image photographs are also possible, but are generally unsatisfactory. If the underwater conditions do not allow photography, sketches based on the screen image may be useful. As previously mentioned, the entire observation and operating procedure should be recorded on a soundtrack. It is very useful, especially if photographic conditions are poor, to set down the television observations by means of an image-recording instrument. Such instruments are now commercially available in manageable sizes and at acceptable prices.

2. UNDERWATER PHOTOGRAPHY

The need for brevity will be even more noticeable in the following sections than in the foregoing pages. The special characteristics of underwater photography and its specific problems require a thorough discussion in themselves in regard to both the optical and physical properties of the medium and the resulting photographic and photochemical modes of procedure. These principles are described in a prototype manner in the book by H. U. Richter, *Underwater photography and television*, Halle (Saale) 1960, which is unfortunately out of print and naturally somewhat out of date. This book also includes an extensive list of relevant bibliographical material (up to approximately 1957).

We must replace the comprehensive introduction by one basic point: anyone who wishes to work with underwater photography for scientific purposes, especially under conditions which are photographically unfavourable or involve objects which are photographically unrewarding (such as marine algae), requires a solid theoretical background and sufficient practical experience to develop his own optimum photographic method in each case, as well as a suitable photographic processing procedure based on his own knowledge. General guide lines may serve as initial directives only.

(a) Photography in the Area of the Water Surface

Photography in the area of the water surface can be of great importance for the marine biologist in carrying out shallow water botanical studies or zoo-ecological and similar studies in coastal waters. Pictures of this kind require a calm water surface and a place for the camera which is not disturbed by reflections. An obvious aid is the combination of a water observation chamber and a camera.

The construction and making of such a device is so simple, in relation to particular requirements, that no extensive description is required. The incorporation of an exposure meter in the underwater casing is useful. If possible, photographs in the region of the water surface should not be taken when the sky is cloudless or if the sun is low, in order to avoid a bluish tint and reflections when taking colour pictures. The best is diffuse light with a thin cloud cover. A polarization filter may be used against reflection, providing a favourable camera angle of approximately 37° (in relation to the total polarization angle of the water of approximately 53°).

(b) Underwater Photography by a Diver

Underwater photography by a diver is the true field of application of this technique. It presupposes a suitable camera contained in a pressure-resistant and waterproof casing. Operation and adjustment must be possible from the outside, i.e., waterproof connections for the operating mechanism and a probe adjusted to the underwater optical conditions are required. Underwater casings of this kind are available for a great variety of cameras and are naturally of very varying quality in regard to the reliability and manageability of the operation while diving. Guides to special constructions are found in pertinent technical literature (see, for example Richter 1960) and the especially in the sports diving publications, such as 'Delphin.'

For scientific purposes we generally do not turn to elementary work of this type, and even the commercially available equipment offers only a few really suitable instruments. An outstanding underwater camera whose superiority has remained unquestioned until now is the 'Rolleimarin' of Franke and Heidecke, which contains – in a well-constructed and easily operated pressure casing – a twin lens mirror reflex camera ('Rolleiflex') with a 6 × 6 cm picture format. A prism conveys the focusing screen picture to a large probe eyepiece which is slanted upwards and allows sharp direct optical focusing. Forelenses for close-ups can be inserted in front of the optical equipment from the outside. The mirror reflex principle makes distance conversion calculations to underwater optics unnecessary and, moreover, does not present any parallax problems. The 6 × 6 cm picture format is also an advantage; anyone who has seen a projection of 6 × 6 cm underwater photographs and a microphotography format next to each other will not dispute this. Even the limited number of 12 pictures per roll of film is an advantage rather than a disadvantage for scientific purposes.

Certain special cameras such as the French 'Calypsophot' and the Japanese 'Nikonos Allweather Camera' (microphotography format) should also be mentioned; these can be used without an underwater casing up to a certain depth. It is my opinion, however, that the quality of the 'Rolleimarin' photographs cannot be attained with these.

A necessary accessory is a reliable exposure meter in the underwater casing. In addition, one will require an underwater flash instrument, preferably independent of the problematical underwater light conditions. Plunger flash instruments are as easy to operate when diving as on land. In the 'Rolleimarin' the reflector is screwed to the underwater casing of the camera by means of a supporting rod. Instead of the often recommended bag for flash bulbs, it is advisable to use a clamp strip made of synthetic material (PVC) and attached to a supporting rod. Flash bulbs often tend to fall out if sacks or nets are used. Finally we should mention that underwater casings are available for many cine-cameras which, for obvious reasons, are larger and therefore less manageable than photographic casings. They are therefore often given a streamlined form and provided with stabilization fins. Films usually require special underwater floodlights; we should also mention the lighting apparatus developed by Rebikoff, the so-called Kino-Torpille.

The filming conditions in the sea are principally determined by the specific properties of the underwater light, familiar to every marine biologist, in regard to quantity (rapid decrease of intensity as the depth increases) and quality (alteration of the spectral composition with the depth). In addition it must be remembered that even under the best visibility conditions, large-scale overall pictures are possible only within certain limitations, since even then the underwater situation still only corresponds to terrestrial conditions in fog with a visibility range of approximately 20 m. It must also be remembered that in a photographic sense, the underwater landscape is basically poorer in contrast than the terrestrial one. This characteristic naturally shows up less when taking close-ups of sharply coloured marine animals than when taking vegetation survey pictures. It is obvious that all these problems are best met by the use of a flash instrument (within the framework of the particular power capacity). Arguments have sometimes arisen as to whether flash instruments provide 'unnatural' light and colour underwater. The marine biologist will not be greatly affected by this discussion; as a rule his purpose will be to represent his objects in the true colours, i.e. in colours which are not altered by the underwater light, and if necessary, he will also be capable of portraying the real submarine light situation.

A further special optical underwater feature is the apparent shortening of the distance, resulting in a slight tele-effect of the normal optics of the camera. The following rule of thumb is used:

actual distance divided by 4×3 = camera focus. One may also use a measuring rod which is $1 \cdot 33$ m long, which then corresponds to a camera focus of 1 m. However, obviously the best solution is the use of a mirror reflex camera.

(c) Deep Sea Photography

Deep sea photography in the narrower sense has been used for a fairly long period. We understand the term to mean the photographic representation of the sea bottom outside the shelf regions and at depths which cannot be reached by divers. It is difficult, but essentially unimportant, to differentiate deep sea photography with diving apparatus from deep sea photography with underwater vessels. Strictly speaking, deep sea cameras consist as a rule of a box-shaped set of metal rods to which two pressure casings are attached in the lighting position which is correct in relation to each other. One casing contains the underwater flash apparatus and the other a special automatic microphotography camera which can take some 500 pictures. The apparatus can be lowered from the research ship to the greatest oceanic depths. The camera can be operated either by a previously set time switch or by a bottom contact switch. The photographing is carried out virtually 'blind.' A certain element of control is provided by the 'pinger', which is nowadays much used in oceanographic technology. This is an underwater supersonic emitter which is attached to the deep-sea camera and whose impulses are traced on a suitable echograph. The depth of the camera in relation to the bottom echo tracing can thus be determined to the meter in each case without it being necessary to know the particular cable curve, which becomes complicated and can no longer be calculated when great depths are involved.

Finally we should include the deep-sea underwater television units (to 2000 m at present) in which the television camera functions as an electronic probe for an automatic, and (in this case) remote-controlled photocamera. This method allows specific but controlled and selective individual pictures.

(d) Photographic Material and its Processing

In regard to the photographic material and its processing it should first be said that the pertinent literature offers a great variety of ideas and that some underwater photographers claim a 'secret process'. It would seem obvious that the most sensitive types of film should be used in order to overcome the poor underwater light conditions. However, in the presence of a flat gradation curve and a low gamma value, these films are too low in contrast to provide satisfactory results. It is best to use moderately sensitive films of

14 to 17 °DIN if no extremely specialized conditions are involved. The usual colour films of 18 °DIN are also well suited for colour pictures. In principle, a flash instrument should be used for such pictures, namely, a blue glass bulb or electronic flash for close-ups to 1 m and a clear glass bulb for distances from 1 to approximately 5 m because of the better colour absorption.

In principle, the underwater photographer should develop and enlarge his own black and white films in order to obtain optimal results. Particularly good are rapid developers or the more highly concentrated fine-grain developers, which are made for line work; high acutance developers are unsuitable because underwater pictures taken without a flash are too poor in contrast. In this regard the otherwise excellent Neofin method in underwater photography is also disputed. The use of flash equipment, however, involves an adjustment to normal photographic conditions in each case and makes special work procedures largely unnecessary. Enlargements are made on white, glossy or highly glossy paper of normal or special grade. Underwater photographs tend to have a certain lack of sharpness in contours, which is further emphasized by matt paper surfaces; chamois papers yield dull pictures.

Once the methods have been worked out and the materials have proven themselves for the particular photographic conditions, it is advisable to retain them; this is the only way of working out an evaluation method for the possibilities and limitations of underwater photography.

REFERENCES

BARNES, H. (1963). 'Underwater Television.' *Oceanogr. Mar. Biol. Ann. Rev.* **1**, 115–128.

BIEBL, R. (1960). 'Unterwasserfotografie in der Botanik.' RICHTER, H. U., *Unterwasser-Fotografie und -Fernsehen.* Halle/Saale.

BRESLAU, L. R., J. M. ZIEGLER, and D. M. OWEN (1962). 'A Self-Contained Portable Tape Recording System for Use by SCUBA Divers.' *Bull. de l'Inst. Océanographique Monaco*, 1235.

CHAPMAN, V. J. (1944). 'Methods of surveying *Laminaria* beds.' *J. Mar. Biol. Assoc.* **26**, 37–60.

DIETRICH, G., und H. HUNGER (1962). 'Gezielte Tiefsee-Beobachtungen: Eine neue Tiefsee-Fernsehkamera mit eingebauter Fotokamera und mit gekoppelten Sammelgeräten.' *Dt. Hydrogr. Z.* **15**, 229–242.

EIBL-EIBESFELDT, I. v. (1960). 'Unterwasserfotografie in der Zoologie.' RICHTER, H.-U., *Unterwasser-Fotografie und -Fernsehen.* Halle/Saale.

ERNST, J. (1959). 'Studien über die Seichtwasservegetation der Sorrentiner Küste.' *Pubbl. Staz. Zool. Napoli* **30**, Suppl., 470–518.

KRIENKE, G. (1964). '*Mit der Kamera im Meer*. Rüschlikon-Zürich-Stuttgart-Wien.

NEUSHUL, M. (1961). 'Diving in Antarctic waters.' *The Polar Record* 10, No. 67.
— (1965). 'SCUBA diving studies of the vertical distribution of benthic marine plants.' Proceedings of the Fifth Marine Biological Symposium. *Botanica Gothoburgensia* 3, 161–176.
OHLE, W. (1960). 'Fernsehen', Photographie und Schallotrung der Sedimentoberflache in Seen.' *Arch. Hydrobiol.* 57.
RICHTER, H.-U. (1960). *Unterwasser-Fotografie und -Fernsehen.* Halle/Saale.
SCHRÖDER, R. (1961). 'Untersuchungen über die Planktonverteilung mit Hilfe der Unterwasser-Fernsehanlage und des Echographen.' *Arch. Hydrobiol.* 25, Suppl., Falkau-Schr. IV.
SCHWENKE, H. (1965). Uber die Anwendung des Unterwasser-Fernsehens in der Meeresbotanik.' *Kieler Meeresforsch.* 21, 101–106.
— (1966). 'Untersuchungen zur marinen Vegetationskunde. I. Über den Aufbau der marinen Benthosvegetation im Westteil der Kieler Bucht (westl. Ostsee).' *Kieler Meeresforsch,* 22 (2), 163–170.

Section IV
CONSERVATION AND CULTURE IN THE LABORATORY

A. Algae
BENTHONIC ALGAE

(P. KORNMANN)

In the last half century our knowledge of the development and propagation of marine algae has been greatly enlarged by the use of laboratory cultures. The sensational discovery of microscopic sexual reproduction of *Saccorhiza* by Sauvageau (1915) was one of the first results of this method. Soon afterwards the heteromorphic alternation of generations was also proved for other families of the Laminariales, by Kylin (1916, 1918) for *Laminaria digitata* and *Chorda filum*, by Kuckuck (1917) for *Laminaria saccharina*. Whereas Kuckuck carried out cultures in pure seawater, Kylin enriched his cultures with nitrate and phosphate. On the basis of nutritional physiology studies with diatom cultures, Schreiber (1927) produced a nutrient medium suitable for the culture of marine algae which contained the above-mentioned minimum substances in suitably formulated proportions in seawater. Schreiber explained the life cycles of *Desmarestia* and *Cladostephus* and demonstrated the genotypical sex determination in *Dictyota* by analysis of their spore tetrads and in *Laminaria* by cultivating the 32 zoospores of one sporangium in a culture experiment.

It was only in 1929 to 1934 that the isomorphic alternation of generations in *Chaetomorpha*, *Cladophora*, *Ulva*, and *Enteromorpha* became known through the work of Hartmann and Föyn. Successful culture of these species was achieved with the use of the nutrient medium introduced by Föyn as 'Erdschreiber solution', which contains, in the Schreiber nutrient solution, the growth-promoting substances of the soil extract discovered by Pringsheim in his studies of freshwater algae. From the large number of studies in life history – Papenfuss reviews the literature published until 1950 – we should like to single out only one report, a surprising discovery by Drew (1949, 1954): the calcium-boring *Conchocelis*, which lives in mollusc shells, is one stage of the life cycle of the flat-*Porphyra*.

The soil extract introduced an unknown factor into the nutrient solution, and numerous nutritional studies set out to identify it. The need for minimum mineral substances, growth hormones and vitamins for the growth of algal cultures was tested under sterile conditions. Provasoli (1963) listed the composition of various artificial nutrient media and the requirements of a series of marine algae.

Nutritional physiology questions will not be discussed in the following sections. The Erdschreiber solution yielded the best results when the aim was principally to culture algae of typical characteristics in the laboratory, which assuredly do not grow more slowly there than at their natural site. The method thus opens a broad field of study, as illustrated by a few examples in Section 2. An understanding of the complete life cycle of an alga is in itself a rewarding study. Taxonomy and systematics can no longer dispense with such studies, which offer the opportunity of further distinguishing forms with slight morphological differences on the basis of their developmental characteristics. The observation of a subject in its natural habitat in comparison with its observation under the defined conditions of the culture experiment leads to an understanding of its ecology. Finally, cultures of genetically uniform algae are rewarding material for study of developmental physiology.

In regard to the title of the report we should like to mention very briefly that conservation of marine algae is only possible to a very limited extent. Algae removed from their original habitat generally cannot be conserved without damage over a protracted period in containers filled with seawater. For physiological experiments it is therefore preferable to use freshly collected material kept under running seawater.

Certain algae show further growth in tanks with running or aerated seawater under suitable light and temperature conditions (Hillis, and others 1965); conservation thus changes directly into culture. *Caulerpa* species can be easily cultivated in this manner. Often various algae readily become established in aquarium tanks with running seawater; in Heligoland *Bryopsis* occurs regularly every spring. *Derbesia* is a typical aquarium alga, and the pink scales of the calcarous alga *Lithothamnion* also thrive on the walls of conservation tanks. A luxuriant growth of these algae requires the chance concurrence of favourable growth conditions; these should be studied in order to yield information which will be useful for a planned culture.

According to personal experience, a group of the larger brown algae shows excellent growth under suitable experimental conditions in the laboratory with running seawater. At a water temperature which fluctuated little around 10°C, *Laminaria* sporophytes grew to a length of 40 cm in three months. *Desmarestia aculeata* grew to a length of 50 cm and was abundantly covered with hair tufts, which were then discarded. *Desmarestia viridis* of entirely normal characteristics (length 30 cm) even became fertile during this period; there is no doubt that the plant attained this stage much more rapidly than under the natural conditions.

1. Culture Technique

In this section we will describe the methods which have proven themselves for the culture of green, brown and red algae after many years of experiences at the Biological Institute of Heligoland.

(a) The Nutrient Solution

Five litres of freshly collected seawater (unfiltered) is mixed with a solution of 0·5 g $NaNO_3$ and 0·1 g $Na_2HPO_4 + 12 H_2O$ in 20 cc distilled water and heated to approximately 90°C. No precipitation occurs during this process. To this Schreiber solution, which is always kept on hand, we add 25 cc soil extract per litre before use. This quantity has been found sufficient. In the literature the use of double this quantity of soil extract is generally reported.

Preparation of the soil extract:

(1) According to the specifications of Föyn (1934), 1 kg good garden soil is boiled in the autoclave for one hour together with 1 kg distilled water. After 2 to 3 days the mixture is decanted and re-sterilized. The brownish-yellow liquid, which has gradually become clear after being kept in the refrigerator for weeks, is re-decanted and nitrate and phosphate are dissolved in it in the quantities specified by Schreiber.

(2) In the Biological Institute of Heligoland the Soxhlet apparatus has been used for the last few years for the soil extraction. A flask containing 65 cc soil is extracted for 1 hour with approximately 750 cc water. The entirely clear, dark brown extract is re-heated to boiling point and is then ready for use. After standing for some time a flocculent precipitate is formed, which is filtered off.

(b) Preparation of the Cultures

Pure species cultures or unialgal cultures contain only one type of alga and no other contamination except bacteria. They are generally obtained from the propagation cells of the algae: immobile spores, swarm spores or zygotes. If no fertile material is available, an impure culture must first be prepared with a piece of thallus which is as free of contamination with foreign algae as possible. In general the algae becomes fertile after a few days. Occasionally the separation of freshly growing thallus ends which are not occupied by epiphytes also produces pure cultures.

The cultures should always be based on single plants which are kept either in the herbarium or, after careful fixation, as permanent preparations. If several species occupy one substrate from which they cannot be isolated, such as endophytes in spongy host plants or in mollusc shells, then the various types of germinating shoots must be isolated early.

In general, swarmers aggregate positively or negatively photo-tactically so that they can be collected with glass capillaries and placed directly in the culture dishes filled with nutrient solution. In many cases it will be useful to transfer the swarmers into fairly large drops until they are settled, and to fill the dish with nutrient solution subsequently. The advantage of this is that the germination stages in the region of the original drops are easy to find. When settled in drops on coverglasses it is easy to make up preparations of microscopic stages. The shoots generally adhere to the substrate solidly enough so as not to get lost during fixation and staining.

(c) Culture Vessels

In our experience, plastic petri dishes are more suitable for the culture of algae than the previously used glass dishes. The growth is entirely uniform, whereas a test series prepared in glass dishes under the same conditions can show considerable differences in growth in the individual containers. The plastic dishes can be re-used after cleaning in diluted chemically pure hydrochloric acid and thorough rinsing in clean water.

In general, dishes of 6 cm diameter, containing approximately 20 cc nutrient solution, are adequate. For example, a single plant of *Enteromorpha prolifera* will grow to a length of approximately 20 cm and become fertile in 4 weeks in one of these dishes without changing the nutrient solution. Fast-growing, tufted algae, which rapidly reach the surface (such as *Derbesia*) are best cultivated in cylindrical glass containers filled with a large quantity of liquid and loosely covered with glass tops or plastic foil.

(d) Culture Rooms, Temperature and Light

As the main culture room and, at the same time, workroom for the care and continuous observation of the algal cultures it is advisable to use a dark room kept at a uniform temperature of 15°C.

The cultures stand on tables provided with a 40 watt daylight fluorescent lamp at a height of 25 cm above the table centre. The light intensity perpendicularly below the centre of the lamp is some 1200 lux. The light intensity on the table surface is adapted to the needs of the individual species. The light is kept on for 14 hours a day.

Because of the seasonal temperature trend in our ocean area, only a few algae go through their complete life cycle at a uniform temperature of 15°C. Consequently, arrangements for cultures in various temperature ranges cannot be dispensed with: culture rooms or incubators, or containers cooled with circulating water. An over-all statement cannot be made in regard to the temperature range optimal for the plant growth or the fertilization of the individual species. According to our experiences an alternation between 15, 10 and 5°C was good for the observation of all the stages of development.

(e) Care

Here too only a few general statements can be made, of course. For the algae to grow well, it is important to limit the number of specimens growing in any one culture dish. It is only desirable to limit the swarming quantity in a beginning culture when some knowledge about its germination rate is already available. Otherwise it is preferable to remove a few plants for further culture from an overly dense initial culture. Single-plant cultures not only ensure good plant development, but are also necessary for an understanding of the sexual relationships.

In general a single cultivated alga, such as *Enteromorpha*, will become fertile in 4–5 weeks without a change of nutrient solution. The swarming discharge of a ripe plant can be triggered by allowing it to remain lying dry in the closed dish for several hours or even by changing the liquid.

If the nutrient solution is not changed, the plants gradually alter their characteristics as the nutrient substances are used up. By transferring small pieces into fresh nutrient solution it is possible to maintain the cultures fresh and for a number of years. At 15°C and with suitably weak light the algae grow so slowly that their renewal need only be carried out at intervals of 3 to 4 months.

2. APPLICATION OF THE CULTURE TECHNIQUE

(a) Developmental History and Taxonomy

(α) *Acrosiphonia* spp.:

Fig. 69 illustrates the heteromorphic life cycle of an *Acrosiphonia* species which forms isogamous gametes (A). A single-cell sporophyte

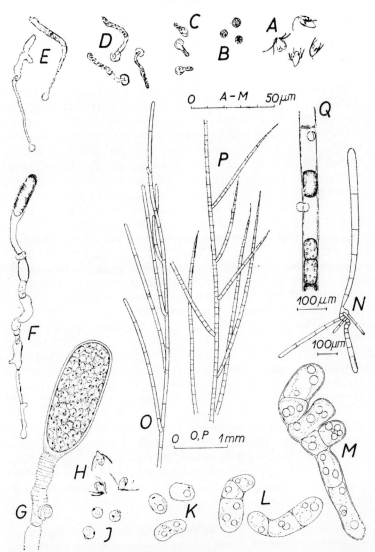

Figure 69. *Acrosiphonia* spp. Heteromorphic alternation of generations: A — G = development of the sporophyte; H — P = development of the gameto-phyte; Q = piece of filament with a ripe gametangium between two empty ones

(C – G) develops from the zygote (B). At a temperature of 15°C the little plants grow rapidly; the green head caps a stalk which is continually elongating and segmenting by means of false septa. At

15°C the sporophytes do not ripen; occasionally swarmers develop, but degenerate into sporangia. The formation of zoospores (G) which are capable of germinating is attained by transferring the 5-week old sporophytes into a temperature of 5°C.

The culture of the gametophytes is best achieved at 15°C; at 5°C the zoospore shoots only grow very slowly. A filamentous branched alga with the characteristics of a *Cladophora* (H – O) develops; it too only becomes fertile at a low temperature. Prior to the formation of the gametangium, the phenotype of the alga also alters in an entirely characteristic manner; the growing branches become pointed (P). The gametophyte is monozoic; the gametangium opening is a circular membraneous aperture which frequently adheres to the empty gametangium, forming a cover (Q).

This species, characterized by its life cycle, has not yet been found growing freely by the investigators of the Heligoland Institute. It occurred by chance in a single specimen of one culture. However, they often found a morphologically very similar form with similar generations; in this form the monozoic sexual plant develops directly from the zygote. The culture experiment carried out without interruption and repeatedly with both species is the basis for the taxonomic separation of these two forms, which are similar in characteristics and very different in their growth period. Several species described according to morphological characteristics are regarded as synonymous. Clarification of this morphological cycle and at the same time clarification of the nomenclature can only be expected from studies of the developmental history based on material obtained from the original sites of the types.

(β) *Urospora Wormskioldii:*

Fig. 70 shows a diagram of the life cycle of a group of *Urospora* species, which includes three entirely different morphological stages. Long, unbranched threads and a single cell *Codiolum* stage reproduce by zoospores bearing four flagella. A dwarf plant generation also forms four-flagellate swarmers, and under special conditions two-flagellate non-sexual swarmers.

In its natural location, *Codiolum* (the shoots grow in dense grass of approximately 1 – 1·5 mm height) becomes fertile in the autumn and winter. In the late winter and spring one finds the long, large-cell *Urospora* threads. The unremarkable dwarf generation was not observed under natural conditions.

According to the laboratory studies, the development tendencies of the four-flagellate swarmers are largely determined by temperature. At 10 and 15°C the *Codiolum* zoospores produce dwarf plants, and at 5°C threads. Their zoospores develop into threads or into dwarf

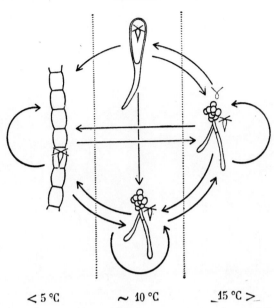

$< 5\ ^\circ C$ $\sim 10\ ^\circ C$ $15\ ^\circ C >$

Figure 70. *Urospora Wormskioldii.* Diagram of the life cycle. Further explanations in text

plants, according to the choice of the experimental conditions. Only at a high temperature do the dwarf plants form – apart from four-flagellate swarmers – two-flagellate ones from which the *Codiolum* generation originates. When cultivated at moderate and lower temperatures, it becomes fertile in as little as 4 to 5 weeks. At 15°C a resting stage in the development process occurs; the *Codiolum* cells only become ripe after being transferred to a lower temperature.

These data obtained in the laboratory and the pattern of the cycle can surely be transferred to the natural conditions; they allow us to understand the seasonally-determined appearance of *Urospora Wormskioldii* and *Codiolum gregarium*. These names are only meant to designate members of a cycle of morphologically similar forms. Their separation and taxonomical processing on the basis of culture studies is not yet completed.

(b) Developmental Physiology

(α) growth and synthesis of *Acrosiphonia:*

Fig. 71 shows the development of an *Acrosiphonia* thallus from a regenerating multi-nuclear special cell. The picture series encompassing five days requires only a brief explanation. The separated

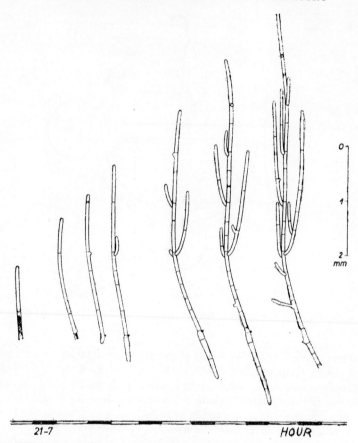

Figure 71. *Acrosiphonia* spp. Regeneration of the thallus in a separated apical cell. Daily light exposure 14 hours, temperature 15°C

apical cell immediately continues to grow. It grows at the same rate during the light and dark periods of the observation time. The cut sub-apical cell heals and grows into a rhizoid, which issues from the membrane sheath as early as on the day after the separation.

At 15°C and 14 hours of exposure to light the apical cell divides once a day, namely at around midnight. The division of the sub-apical cell and the formation of side branches at its upper end also occurs with complete regularity. Like the main axis, the branches increase by one cell a day. The growth rate of the individual axes varies, however, and consequently so does the length of the branched segments. For example, the younger of the opposing

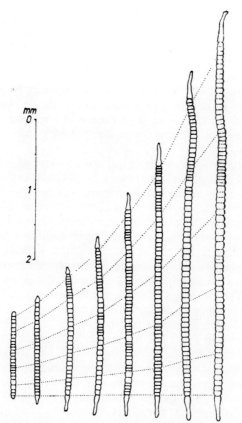

Figure 72. Urospora wormskioldii. Intercalary growth of a piece of filament at 5°C and daily light exposure of 14 hours

branches (on the three right-hand pictures) shows significantly slower growth than the older one, whose own side branch is even slower. It has been found that the size and growth rate of each apical cell is related to the number of its nuclei, which are irregularly distributed in the new cells at the time of each separation (Fig. 73).

(β) The intercalary growth of *Urospora:*

Fig. 72 illustrates an entirely different type of growth, an example of intercalary growth. It shows the change of a cell series of 1·3 mm length cut out from a thread of *Urospora wormskioldii* at intervals of days. All the cells constantly grow in length and in breadth; having attained a suitable length, they divide at a 4-day rhythm.

The result is a uniformly accelerated growth process, which stops when fertilization of the cells begins. As a rule fertilization progresses from the tips to the base; the upper end of a thread may be fertile and may already have swarmed while cell division is still taking place at the base.

The section cut out from the filament continues to grow in the same manner as it would have done in the sheath. Only the undamaged end cells lengthen in a rhizoid-like manner. The upper tubular cell is even re-incorporated into the division.

(γ) The division of the apical cell of *Acrosiphonia:*

The apical cells of a plant growing at 15°C and exposed to light for 14 hours a day divide simultaneously. The process, observed on the living specimen, shows one peculiarity: between 8 and 9 p.m. a hyaline ring forms at a certain distance from the tip. The ring does not change its position at first, but the cell continues to grow uniformly at the tip. Shortly after midnight a delicate separation membrane can be seen, positioned in the centre between a broad hyaline ring in the new apical cell and a much flatter one in the subapical cell. The two rings slowly separate further and further from each other and diminish during this process.

A series of stained preparations reveals the activity of the nuclei during this process (Fig. 73). All the nuclei from the tip of the cell as well as some from the underlying section (A, B) collect in the hyaline plasma ring. All the nuclei divide simultaneously (C). When the wall becomes visible, the nuclei with the hyaline rings have already progressed into the new cells (D). Their distribution into the two cells is irregular; the apical cell receives most of the nuclei.

The apical cell of *Acrosiphonia*, with its daily growth increment of approximately 1 mm, is certainly a rewarding object for gaining greater insight into membrane growth and differentiation processes of the cell, with the help of cinematography and the electron microscope.

3. FINAL COMMENTS

Growth processes and developmental processes are basic phenomena of life, whose study is of fundamental importance. Although the life cycles of many marine benthonic algae have been studied, until now they have been the subject of developmental physiology investigations to only a limited extent. Cultivated algae are particularly ideal subjects for observing the growth and development processes in relation to the test conditions and for attaining some understanding of the phenomena of the form changes. Typical forms

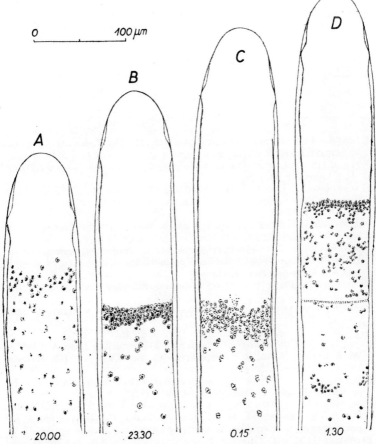

Figure 73.   Acrosiphonia spp. Activity of the nuclei in the dividing apical cell

are easily cultivated in the laboratory; they grow rapidly and are completely responsive. A special advantage is their simple structure; in monosiphonic unbranched or branched filaments the observed reactions can be recorded easily and evaluated quantitatively. Under certain test conditions the functions can be given a rhythmical pattern, an essential prerequisite for the study of division processes and differentiation processes in the cell.

It is certainly true that in the above paper the subject has been treated in a one-sided manner, the area of interest of the author having been given the first place. With an ecological orientation, V. Stosch (1962) has discussed the use of the culture experiment. A unique subject for cell physiology and biochemical studies are the single-nucleus giant cells of *Acetabularia*, whose morphogenetic processes in relation to the material conditions were studied by Hämmerling (1962) and co-workers. Further results of developmental physiology studies were collected by Lang (1965). The list of references also contains some studies which were not mentioned in the text.

REFERENCES

DREW, K. M. (1954). 'Studies in the Bangioideae. III. The life-history of *Porphyra umbilicalis* (L.) Kütz. var. *laciniata* (Lightf.).' *J. Ag. Ann. Bot.*, *N.S.* **18**, 183–211.

FÖYN, B. (1934). 'Lebenszyklus der Chlorophycee *Cladophora Suhriana* Kützing.' *Arch. Protistenk.* **83**, 1–56.

HÄMMERLING, J. (1962). 'Neuere physiologische und biochemische Ergebnisse über die Beziehungen Kern – Plasma bei *Acetabularia*.' *Vortr. Gesamtgeb. Bot.*, *N. F.* **1**, 20–28.

HILLIS (Colinvaux), L. K. M. WILBUR, and N. WATABE (1965). 'Tropical marine algae: Growth in laboratory culture.' *J. Phycol.* **1**, 69–78.

KOCH, W. (1965). 'Cyanophyceenkulturen. Anreicherungs- und Isolierverfahren.' H. G. SCHLEGEL (Hrsg.), *Anreicherungskultur und Mutantenauslese. Supplementheft 1 zum Zbl. Bakt.* 1, 415–431.

KORNMANN, P. (1965). 'Was ist *Acrosiphonia arcta*?' *Helgol. wiss. Meeresunters.* **12**, 40–51.

— (1965). 'Zur Analyse des Wachstums und des Aufbaus von *Acrosiphonia*.' *Ibid.* **12**, 219–238.

— (1966). 'Wachstum und Zellteilung bei *Urospora*.' *Ibid.* **13**, 73–83.

LANG, A. (1965). 'Physiology of growth and development in algae. A synopsis.' W. RUHLAND, *Handb. d. Pflanzenphysiologie.* Bd. XV/1, 680–715. Springer-Verlag, Berlin-Heidelberg-New York.

MÜLLER, D. (1962). 'Über jahres- und lunarperiodische Erscheinungen bei einigen Braunalgen.' *Botanica Marina* **4**, 140–155.

PAPENFUSS, G. F. (1960). 'Culturing of marine algae in relation to problems in morphology.' BRUNEL, J., G. W. PRESCOTT, and L. K. TIFFANY (eds.), *The culturing of algae.* A symposium. Kettering Foundation, Yellow Springs 77–95.

PROVASOLI, L. (1964). 'Growing marine seaweeds.' DE VIRVILLE, A. D., and J. FELDMANN (eds.), *Proc. 4. Intern. Seaweed Symposium*, 9–17. Pergamon Press, Oxford–London–New York–Paris.

SCHREIBER, E. (1927). 'Die Rheinkultur von marinem Phytoplankton und deren Bedeutung für die Erforschung der Produktionsfahigkeit des Meerwassers.' *Wiss. Meeresunters., Abt. Helgoland, N. F.* **16**/10, 1–34.

— (1930). 'Untersuchungen über Parthenogenesis, Geschlechtsbestimmung und Bastardierungsvermögen bei Laminarien.' *Planta* **12**, 331–353.

V. STOSCH, H. A. (1962). 'Kulturexperiment und Ökologie bei Algen.' *Kieler Meeresforsch.* **18**, Sonderheft, 13–27.

PLANKTONIC ALGAE

G. DREBES

The laboratory culture of planktonic algae has assumed special importance among marine biology working methods at present. This method makes it possible to clarify problems of form change or nutrition, as has already been shown by some examples in the section on the benthonic algae. Moreover, field observation in marine ecology has revealed a quantity of problems which cannot be solved as long as the culture experiment is not introduced. Thus, an analysis of the horizontal and vertical distribution pattern of algal populations, of the successions of plankton associations,

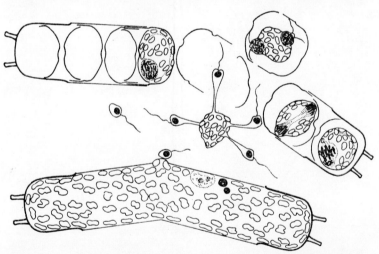

Figure 74. Diagram of the sexual organs of *Stephanopyxis turris.* Top: open spermatogonangium; only the husks of the first three spermatogonia from left shown; a cell in the pachytene phase follows. In the other half spermatogonangium, from right to left: metaphase and anaphase I, interchinesis with the flagella developing in pairs, and finally, already free of the husk, a tetrad with the sperm being loosened from the residual body of cytoplasm, which retains the plastids. Bottom: open oogonium ready for fertilization, with egg nucleus and two pycnotic nuclei as well as a spermatozoon on the open cytoplasmic surface (according to von Stosch and Drebes 1964)

of the annual cycles of certain species, is only possible if we know the reactions of each species or group of species within the complex environmental system. Since the algae occupy an important position in marine biological research in their role as primary producers, an understanding of their biology has become a pressing necessity. A few examples will illustrate the successful use of the culture process in planktonic algae.

The complete form change of the centric diatom *Stephanopyxis turris* (parts of the sexual cycle in Fig. 74) was investigated and studied in relation to environmental factors in the culture test (von Stosch and Drebes 1964). This plankton diatom can be used as a didactic example for the study of plant cell division, formation and germination of resting spores, differentiation of male and female gametes, fertilization and auxospore formation.

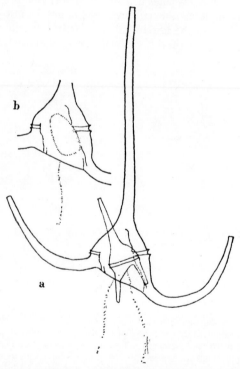

Figure 75. Ceratium horridum. (a) Early copulation stage of a small male gamete with a female gamete indistinguishable from a plant cell; (b) the same 2·5 hours later; the male gamete has fused into an oval body (drawn from life). (According to von Stosch 1964)

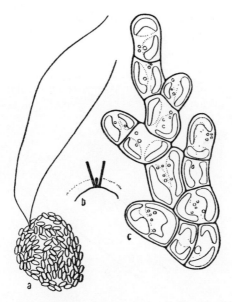

Figure 76. *Cricosphaera carterae* (Carter) Braarud (list of strains). (a) monad phase, only the small calcareous discs of the sheath and the flagella are shown; (b) flagella and haptonema, diagram, the coccolith-carrying jelly probably reaches to the dotted line; (c) *Apistonema*-phase, the pyrenoids are only shown at the lower side branch on the right. 750 X (according to von Stosch 1962)

The culture experiment also led to the first well-founded results relating to the question of the sexuality of dinoflagellates. In particular, the earlier literature to 1930 (reviewed by von Stosch 1964) yielded an abundance of morphological and cytological data on the family *Ceratium* which, however, was all found to be incorrect. By culture trials with *Ceratium horridum*, von Stosch (1964) established the described juvenile forms, temporal variations, secondary forms as *male gametes* and the amitotic budding as *gamete fusion* (Fig. 75).

A surprising discovery made by von Stosch (1955; also von Stosch and co-workers, unpublished) was the heteromorphic generation alternation (Fig. 76) between a highly differentiated diploid calcareous flagellate (*Cricosphaera*) and a haploid flat trichal alga (*Apistonema*). In the same year culture experiments by Kornmann (cited by von Stosch 1962) established that the life cycle of *Phaeocystis* includes a freely motile monadal phase and a palmelloid colony stage.

In regard to the nutritional requirements, it is surprising to discover through the preparation of absolutely pure (abacterial) cultures that the ideal autotrophic organism is not the rule under natural conditions among the algae. Numerous microscopic and macroscopic forms require supplementary organic substances – above all vitamins (for the literature see Lewin 1961). Figure 77 shows a linear correlation of the growth (cell number) of the chrysomonad *Monochrysis lutheri* with the vitamin B_{12} concentration in the medium (Droop 1961).

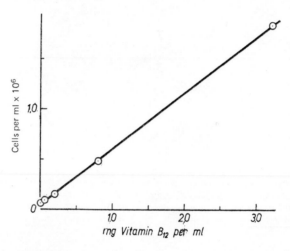

Figure 77. Dependence of the vegetative cell multiplication of the chrysomonad *Monochrysis lutheri* on the vitamin B_{12} concentration in the medium. Culture experiment at 15°C and 2000 lux. (According to Droop 1961)

We are indebted in particular to T. Braarud and his students in Norway for the use of the culture study as a means of answering ecological questions, such as the behaviour of a population of algae in relation to the classical environmental factors: salinity, temperature, light, and inorganic nutrient substances. Although these few experimental studies certainly cannot keep pace as yet with the data recorded from nature, their outstanding importance has been emphasized (see Braarud 1961). Figure 78 shows the dependence of the maximum growth rate on the temperature in a group of dinoflagellates representative for the North Sea.

Before describing the culture method – in this brief report we can only cast a rapid glance at the practice – we should like to refer to the fundamental work of Pringsheim (1954), which discusses in detail the preparation and maintenance of pure algal cultures.

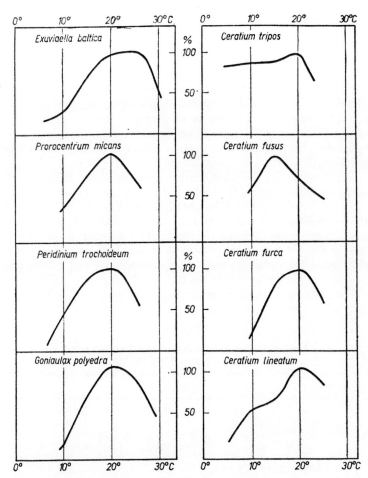

Figure 78. Dinoflagellates: growth at various temperatures as a percentage of the maximum growth rate (Bakken, unpublished; Nordli 1957; from Braarud 1961)

1. CULTURE METHOD

(a) Nutrient Solution

Difficulties with the culture may be due either to special require-ments of the algae or to particular sensitivity on the part of the subjects. The nutrient solutions proven in practice range from the Erdschreiber solution through semi-synthetic solutions to fully synthetic solutions. Only the latter can claim to be exactly definable and consequently reproducible. In many cases the choice of the nutrient medium is a compromise with the experimental expediency. Thus, a semi-synthetic solution is adequate when problems of taxonomical-life history or developmental physiology are involved. Those who only cultivate the algae for feeding purposes generally use the loosely defined Erdschreiber solution. In contrast, nutritional physiology studies always require entirely synthetic nutrient solutions and the simultaneous exclusion of bacteria. Such sterile cultures are known as *axenic cultures* or *absolute pure cultures*. For cultures containing bacteria the name *unialgal culture* or *pure species culture* has been adopted.

The formula for the Erdschreiber solution (Föyn 1934) has already been described in the foregoing section relating to the culture of benthonic algae (cf. p. 192). We should now like to discuss briefly the composition of a semi-synthetic nutrient solution. It is a nutrient medium which has been successfully used for years in developmental history and developmental physiology studies with marine plankton diatoms at the Botanical Institute of Marburg by von Stosch and now by the author himself in the Biological Institute of Heligoland. Table 8 shows the culture medium for the diatom *Stephanopyxis*

TABLE 8

Culture medium for the diatom *Stephanopyxis turris*
(according to von Stosch and Drebes 1964)

Seawater	1·02 kg			
NaNO$_3$............	42·5	mg	70,000	μg nitrogen
Na$_2$HPO$_4$.12 H$_2$O ...	10·75	mg	930	μg phosphorus
FeSO$_4$.7 H$_2$O	0·278	mg	55·8	μg iron
MnCl$_2$.4 H$_2$O	0·0198	mg	5·9	μg manganese
SiO$_2$................	12	mg	5·6	μg silicon
Na$_2$EDTA.2 H$_2$O....	3·72	mg		
Vitamin B$_{12}$	0·0007	mg		

turris. In contrast to an entirely synthetic solution (Table 9), this semi-synthetic nutrient solution is based on natural seawater. It represents a modification of the Schreiber formula (1927), since in addition to nitrate and phosphate, the natural seawater has had iron, manganese and SiO$_2$-Sol added to it. The metal chelator EDTA and the vitamin will be discussed later. In the waters of the North Sea nitrate and phosphate act as the two primary minimum substances for diatoms (Schreiber 1927). The absence of these substances leads to inhibition of the division process and bleaching of the chromatophores; since photosynthesis continues, the reserve substance *leucosin* (= chrysolaminarin) accumulates, and fat is stored in the form of droplets. The iron requirements of algae vary (see the literature cited by von Stosch and Drebes 1964). On the whole the problem of the iron supply in seawater nutrient solutions, particularly in the presence of chelators, is a complex one. Deficiency of this element also leads to bleaching of the cells. Manganese must be added to the solution because some natural seawaters produce signs of manganese deficiency in diatoms. In *Achnanthes longpipes* (von Stosch 1942) and in *Stephanopyxis turris* (von Stosch and Drebes 1964) division was inhibited shortly after the occurrence of a manganese deficiency, while plasma growth continued. The quantities of silicic acid present in the seawater and the additional quantities given up by the glass are not sufficient to meet the requirements of the diatoms; we therefore add it in the form of $0 \cdot 6\%$ brine. The addition of sodium silicate would have an appreciable effect on the total alkalinity and on the hydrogen ion concentration. The brine is prepared by hydrolysis of $SiCL_4$ or, more simply,

TABLE 9

ASP 2-medium (an entirely synthetic nutrient solution from the Haskins Laboratories)
(see Provasoli, McLaughlin and Droop 1957)

NaCl	1,800	mg
$MgSO_4 . 7 H_2O$	500	mg
KCl	60	mg
Ca (as Cl^-)	10	mg
$NaNO_3$	5	mg
K_2HPO_4	$0 \cdot 5$	mg
$Na_2SiO_3 \cdot 9 H_2O$	15	mg
'TRIS'*	100	mg
B_{12}	$0 \cdot 0002$	mg
Vitamin Mix S 3†	$1 \cdot 0$	ml
Na_2EDTA	3	mg
Fe (as Cl^-)	$0 \cdot 08$	mg
Zn (as Cl^-)	$0 \cdot 015$	mg
Mn (as Cl^-)	$0 \cdot 12$	mg
Co (as Cl^-)	$0 \cdot 0003$	mg
Cu (as Cl^-)	$0 \cdot 00012$	mg
B (as H_3BO_3)	$0 \cdot 6$	mg
H_2O	100	ml
pH	$7 \cdot 6 – 7 \cdot 8$	

* 'TRIS' = Tris (hydroxymethyl) aminomethane (buffer substance).
† 1 ml Vitamin Mix S 3 contains: thiamine.HCl, $0 \cdot 05$ mg; nicotinic acid, $0 \cdot 01$ mg; Ca-pantothenate, $0 \cdot 01$ mg; p-aminobenzoic acid, $1 \cdot 0$ μg; biotin, $0 \cdot 1$ μg; inosit, $0 \cdot 5$ mg; folic acid, $0 \cdot 2$ μg; thymine, $0 \cdot 3$ mg.

from sodium silicate by exchanger according to the Pramer technique (1957). The diatoms require the silicic acid to build up their shells. In addition, however, silicic acid appears to be involved at another point in the metabolism of the cell, since it is frequently the cell division which is first inhibited in the presence of a silicic acid deficiency, whereas plasma growth and chromosynthesis continue more or less vigorously (von Stosch 1942; von Stosch and Drebes 1964). This sign of SiO_2 deficiency can be used as a method for altering the cell size by vegetative means in a number of diatoms (von Stosch 1965).

Other culture problems can be attributed to a special sensitivity on the part of the subject. Thus, according to Hutner and others (1950) a substance such as ethylene diamine tetra-acetic acid (EDTA) should be added to the seawater medium in order to maintain growth after the first division in some algae – *Stephanopyxis turris*, for example. The EDTA was introduced in the form of metal buffer in order to administer the quantities of heavy metals required for a high production without causing toxicity symptoms. To a certain extent the soil extract of the Erdschreiber solution, which contains humic acid of equally complex-building properties, may replace EDTA. However, the synthetic product is preferable to the soil extract, as the latter has a toxic effect on some algae and varies in its composition, which is not exact in any case.

It was discovered with surprise in the two last decades that a large number of algae are not entirely autotrophic; their synthesis apparatus cannot produce certain organic substances, primarily *vitamins*. These must be obtained externally, a fact known as *auxotrophism*. Vitamin B_{12} (cobalamine) is needed especially often, and in addition vitamin B_1 (thiamine) and biotin (extensive discussion in Lewin 1961). According to a rough estimate, some 70% of the plankton algae have an auxotrophism for vitamin B_{12} (Fogg 1965).

The development of artificial media for marine algae has been extensively discussed by Provasoli, McLaughlin and Droop (1957). These and other investigatiors prepared a series of different fully synthetic nutrient solutions and also established the special requirements of certain types of algae in axenic culture in regard to nutrient substances. Table 2 shows the composition of one fully synthetic nutrient solution, known as *ASP 2-medium*. This solution can be used equally for unialgal and axenic cultures and has been found very good for a number of diatoms, chrysomonads, dinoflagellates, blue algae, and green algae.

To summarize, the significant advances in the culture of algae consist of the introduction of chelators (metal buffer) and in clarification of the vitamin requirements. However, the properties of the metal chelators and of the other synthetic products used as buffers (TRIS, Glycyl-glycine, etc.) should not be overestimated, since they may produce toxic effects in some algae. Taylor (1964) has been attempting to produce media without chelators.

(b) Culture Vessels, Temperature and Lighting

The culture vessel consists of quartz, glass or synthetic material. We use small petri dishes (diameter: 5, 6 or 10 cm) made of Jena glass (G 20). In Britain 'Pyrex' reagent tubes and bottles are used. Recently pre-sterilized plastic dishes which are discarded after one use have become commercially available. The breeding of algae as foodstuffs requires larger containers. We culture the green alga *Dunaliella*, for feeding to the brine shrimp *Artemia salina*, in 5-litre bottles made of Jena glass. As containers in the widest sense we should also mention the apparatus known as a *chemostat* and *turbidostat*, used for *continuous* mass cultures (for description, see Fogg 1965).

The culture glassware is best cleaned with chlorate sulphuric acid (not chromosulphuric acid!) and subsequently dry-sterilized at 120°C.

The cultures are kept in windowless rooms at a constant temperature. An adequate number of rooms with a constant temperature is often not available, but these can be successfully supplemented by temperature-controlled cupboards. A temperature range of 0-30°C is sufficient for culture experiments. The cultures stand on a white base and are exposed to light from fluorescent light tubes. In most cases the light intensity need not exceed 5,000 lux. Usually 14:10 hours is selected as the light-dark rhythm.

(c) Isolation Technique and Care of the Algal Culture.

For the isolation of one alga from a plankton sample we use a *mouth pipette*. At its broad end this pipette is provided with a thin rubber tube for drawing up and blowing out, and has a small opening with a diameter which should not exceed by more than 2 to 3 times the cells to be pipetted. A preparation microscope is used for optical aid (magnification 10 to 40 times, if possible with an additional dark field device). The algae are gradually washed free of flagellates by being transferred repeatedly into solid watch-glasses filled with seawater. Distilled water and a reflection burner stand ready for sterilization of the mouth pipette. The purity test of a freshly prepared culture or one made up a few days previously is best carried out under dark field illumination; only by this method can any contamination, such as minute flagellates, be identified as bright, flashing, mobile dots. The algae can be freed of bacteria by spreading on agar nutrient substrate. A further possibility is offered by centrifugation of the algae together with some culture fluid. The supernatant fluid in the centrifugal glass is repeatedly replaced by sterile media during this process. Short-term treatment with antibiotics can further improve these methods.

Cultures of North Sea algae can be kept for several months at 15°C on shelves in the shade (100-400 lux) without re-inoculation. It is advisable to replace the evaporated water (weigh!) in the cultures from time to time. The cultures are re-inoculated by the mouth pipette. It is best to allow freshly prepared cultures to begin growing vigorously under strong light exposure before placing them in the shadows. A great deal more could be said regarding the activity and treatment of members of various classes of algae, as well as about the information yielded by culture studies on the modalities under which external factors – especially light, temperature and nutrition – control the phase changes of the algae. We would like to terminate this brief report, which has at its core the detailed discussion of a semi-synthetic nutrient solution, by reproducing (Fig. 79) a growth diagram by Fogg (1965). This diagram is characteristic for a vegetatively growing unicellular alga in a culture with

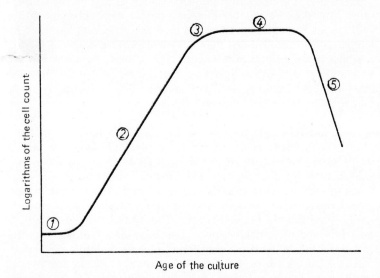

Figure 79. Characteristic growth diagram of a one-cell alga in a culture with a limited volume. 1 = resting phase, 2 = exponential (logarithmic) phase, 3 = phase of the decreasing relative growth rate, 4 = stationary phase, 5 = declining phase. (According to Fogg)

a limited volume. The individual growth phases are discussed in detail by Fogg (1965).

REFERENCES

BRAARUD, T. (1961). 'Cultivation of marine organisms as a means of understanding environmental influences on populations.' *Intern. Oceanogr. Congress New York 1959. Amer. Assoc. Adv. Sci.*, Washington, No. Publ. 67, 271–298.

DROOP, M. (1961). 'Vitamin B_{12} and Marine Ecology: the Response of *Monochrysis lutheri.' J. Mar. biol. Assoc. U.K.* **41**, 69-76.

FOGG, G. E. (1965). *Algal Cultures and Phytoplankton Ecology.* 126 S. London.

FÖYN, B. (1934). 'Lebenszyklus, Cytologie und Sexualität der Chlorophycee *Cladophora Suhriana* Kützing.' *Arch. Protistenkd.* **83**, 1–56.

HUTNER, S. H., L. PROVASOLI, A. SCHARTZ, and C. P. HASKINS (1950). 'Some approaches to the study of the role of metals in the metabolism of microorganisms.' *Proc. Amer. phil. Soc.* **94**, 152–170.

LEWIN, R. A. (1961). 'Phytoflagellates und Algae.' *Handb. d. Pflanzenphysiol.* **14**, 400–417. Springer, Berlin.

PRAMER, D. (1957). 'The influence of physical and chemical factors on the preparation of silica gel media.' *Appl. Microbiol.* **5**, 392–395.

PRINGSHEIM, E. G. (1954). *Algenreinkulturen, ihre Herstellung und Erhaltung* 108 S. Fischer, Jena.

PROVASOLI, L., J. J. A. MCLAUGHLIN, and M. DROOP. (1957). 'The development of artificial media for marine algae.' *Arch. Microbiol.* **25**, 392–428.

SCHREIBER, E. (1927). 'Die Reinkultur von marinem Phytoplankton und deren Bedeutung für die Erforschung der Produktionsfähigkeit des Meerwassers.' *Wiss. Meeresunters., Abt., Helgoland*, N. F. **16**, 10, 1–34.

STOSCH, H. A. v. (1942). 'Form und Formwechsel der Diatomee *Achnanthes longipes* in Abhängigkeit von der Ernährung. Mit besonderer Berücksichtgung der Spurenstoffe.' *Ber. dt. bot. Ges.* **60**, 2–15.

— (1955).'Ein morphologischer Phasenwechsel bie einer Coccolithophoride.' *Naturwiss.* **42**, 423.

— (1962). 'Kulturexperiment und Ökologie bei Algen.' *Kieler Meeresforsch.* **18**, Sonderheft, 13–27.

— (1964). 'Zum Problem der sexuellen Fortpflanzung in der Peridineengattung *Ceratium.' Helgol Wiss. Meeresunters*, **10**, 140–152.

— (1965). 'Manipulierung der Zellgrösse von Diatomeen im Experiment.' *Phycologia* **5**, 21–44.

— und G. DREBES (1964). 'Entwicklungsgeschichtliche Untersuchungen an zentrischen Diatomeen. IV. Die Planktondiatomee *Stephanopyxis turris*—ihre Behandlung und Entwicklungsgeschichte.' *Helgol. Wiss. Meeresunters*, **11**, 209–257.

TAYLOR, W. R. (1964). 'Inorganic nutrient requirements for marine phytoplankton organisms.' *Proc. Symp. Experim. Mar. Ecol.*, Occasional Publ. No. 2, 17–24.

B. Animals

INVERTEBRATES

C. HAUENSCHILD

In virtually all scientific studies of marine invertebrates the test animals must be kept in the laboratory if continuous live observation is required. We distinguish two methods in this context: for certain physiological and behavioural studies '*maintenance*' is sufficient; the animals are kept in aquaria or similar containers and the aim is to keep the specimens – which have been captured under free conditions – alive for the duration of the experiment and under the best possible conditions; there is no attempt at regular propagation or at continued breeding of the offspring. This method also includes the raising of a test animal in captivity from the egg to sexual maturity, i.e., care extending only to one generation. For many developmental physiology studies and all genetic investigations it is indispensable, however, to set up a *rearing trial* (*=culture*) in the laboratory, in which one type of animal is continuously cultivated over a number of generations; of primary importance here is not so much the time-limited conservation of the individual animals, but rather the regular alternation of rearing and propagation, i.e., the maintenance of a stock, theoretically without time limit (to some extent an artificial population). The continuous culture of one animal species generally requires a greater investment of work than does preservation; however, its advantage is that irrespective of the station, weather and season, various developmental stages of the particular organism are always available and its entire life cycle can be followed repeatedly under controllable conditions. Since the requirements for the preservation method and the rearing method differ greatly, the two will be discussed separately below.

1. MAINTENANCE

Maintenance, i.e., the temporary care of living marine animals in the laboratory, is a possibility above all for large-size invertebrates which can scarcely be reared if only because of their body mass, and which can only be kept in large aquaria for a short time. In contrast, if small forms are involved (< 1-2 cm), if possible one should attempt to prepare definite cultures if these are frequently required for study or demonstration purposes (p.223); for in the immensely complex biocenosis of a 'normal' aquarium such animals generally tend to disappear again as suddenly as they appeared.

If we survey the limited literature which has been published on the subject until now, it would appear that on the whole the care of marine invertebrates in the aquarium is a subject which has received little study in the past. Accordingly, only a very few points of reference are available in regard to the requirements of certain species and these are generally limited to a few standard animals; in most cases one must begin the study of every new animal species by collecting one's own data. General information on the equipping and operating of seawater aquaria is given by Müllegger (1955) and Wickler (1962); this pamphlet, as well as the work of Riedl (1963) also contains some brief information on the conservation of marine invertebrates. Hückstedt (1963a, 1963b) and Geisler (1964) deal in particular with the chemical properties of seawater. All the data in these reports naturally relates to indoor aquaria or exhibition tanks. If marine animals are kept solely for experimental purposes, a somewhat different arrangement may be adopted, the aesthetic aspects being left out; only a few important points will be brought out here.

The *container* and all other parts which come into contact with the seawater (such as pumps, pipes, tubes, regulating taps, heating elements, filtering devices, etc.) must consist of a material with high-grade resistance to seawater; in practice these are mainly glass, earthenware, asbestos cement and a number of synthetics (Plexiglass ('Perspex'), polystyrol, PVC, perlon, etc.). Contact between the chemically very aggressive seawater and any metal object – even if only briefly – must be avoided at all costs, since this dissolves out more or less toxic substances. Consequently, the most suitable aquaria are solid glass tanks (dimensions limited for static reasons, visibility somewhat hindered by irregularities and streaks in the glass) and Eternit containers (wet repeatedly before use!) or those made of fireclay, which have a large glass plate set into the long side walls; framed aquaria should only be used after very careful coating of all metal parts and cement joints with a suitable isolation mass such as Icosit (for coating: K 240; for sealing the joints: K 250). A solidly fastened, thick glass coverplate should be used to protect against evaporation and water splashes, as well as against the escape of freely motile animals; the undesirable crystallization of salt at the top edge of the aquarium can be avoided by laying the cover

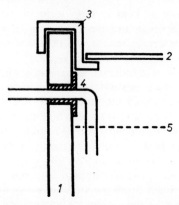

Figure 80.   Arrangement of the cover and pipes on an Eternit tank (drawing by D. Hofmann). 1 = side wall of the aquarium, 2 = cover plate, 3 = PVC holding clamp, 4 = inlet tube introduced through a laterally bored hole, with seal (cross hatched), 5 = water level

plate on a PVC frame and lowering it to a depth of approximately 2 cm into the top opening of the tank (see Fig. 80). When equipping an aquarium it is very helpful if all the pipes, tubes (preferably made of silicon) and cables (with plastic isolation) can be introduced through precisely fitting lateral openings below the coverplate; when Eternit tanks are used, they can be readily drilled for this purpose. It should be possible to expose each aquarium to light at a different intensity and for a different period (time switch); fluorescent tubes have been found useful for this purpose; they are screened off at the top and sides by a reflector and their light intensity can be reduced at will by varying the thickness of the paper placed on the coverplate. Intensive exposure to sunlight is usually undesirable, especially because of the associated heating.

For scientific purposes it is generally useful to set up as many tanks as possible, not too large in size (50 to 200 litres), so that a variety of animals can be accommodated separately. It is best to connect the individual aquaria to a common seawater circulation system, either in a parallel sequence or one after the other. In order to achieve complete water rotation, each tank must be provided with an inlet and an outlet on opposite sides, the one being at the top and the other at the bottom; in aquaria connected one after the other all the communication pipes must be screened on both sides with a perlon gauze of suitable mesh width so as to prevent the animals from wandering from one container into the next. The water circulation can be actuated by centrifugal pumps (seawater-proof model only, however), peristaltic pumps or mammoth pump outfits; if the water flow and water rotation are sufficiently fast, additional aeration of the individual tanks is generally unnecessary. In that case the temperature for all the aquaria connected to the circuit can also be regulated centrally by means of a contact thermometer and a relais. Depending on the temperature required, glass heating elements of suitable capacity and/or a cooling unit with immersion cooler are used; but the latter, which is generally made of metal, should not come into direct contact with the seawater. If animals from the North Sea, Mediterranean, and tropical waters are to be accommodated simultaneously, three separate seawater circuits must be installed and adjusted to different temperatures (approximately 15°, 21°, and 27°C). The internal arrangement of the aquaria is kept to the barest minimum so that all the animals can easily be found and all food remains and excrement can be removed completely. A bottom cover can generally be omitted; only those animals which dig themselves into a sandy or muddy bottom as part of their essential living requirements should be given a suitable substrate. Hiding places can consist of inverted flower pots; these are easier to check than massive constructions made of stones or coral blocks. Special problems are presented by rapidly moving animals who soon damage themselves more or less severely by hitting continuously against the walls of the aquarium. The conservation of such invertebrates can sometimes be improved by placing them in a round container with a circular water flow. It may also be useful to pad the walls with a very soft foam material.

Of decisive importance for the conservation is the quality of the seawater and its maintenance. For institutes which are located directly by the sea, an open circuit in which fresh seawater is continually being drawn in and let out again after passing through the aquaria once is naturally the ideal solution; the water at the withdrawal site must of course be free of harmful contamination (such as oil, sewage) at all times. Before entering the aquarium installation the water should run through a fine perlon sieve in order to prevent the introduction of any undesirable macro-organisms. Even when using an open circuit it is advisable to avoid as far as possible any contact between the seawater and metal parts throughout the pump and tubing system. The only disadvantage of the open system is the relatively high cost for any required heating or cooling. For inland systems a closed circuit only comes into consideration, the seawater being available in a limited quantity; after running through the aquarium chain this water is purified in a filter and then returned, possibly through a fairly large reservoir, into the circuit. Since in this system the slightest contact between the seawater and the metal is particularly damaging because of the progressive concentration of the harmful ions, a closed circuit aquarium unit must be in absolutely perfect condition in this respect. In a plant of this kind synthetic seawater will generally be used for reasons of economy, but it is advisable to add at least some natural seawater (as for example one-fourth); the organic constituents of natural seawater, which have not yet been analysed precisely individually and which are absent from the synthetic medium, do seem to be of some importance for many animals. The analysis of seawater shown in Table 1 (p.2) can serve as a basis for the composition of artificial seawater. The following ten ions are essential constituents: Na^+, Mg^+, Ca^+, K^+, Sr^+, Cl^-, SO_4^-, HCO_3^-, Br^- and I^-; this corresponds to the generally used seawater formula shown in Table 10a (p.225). It is logical to add the numerous trace elements in weighed quantities naturally only when highly purified substances are used for all the principal constituents (especially $NaCl$); the addition of a complex builder (such as 20 mg Na_2EDTA per litre) is sometimes useful, and is particularly important in the presence of an overdosage of trace elements. However, with many animals such an expenditure is certainly not necessary, particularly if a quantity of natural seawater is mixed in, and it may be sufficient to introduce numerous trace elements (in quantities which cannot be regulated, of course) into the synthetic medium by using a little purified $NaCl$.

The density of the seawater ($1 \cdot 02$ to $1 \cdot 03$) must be continuously checked with an hydrometer and must be maintained approximately uniform by the timely addition of distilled water.

A very wide range of materials is in use for filtration of the water; all must be washed out, regenerated or renewed at regular intervals (see Fig. 81). Sand and perlon wadding by nature retain only the coarser suspended matter mechanically; nevertheless certain colloid layers may develop on them, as well as on the basalt chips which are frequently recommended as a 'biological filter'. These layers absorb proteins and their degradation products, and lead to rapid bacteriological decomposition. The same purpose is served in a somewhat different manner by filtration through active carbon or through adsorptive artificial resins (such as 'N-Ex'); apart from the fact that under certain conditions both adsorb important constituents of the seawater, the latter in particular are not entirely acceptable because of a possible release of phenol. The filtration process can be combined with a buffering through regenerating ion exchangers (such as Lewatit MP 60, pH $8 \cdot 1$ to $8 \cdot 4$) and with an ozone treatment which oxidizes the metabolic products in the water. All of these measures are aimed at maintaining the composition of the seawater within physiologically tolerable limits as long as possible,

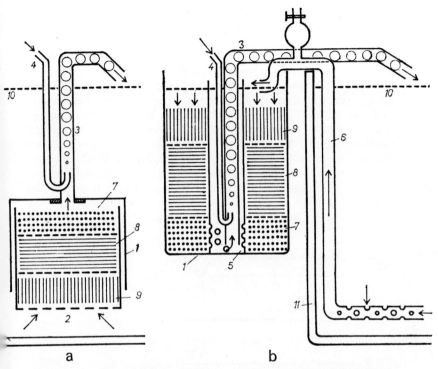

Figure 81. Aquarium filter (according to Hückstedt 1963a; drawing by D. Hofmann). (a) internal filter, (b) external filter. 1 = filter casing (glass or synthetic material), 2 = filter plate, 3 = ascension tube, 4 = air inlet, 5 = cylinder into which the filtered water is introduced from below, 6 = siphon tube, 7 = gravel, 8 = activated carbon (or other filtering material), 9 = perlon wadding, 10 = water level, 11 = aquarium wall

despite the protein metabolism products continuously accumulating in the aquaria. In tanks with a normal quantity of animals, however, even with the most careful care of the water it is not possible to prevent the gradual accumulation of nitrate as end product of protein in ever increasing quantities, finally reaching a NO_3 concentration critical for the animals. Certainly, no nitrate tolerance value valid for all marine invertebrates can be given; in general, however, the NO_3 content will not be allowed to rise above 0.3-0.5 g per litre, except for polysaprobic types. As soon as the nitrate content has risen to a value in this order of magnitude, the old seawater must be replaced by fresh water, it is best to do this by renewing a portion of the water at fairly short intervals, rather than by replacing all of the water from time to time. The intervals between two changes of water naturally depend entirely on the population density in the aquaria or on the relationship between food metabolism per time unit (see pp. 232-3) and the volume of seawater circulating through the whole installation; in general the intervals will range between a few weeks and a half year. In a large plant it is entirely possible to do without water changes by letting the accumulating nitrate be assimilated by green algae; however, the expense involved in such an arrangement should not be underestimated (see Fig. 82). To arrest the nitrate build-up entirely, the algal excess (with regard to the N content) which develops in the seawater circuit and must be removed from it continually would have to correspond to the quantity of food added during approximately the same period. A biological balance of this kind can only be attained when a relatively very large shallow tank is introduced into the closed seawater circuit. A rapid-growth green alga (such as *Ulva*) is

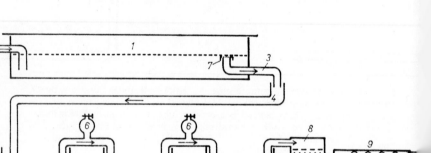

Figure 82. Basic model of a seawater aquarium unit with algae breeding containers for binding the N-containing degradation products (drawing by D. Hofmann). 1 = large, shallow algae breeding tank (water level approx. 10 cm) with pure culture of green algae, 2 = fluorescent light tube unit, 3 = drainage tube, 4 = intermediate tube for protection against droplet infection, 5 = aquarium with animals, 6 = siphon tube (may be filled through suction nozzles with glass stop cock), 7 = perlon sieve, 8 = filter, 9 = arrangement for sterilization of the circulating seawater by UV light or heating to 80°C, 10 = seawaterproof pump

grown in this tank in a pure culture at a high light intensity; this presupposes that the water coming from the aquaria is sterilized before its entry into the algae tank (by UV light, for example) and that the algae culture is started with cleaned zygotes or zoospores. Since regular renewal of the seawater involves a comparatively much smaller expense, this kind of solution to the nitrate problem is probably more of interest on principle and for scientific rather than practical reasons.

2. REARING

A rearing trial must meet the following requirements:

1. The animals must reproduce under laboratory conditions in a chronologically unlimited manner, i.e., over as many generations as required, vegetatively and/or sexually; a surplus of offspring must always be available, allowing continuous removal of individuals for experimental purposes, etc., without impairing the stock.

2. The animals must live under precisely defined and controllable conditions and must be accessible to continuous observation; all the components of a culture must be easy to observe and must be reproducible at any time.

Since as a rule only small species (especially those of the order of magnitude of millimetres) are suitable for a rearing trial, the conditions listed in (2) can usually only be realized in small, manageable culture vessels which can be inspected in their entirety under a binocular microscope; they are virtually impossible to realize in normal aquaria. It is therefore necessary to limit the life space to the barest minimum in a rearing trial, in contrast to a conservation procedure.

(a) Breeding Containers

Neutral glass is the most suitable material for the breeding containers. Transparent synthetic materials such as Plexiglass or polystyrol can also be used for many purposes; however, certain sessile animals with a very strong substrate contact may not tolerate these substances well. To facilitate inspection and to allow unhindered microscopic observation of the culture, the smallest possible container permitting the establishment and maintenance of stable breeding conditions should always be selected, even if this involves attention at intervals of a few days (see Fig. 83). The 'Boveri' dish (diameter approximately 60 mm, capacity 10 to 15 ml), which can be examined rapidly under a binocular microscope, has been found useful for many small animals (< 5 mm); for somewhat larger species

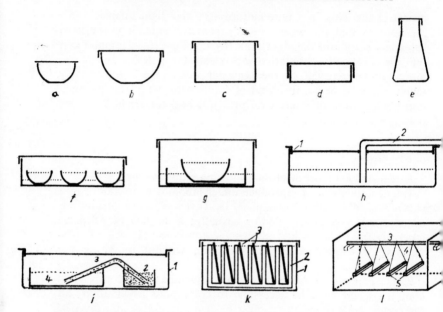

Figure 83. Breeding containers (drawing by D. Hofmann). (*a*) Boveri dish with cover made of window glass; (*b*) semi-spherical crystallizing dish with half a petri dish as cover; (*c*) cylindrical crystallizing dish; (*d*) petri dish; (*e*) wide-neck Erlenmeyer flask with small half petri disk as cover; (*f*) set of three dishes set one within the other (the innermost dish contains animals which frequently crawl out of the water); (*g*) small open crystallizing dish with a large petri dish acting as a moist chamber; (*h*) aerated polystyrene box: 1 = cover with groove for splash water and centrally drilled hole, 2 = aeration tube; (*i*) arrangement for continuous feeding of hydroids: 1 = polystyrene box, 2 = petri dish with *Artemia* larvae, 3 = siphon tube, 4 = petri dish with hydroid polyps; (*k*) culture of sessile animals on slides: 1 = cylindrical crystallizing dish, 2 = glass rod stand, 3 = slide populated with animals; (*l*) culture of sessile animals on glass rods: 1 = polystyrene box, 2 = mounting, 3 = glass rod, 4 = perlon threads, 5 = glass rods with animals

a similarly formed, larger crystallizing dish (diameter approximately 95 mm, capacity 100 to 125 ml) may be considered. For individual cultures or very small organisms, small crystallizing dishes (diameter 40 mm) save space and are easy to examine. Eight of these without individual covers are set into a large, covered petri dish (diameter 150 mm). Animals which must be kept at a very low water level because of their O_2 requirements are best accommodated in petri dishes of suitable size. For species which are easily visible with the naked eye crystallizing dishes with a flat base are suitable, as are the rectangular polystyrene containers sold as refrigerator boxes (10 × 10 or 20 × 20 cm bottom surface, 6 cm height). These boxes of synthetic material are easily ventilated by drilling through their cover, while the inside flange of the cover prevents the water from splashing out; cultures of sessile animals on slides or glass rods can easily be set up or suspended in them. Certain animals often crawl out of the water and then dry up in small culture dishes on the cover. This can largely be prevented by placing two open, shallow glass dishes of different diameters within each other and setting these two into one larger, covered dish; the three dishes are then filled with seawater to the same level and the animals are placed in the inner dish; from time to time the strays can be pipetted back from the outer dish into the inner dish without damage.

It is important to clean the breeding containers thoroughly before every use, preferably under running hot water (do not use rinsing agents!) in special cases it may also be necessary to dry-sterilize these afterwards. All covers should always be marked on the same upturned side so that none of the more or less damaging substances which make up glass pencils (i.e., pencils which write on glass) will enter the culture fluid. Containers which have once contained a toxic fluid (formol, etc.) cannot be used for breeding purposes on principle.

(b) Rearing Medium

It is best to use natural seawater collected from the sea at a favourable site away from metal, as is the case at the Biological Institute of Heligoland. For many purposes it is also possible to mix the natural water in a 1:1 ratio or 1:2 ratio with artificial seawater (Table 10a) without further ado; on the other hand, only relatively insensitive animals can be reared successfully and over a long period in a purely synthetic medium. Large polyethylene bottles

TABLE 10a

Synthetic seawater according to Wiedemann and Kramer, modified by Hückstedt (1963b)

NaCl	27·65	g
$MgSO_4$ (+ 7 H_2O)	6·92	g
$MgCl_2$ (+ 6 H_2O)	5·51	g
$CaSO_4$ (+ 2 H_2O)	1·6	g (dissolve separately)
KCl	0·65	g
$NaHCO_3$	0·25	g (dissolve separately and add at the end)
NaBr	0·1	g
KJ	0·005	g
$SrCl_2$	0·015	g (dissolve separately)
H_2O dist. ....1 litre		

TABLE 10b

Composition of a synthetic culture medium for fully autotrophic marine organisms: seawater with a reduced Mg content

Density .... 1·021 pH.............. 7·6	Constituents in 1 litre distilled water		
	NaCl	26·7	g
	MgSO$_4$(7 H$_2$O)	3·2	g
	MgCl$_2$(6 H$_2$O)	2·2	g
	CaSO$_4$(2 H$_2$O)	1·6	g
	KCl	0·7	g
	NaHCO$_3$	0·2	g
Nutrient salts:	NaNO$_3$	0·1	g
	Na$_2$HPO$_4$	0·02	g
Complex builders:	Na$_2$EDTA	0·02	g
Trace elements 1. Cations:	LiCl	0·006	mg
	RbCl	0·06	mg
	SrCl$_2$	3·8	mg
	AlCl$_3$	0·03	mg
	FeCl$_3$ (in EDTA)	0·2	mg
	ZnSO$_4$	2·3	mg
	MnSO$_4$	0·6	mg
	CoSO$_4$	0·006	mg
	CuSO$_4$	0·001	mg
2. Anions:	KBr	22	mg
	KJ	0·02	mg
	Na-Silicate	20	mg
	H$_3$BO$_3$	2	mg
	Na$_2$MoO$_4$	0·2	mg

(60 l) are very practical for the transportation of the seawater, but if the water is to be conserved for a fairly long period, it is advisable to transfer it to spherical glass flasks. Fairly small differences in the salt concentration of various seas are of no importance for most animals; for example, animals from the Mediterranean can generally be bred in North Sea water without difficulty. However, for animals from waters with a very low salt content (such as the Baltic) suitable dilution of the concentrated water with distilled water (or standard fresh water) is required. Before use the seawater must be freed of the coarser suspended substances by filtering or decanting it. It must then be sterilized by heating for approximately 30 minutes at 80°C; since we do not strive for the absence of bacteria from animal cultures, this procedure is adequate for killing

all other possible organisms. Regular renewal of the water in the cultures is indispensable (cf. p. 232) because harmful metabolic products accumulate fairly rapidly in the very small seawater volumes used to rear most animals. The frequency with which the culture medium must be replaced by fresh seawater depends on the nature and level of the metabolism; generally purely carnivorous animals require fresh water more often (after every feeding, i.e., once to three times a week) than purely herbivorous animals (perhaps twice a month).

(c) Selection of Suitable Animal Species

In general the marine invertebrates from the rocky littoral zone are best suited for culture because they are by nature relatively tolerant to fluctuations in the environmental conditions; in particular, species which exist in tidal pools are usually easily reared. On the other hand, planktonic forms, those from the greater depths or those of very special biotopes, which are as a rule adapted to very stable or very specific conditions, will usually present fairly great difficulties. Within an animal group it is usually the smallest species, requiring little space and usually also with the shortest generation span, which are best suited to a rearing trial. Poorly suited from the start are forms which have extreme requirements of any kind, or those which show high-grade incompatibility to each other within a small space. In practice, the species to be reared can be selected in one of two ways: a number of species can be isolated specifically from the particular animal group obtained under natural conditions, the method of climate deterioration (see p .126) being useful for many small forms in this regard (for example, in a dish filled with algal material the animals collect at the water surface after some time because of O_2 deficiency); since the ecological requirements for all species are established empirically by culture experiments, the species best suited for rearing can be determined. This procedure is often very lengthy because a suitable food organism must first be found and reared in each individual case, and an adequate culture method must be developed by trial and error. But it is also possible to proceed inversely, that is, to obtain by selection from the marine samples those animal species which can be reared by one of the proven standard methods. For this purpose the selected material (especially substrate sediment, algae and other growth on rocks or pilings, *Posidonia*-rhizomes, etc.) is distributed in small portions in many dishes filled with seawater. The dishes are then exposed to various standard conditions; for example, the animals in some of the dishes are regularly fed with a flagellate suspension, while others receive *Artemia* nauplii; in the

TABLE 11

Marine invertebrate metazoa which have been successfully reared in the laboratory; according to the following authors: G. Bacci (1), P. Emschermann (2), A. Fischer (3), B. Föyn (4), K. G. Grell (5), M. Hartmann (6), C. Hauenschild (7), D. Neumann (8), H. Thiel (9), and B. Werner (10)

Animal species	Method of reproduction	Food	Author
Fungi			
Clathrina spec...........	veget.	Bacteria on *Dunaliella**	7
Hydrozoa			
Aequorea aequorea	veget.	*Artemia*	10
Bougainvillia superciliaris..	veget.	*Artemia*	10
Campanulina lacerata	veget.	*Artemia*	2
Chrysaora hysoscella	veget.	*Artemia*	10
Cladonema radiatum	veget.	*Artemia*	7, 10
Clava multicornis	veget.	*Artemia*	2, 4
Clavopsella quadranularia .	veget.	*Artemia*	2, 9
Clythia johnstoni	veget. + sexual	*Artemia*	7
Coryne tubulosa..........	veget.	*Artemia*	10
Eirene viridula	veget.	*Artemia*	5, 10
Eleutheria dichotoma	veget. + sexual	*Artemia*	7
Eucheilota maculata	veget. + sexual	*Artemia*	10
Eutonina indicans	veget.	*Artemia*	10
Gonionemus vertens	veget.	*Artemia*	2, 5, 10
Gonothyraea loveni	veget.	*Artemia*	7
Halitholus cirratus	veget. + sexual	*Artemia*	10
Hydractinia echinata......	veget. + sexual	*Artemia*	7
Laomedea flexuosa	veget. + sexual	*Artemia*	7
Leuckartiara octona	veget.	*Artemia*	10
Margelopsis haeckeli......	veget. + parthen.	*Artemia*	10
Perigonimus spec.	veget.	*Artemia*	2
Podocoryne spec.	veget.	*Artemia*	7
Rathkea octopunctata	veget.	*Artemia*	10
Staurocladia spec.	veget.	*Artemia*	7
Scyphozoa			
Aurelia aurita	veget.	*Artemia*	2, 9, 10
Anthozoa			
Aiptasia diaphana	veget.	*Artemia*	5, 7
Kamptozoa			
Barentsia gracilis			
Barentsia benedeni			
Barentsia laxa	veget. +sexual.	*Cryptomonas*	2
Loxosomella antedonis	veget.	*Cryptomonas*	2
Pedicellina echinata			
Pedicellina nutans	veget. + sexual	*Cryptomonas*	2
Nemertea			
Lineus spec..............	veget.	*Artemia**	2
(an unidentified species)...	veget.	*Artemia*	7

* = Food organism killed before use by heating to 50°C.

Table 11 (Continued)

Animal species	Method of reproduction	Food	Author
Nematodes			
Epsilonema spec.	sexual	*Artemia**	2
Gastrotrichia			
(an unidentified species) ..	sexual	Bacteria	2, 7
Rotatoria			
(an unidentified species) ..	sexual	Bacteria	7
Polychaetes			
Autolytus prolifer	veget. + sexual	*Laomedea*	7
Brania clavata	sexual	*Oxyrrhis* and	
pusilla...........	parthen.	*Platymonas*	7
Capitomastus minimus	sexual	*Dunaliella**	7
Ctenodrilus serratus	veget.	*Dunaliella**	7
Dinophilus gyrociliatus ...	sexual	*Dunaliella**	5, 7
Dodecaceria concharum ...	veget.	*Cryptomonas*	2
Exogone gemmifera.......	sexual	*Oxyrrhis* and	7
		Platymonas	
Micronereis spec.	sexual	*Cladophora* and	7
		Ceramium	
Ophryotrocha (various species)	sexual	*Dunaliella, Chlorella* and *Artemia**	1, 5, 6, 7
Platynereis dumerilii	sexual	*Platymonas*, Vitawil,	
massiliensis....	sexual	*Oxyrrhis*	7
Pomatoceros triqueter	sexual	*Chlamydomonas*	4
Crustacea			
Tanais cavolinii	sexual	*Chlorella*	5
Tisbe spec..............	sexual	*Platymonas* and *Cryptomonas*	7
Insects			
Clunio marinus	sexual	Stinging-nettle powder Diaton. and Cyanophyc.	8
Gastropods			
Crepidula fornicata	sexual	Unfiltered seawater	10
Haminea hydatis	sexual	Diatoms	5
Bryozoa			
Aetea spec...............	veget.	*Cryptomonas*	2
Bugula avicularia........	veget. + sexual	*Oxyrrhis*	5
Valkeria uva	veget.	*Cryptomonas*	2
Victorella pavida	veget.	*Cryptomonas*	2
Ascidia			
Clavelina lepadiformis	veget. + sexual	*Oxyrrhis* and *Cryptomonas*	3

* = Food organism killed before use by heating to 50°C.

same way it is possible, by varying the treatment of the individual containers, to select the organisms in regard to other environmental factors (such as requirements concerning light, temperature, O_2 and water content; tolerance to water pollution, hunger or over-feeding, etc.). In each dish those of the animal species which can live under the given conditions will develop and reproduce; after some time they can be isolated from the impure cultures and reared further, using the method employed for the selection. This procedure requires relatively little time and has the additional advantage that even animals which are very difficult to find in the collected material because of their small size and number of individuals can be readily isolated after suitable concentration in the impure culture. However, the selection method is only possible for animals which are very small and have few requirements.

Although large-scale rearing of marine invertebrates has only begun recently and has not taken place at more than a few institutes only until now, a whole series of species from the most varied metazoan strains has already been bred successfully, as shown by the survey in Table 11.

(d) Monoculture and Association of Organisms

It is absolutely necessary first to entirely isolate the animal species intended for rearing. For this purpose we begin with the developmental stage which is easiest to free of any adhering foreign organisms during several passages through sterile seawater (such as eggs, free-swimming larvae). A true monoculture, which can be established and is generally useful when dealing with autotrophic organisms, normally does not come into question for animals, however, even if the bacteria are not taken into account. When dealing with carnivorous animals, which can easily be bred in a monoculture in theory, this is almost always found to be disadvantageous in practice because of the excessive accumulation of food residues, excrement and bacteria. In most cases it is much better in the long run to give up the monoculture in favour of a defined organism association and to establish certain proven detritus-, excrement- and carrion-consumers (such as *Dinophilus, Tisbe*) as well as certain bacteria consumers (such as *Oxyrrhis*, some ciliates and rotatoria) in the culture; such culture partners protect sessile animals in particular from being covered and suffocated by a layer of excrement and food residue. However, this by no means suggests that sterilization of the breeding containers, of the pipettes and of the seawater, as well as the most painstaking cleanliness in handling the cultures under these conditions are unnecessary. Most of the organisms which are involuntarily brought in from the natural seawater if this safety measure is not observed are found to be markedly damaging in the culture because of their mass reproduction; such epidemic forms include, above all, many amoeba, ciliates, diatoms and cyanophycea. The cycle of organisms allowed in a single culture must therefore be tested, well defined and continuously observed in each case. Herbivorous animals can sometimes be reared together with only a single auto-trophic food organism, but here too is it often advantageous to introduce into the culture one or several further organisms which both consume the bacterial excess and can serve as an additional food source. It also happens that two species which, in principle, can be cultivated in a completely isolated manner will flourish much better together in a mixed culture because the conditions produced by one partner are particularly favourable to the other one, and vice versa.

(e) Temperature and Light

In the interest of rapid multiplication of the culture organism, we should aim at the special temperature optimum of the latter if possible. However, in most cases animals of various origins are bred in the same room; if animals originate from the littoral zones of the North Sea, of the Mediterranean and of comparable areas, a compromise temperature of 19° to 20°C has been found acceptable for most species; at this temperature the North Sea animals still grow well and those of the Mediterranean also grow sufficiently rapidly. The top temperature limit for long-term tolerance can be set at 22°C for North Sea animals, 30°C for Mediterranean animals and 38°C for those from tropical seas, insofar as inhabitants of the tidal areas are concerned. However, it is highly advisable to avoid such temperature extremes in the cultures by setting up climatic equipment in the rearing room; the cooling and heating devices combined in this equipment can maintain the air temperature uniformly at $\pm 0\cdot 5$°C by the control of two contact thermometers, if the apparatus has the correct dimensions and the room is adequately sealed off.

No direct sunlight should reach cultures under any circumstances, especially because of the associated heating; if necessary the culture room must be entirely darkened and lit artificially with daylight-fluorescent tubes. Light exposure for 16 hours a day (set by time switch) will generally meet the requirements of both the animals and of the autotrophic food organisms. The optimal brightness must be tried out for each organism; but many of the marine invertebrates within the order of magnitude involved behave more of less indifferently in regard to the light intensity. In general, cultures containing an autotrophic food organism will be kept somewhat lighter than hydroid cultures, i.e., will be set closer to the light source because of the O_2 supply; undesirable diatoms and flagellates are always brought into hydroid cultures with the *Artemia* larvae, and their development should be inhibited by keeping the cultures in a darker place. If the culture is not too crowded, a change in the light intensity sometimes also has a stabilizing influence on the balance between an animal and its food plant.

(f) Feeding

It is simplest to rear those animal species which can be fed directly with autotrophic organisms. Of these, the following in particular are a possibility: autotrophic flagellates (such as *Dunaliella, Platymonas, Cryptomonas*), diatoms (such as *Phaeodactylum, Thalassosiria*) or chlorophycea (such as *Ulva*). If the food organism is fully autotrophic, it can be grown rapidly and in a large quantity as a pure culture in a synthetic medium under fluorescent light. Table 10 shows the composition of a culture medium of this kind; many species may require the addition of vitamins (such as B_{12}). In comparison with the so-called Erdschreiber solution (see p.251), which is still much used today, this medium has the advantage that bacteria scarcely develop in it and that it always has the same controllable composition. Freely swimming food organisms are reared in Erlenmeyer flasks, and sessile ones in crystallizing dishes. The colourless, phagotrophic dinoflagellate *Oxyrrhis marina* has also often been found useful as a food; according to a method by Grell, it can in turn be fed with *Dunaliella* (three-part culture procedure). Some detritus consumers and bacteria consumers to whom live flagellates are not accessible can be cultivated by giving *Dunaleilla* which have been killed by heating to 50°C as a food basis. To supplement the plant food, the use of small quantities of suitably granulated dry fish food may be advantageous for many animals which reject the bottom deposit. For pure carnivores, *Artemia salina* is available as standard food; if seawater is poured over the commercially available dry eggs, nauplii can be obtained at any time (aeration is advisable); as the fat content of freshly hatched larvae is fairly great and poorly tolerated by many animals in the long run, it is best to use the Artemia two to three days after a hatching. The Artemia can either be placed in the culture in a surplus (collective feeding) or can be given to the animals (such as hydroid polyps); individually with a pipette (individual feeding); when using the latter method it is also possible, if necessary, to feed larger or smaller pieces of cut up nauplii. Animals which predominantly eat carrion can be fed with Artemia which have been killed by heating (50°C). Apart from *Artemia*, animals which come under consideration as food animals are certain Harpacticoida (such as *Tisbe* species), which can easily be bred with autotrophic flagellates. Finely cut-up mollusca or *Tubifex* can also be used for individual feeding, but this does entail the risk of great contamination of the culture because of the decaying tissue portions.

The optimal quantities of food to be given must be worked out individually for each species reared and always requires a certain delicacy of judgement; in general, it is more harmful to give too much food than to give too little. Whereas the herbivorous animals can be supplied more or less continuously with food, carnivorous animals (except for carrion eaters) are generally limited to individual meals given at intervals of 2 to 7 days; it is best to renew the water 1 to 2 hours after each feeding and to remove all food residues at the same time. However, there are some animals (such as some hydroids) which only grow well if food is continuously available; if Cnidaria are being reared, these cannot be supplied with Artemia for several days ahead of time because they immediately kill all surplus nauplii but do not consume the dead animals later. In such cases the *Artemia* can be accommodated in a separate supply container which communicates through a waterfilled glass tube with the culture dish placed at the same level; if the siphon tube is of suitable dimensions and the Artemia suspension is of a certain concentration, this method makes it possible to supply a culture automatically and fairly regularly for several days with living *Artemia* larvae which migrate through the communication tube (see Fig. 83i).

(g) Maintenance of the Culture Conditions

In every culture the living space is necessarily reduced to a minimum. It is therefore virtually impossible to set up a culture in such a manner that it develops a permanent, self-regulating balance; one of the numerous prerequisites for this would be a virtually unlimited substrate and medium, as are only present under free natural conditions. One therefore cannot avoid continuously interfering in a variety of ways in every rearing experiment in order to maintain the conditions favourable for the development of an animal species. In practice it is impossible to observe all cultures continuously night and day; every culture requires specific arrangements which limit inspection and maintenance to practicable intervals and stabilize the living conditions for the organism involved from the time of one maintenance check to the next in such a manner that deviations from the optimum during this interval can be kept within tolerable limits. For example, herbivorous animals

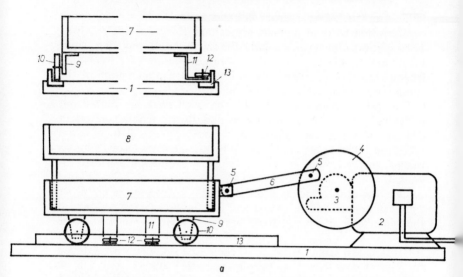

Figure 84. Slowly operating shaking apparatus (drawing by D. Hofmann). (*a*) lateral aspect, (*b*) cross section (left at the level of the ball bearings, right at the level of the guide wheel). 1 = base plate, 2 = electric motor (1360 R/min, 50 W), 3 = drive axis of the transmission unit (55:1), 4 = pulley wheel (diameter 10 cm), 5 = bearing pivot, 6 = crank rod, 7 = carriage, 8 = removable attachment (several tiers possible), 9 = support for ball-bearing axis, 10 = ball-bearing, 11 = support for the axis of the guide wheel, 12 = guide wheel, 13 = steel runner

must be provided with a food supply which will suffice until the next supply is provided; however, since the animals may easily suffocate during the dark period in the presence of a surplus of autotrophic organisms which consume O_2 and produce CO_2 at night, gas exchange must be promoted in densely populated cultures of this type. This is also true, of course, for all animals with an especially large O_2 requirement, irrespective of the mode of feeding. If a very low water level (<5 mm) does not produce sufficient gas diffusion, continuous mixing of the water is best. Two methods have been found good for this: if large cultures kept in polystyrol containers are involved, the water movement can be produced by large air bubbles which issue from a glass tube near the bottom in the middle of the container; however, this only comes into consideration for animals which tolerate fairly strong turbulence without damage to their general condition. Another procedure used above all for smaller breeding dishes, for mechanically sensitive animals and especially for sessile animals is placement of the dish on a slowly operating shaking apparatus (see Fig. 84); this consists of a tray-like arrangement (sometimes multi-storied) which runs on four ball bearings in two runners and weaves back and forth continuously 20 to 30 times a minute with an amplitude of 5 to 10 cm; it is actuated by a transmission motor through an eccentrically positioned sliding rod. The resultant mixing of the water layers depends in its extent on the size, form and filling level of the breeding containers; it promotes economization of the gas exchange and in sessile animals leads to contamination by their own excrement, especially when these animals are not settled at the bottom of the dish, but adhere to vertically standing slides or to glass rods which are freely suspended on perlon threads (see Fig. 83, *k*, *l*).

However the measures listed can only maintain the rearing conditions for a limited time (days, or at most weeks) within the tolerance limits; after this, intervention is necessary to re-establish the initial situation. The principal measures of this kind are: the stirring up and removal of sediments with a pipette and the renewal of the water with its concentration of metabolic products, either by pouring it off or by transferring the animals into a clean culture dish; the supplying of food; the removal of excess numbers when over-population begins; the re-establishment of the optimal qualitative and quantitative composition of an organism association; the isolation of ripe breeding animals and the separate rearing of the young. Naturally, the favourable intervals for the various measures can only be determined empirically for every individual animal species.

REFERENCES

GEISLER, R. (1964). *Wasserkunde für die aquaristische Praxis.* A.-Kernen-Verlag, Stuttgart.

HAUENSCHILD, C. (1962). 'Die Zucht mariner Wirbelloser im Laboratorium (Methoden und Anwendung).' *Kieler Meeresforsch.* 18, 28–37.

HÜCKSTEDT, G. (1963 a). *Aquarientechnik,* 91 Franckhsche Verlagshandlung, Stuttgart.

— (1963 b). *Aquarienchemie.* 88 S. Franckhsche Verlagshandlung, Stuttgart.

LOOSANOFF, V. L., and H. C. DAVIS (1963). 'Rearing of bivalve mollusks.' *Advances in Marine Biology,* ed. by F. S. RUSSEL. Vol. 1, 1–136.

MÜLLEGGER, S. (1955). *Das Seeaquarium.* 2. Aufl., 136 A.-Kernen-Verlag, Stuttgart.

NEEDHAM, J. G. et al. (1959). *Culture methods for invertebrate animals.* Dover-Publ., New York.

RIEDL, R. (1963). *Fauna und Flora der Adria.* P.-Parey-Verlag, Hamburg und Berlin.

WICKLER, W. (1962). *Das Meeresaquarium.* Franckhsche Verlagshandlung, Stuttgart.

FISH

J. FLÜCHTER

It is almost impossible to present concisely general maintenance and rearing instructions for so heterogeneous an animal group as the 'sea fish'. Consequently the following are only intended as guide lines; the special points relating to each species must be taken from the literature, which, in the selection offered here, should act as an indicator to further reading.

The question of the special requirements of marine fish cannot be answered in a general way either. The normal salt content of seawater is definitely not a vital prerequisite in all cases.

However, three groups may be distinguished in a simplified way:

1. We first list the species which are found near the shore and which are easy to maintain even for untrained individuals, with the help of the usual aquarium techniques. We are thinking particularly of the various gobies (*Gobius*), etc.

2. The true inhabitants of the open sea, which are sometimes shown in public display aquaria, present fewer difficulties – apart from the expense involved in keeping them – than the third group.

3. The larvae of marine fish, which as a whole have very uniform requirements, have little in common with their parents. In the discussion which follows, their rearing – which is not entirely simple – will take first place.

Attempts to rear marine fish spawn, which have often been repeated over a number of decades, were stimulated by the well known fact that the natural mortality in the youngest larval stages of marine fish is incredibly high and decisively determines the fisheries yield of commercial fish (Berverton and Holt 1957, Hempel 1963). Whereas the older procedure – replenishment of the stocks thinned by commercial fishing by rearing large quantities of fish brood to the young fish stage – first failed, rearing trials under controlled conditions in the laboratory yielded an abundance of data relevant to the high natural mortality of the larvae. The old subject of commercially oriented rearing of marine fish brood has recently come under discussion again, but now with the emphasis on the extent to which commercial marine fish may be suitable for pond cultures (discussed in detail, with a very extensive historical review, by Shelbourne 1962). Although the rearing trials carried out until now have not yet yielded a complete picture, some basic principles necessary for the growth of marine fish broods are already apparent.

1. Capture and Transport

In regard to the forms living in the coastal areas we refer the reader to the literature on freshwater fish and on ornamental fish. Most of the time, however, selection will be limited to the specimens in a trawl net catch. If possible, brief trawling periods of some ten minutes each should be used. When the catch has arrived on deck, the animals should be placed in containers immediately, with running seawater if possible. Even fish with a swim bladder – which are generally over-expanded because of their being brought up from the depth, causing the fish to lose their balance – often recover after some time (cf. Flüchter 1966). Heating of the water must be avoided at all costs. For transport on land, polyethylene foil bags have recently been found useful, the best thickness being 0·1 mm. The bags are filled to a maximum of one-third with water and the remaining two-thirds with air, or, better – if available – with pure oxygen. This procedure is especially good for small forms.

2. Obtaining the Spawn

When a species spawns or breeds in the aquarium, the spawn is easily obtained. Occasionally one also finds spawn clumps which have been washed on to the shore (determinable according to Ehrenbaum 1904). When dealing with the inhabitants of the open sea, which – with the exception of the herring – possess pelagically floating eggs, it generally happens that animals ready for spawning are captured at sea. As a result of the pressure incurred during capture, the sexual products generally escape by themselves. The ripe gonads are largely emptied by gently stroking the ventral side of the fish from front to back. Eggs and seminal fluid can be placed together in a dish and suspended in water after careful stirring, or they can be scraped into a container with water one after the other. In principle, the artificial fertilization procedures commonly used in the lake and river fishing industry (cf. Lindroth 1946) can be transferred to marine conditions. The fertilized eggs, which can be identified by the fact that they remain clear and float at the surface, must be transferred into fresh water immediately, and this water must be changed repeatedly. The dependence of embryonic development on the temperature was early noted by researchers (such as Apstein 1909). The eggs of fish from the open sea, which are pelagically floating by nature, are best transported with abundant water in thermos bottles, or if available on board, in the refrigerator.

When working with herring it is advantageous to drop the eggs (which adhere to the substrate by nature) uniformly in one layer onto the rough side of a ground glass plate, where they will stick firmly within a few seconds. It has been found best not to keep them in water while being transported, but simply to wrap them in a wet cloth. Also the similarly relatively robust eggs of the other substrate spawners are successfully transported by this moist procedure; this is because the oxygen supply is better, bacteria do not settle on the spawn as rapidly, and finally, the transportation weight is reduced.

In general, it will be best to carry out fertilization right away and to take the material to the determination site immediately; the oxygen requirements of the egg are very low at first, as is its sensitivity to shocks, so that it can tolerate even several hours of transportation by car during the first two days.

If artificial fertilization cannot be carried out immediately, the prepared gonads should be kept as cool as possible (at approx. 4°C); a satisfactory fertilization rate will still be possible later, but preferably within the first twenty-four hours. A small percentage of the eggs and sperm even remains capable of fertilization for several days.

The male gonads yield entirely viable sperm after preservation for six months, if treated with 12% glycerin solution, cooled rapidly to −79°C (dry ice) and kept at these low temperatures. In this manner herring races which spawn at different seasons and those originating from various spawning sites have been crossed with each other repeatedly. A procedure for long-term living conservation of unfertilized fish eggs has not yet been developed. (Discussed in detail by Blaxter 1955).

Eggs collected with plankton nets can also be used for rearing trials; the suitable eggs are determined by measuring the diameter (Heincke and Ehrenbaum 1900). Occasionally fish kept in large tanks will spawn spontaneously. These eggs can be fished out with narrow-mesh hand nets, but one can also produce strong air circulation in the centre of the tank, close off the corners with a partition, and leave a free space near the bottom. The eggs circulating with the current then gradually collect in these calm corners.

It has recently been successfully attempted to stimulate the spawning of almost ripe freshwater fish by injecting them with pituitary extracts. Although this procedure has not yet been applied to marine fish, logically it would appear to be transferable (see Atz and Pickford 1964).

3. CONSERVATION VESSELS

The material of containers in which marine animals are kept should not release toxic substances into the seawater and must furthermore withstand the chemical attack of the latter. Some of the modern synthetics meet these requirements. Large concrete or Eternit tanks are coated with chlorinated rubber varnish*; small containers should be made of glass or Trovidur†. The usual aquaria can be used for littoral forms (solid glass tanks are preferable because of the risk of corrosion) but in principle dark-walled containers should be used for brood rearing. The black 'mortar buckets' and pails made of Lupolen‡ have been found useful, and ceramic§ products can also be used. The pipe lines can be made of Trovidur; silicone rubber is outstanding as tubing material. Tubes made of polyvinyl chloride (soft) are also chemically inert, but become stiff and unmanageable at lower temperatures; rubber tubes are entirely unsuitable.

4. TREATMENT OF THE WATER

The availability of natural seawater is assumed in the discussion which follows. The usefulness of the seawater for our purposes is not independent of its previous history; when working in research stations one is consequently guided by the nature of the seawater supply (cf. also Communications Ier Congrés International d' Aquariologie, Monaco 1962). The common practice of conserving seawater in large dark tanks appears to have an unfavourable effect on the rearing of fish. The water improves in quality when it is continuously aerated and constantly filtered. The construction of our unit always depends on whether we use running seawater or whether we are working with closed water bodies. Especially when rearing fairly large fish, a continuous amount of fresh water is simply allowed to flow through the tanks. Any desired flow can be obtained by suitably arranging the water supply. If the tank wall can be perforated by drilling, it is very simple to set in the outlet. For safety purposes a sieve is placed in front of the *outlet tube;* it consists of a short piece of tubing of approximately 8 cm diameter,

---

* Ikosit A 20 30, Lechler.
† Polyvinylchloride resin, Dynamit Nobel A.G., Troisdorf.
‡ Polyethylene BASF, Ludwigshafen.
§ So-called 'clay pipes', Deutsche Steinzeugwerke Mannheim.

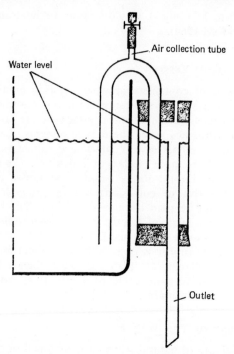

Figure 85.   Overflow lever with level regulator

which is closed on both sides by coarse gauze and should be located laterally on the outlet tube (Fig. 85). This ensures that sudden occlusion of the outlet and the escape of the inhabitants can be avoided. However, in glass aquaria it is difficult to drill through the walls; a so-called overflow pipe can be of help here (Fig. 86); this pipe consists in essence of a water-filled U-tube of at least 10 mm internal diameter, which connects two aquaria with each other. After the system has been operating for some time, air accumulates at the highest point of this tube and blocks further circulation; a transparent air chamber is therefore installed at this point. It is important to have an internal diameter of no less than 6 mm for the air tube. From time to time the accumulated air must be siphoned off. As many aquaria as desired can be connected in a series in this manner. The aquarium series should terminate with a 'level regulator', which produces a uniform water level in all the tanks (Fig. 85). The advantage of rapid removal of organic excreta when using running water is countered by the danger of introducing parasites and diseases, the disadvantage of virtually unavoidable temperature

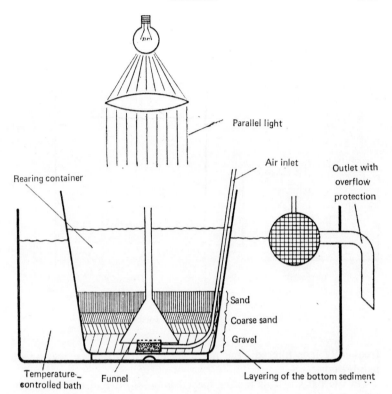

Parallel light

Air inlet

Outlet with overflow protection

Rearing container

Sand

Coarse sand

Gravel

Temperature-controlled bath   Funnel   Layering of the bottom sediment

Figure 86. Rearing container with internal filter set into a controlled-temperature bath

fluctuations, and – because of the high water consumption – the need for the investigator's presence at the marine research stations. If one wishes to use running water for the rearing of the fish brood despite the above-mentioned disadvantages, containers with an outlet near the bottom should be employed. The bottom is covered with gravel, coarse sand and fine sand (Fig. 87), and the water level in the container is regulated by means of an ascending tube attached to the outlet. The layering of the bottom sediment takes care to some extent of fish larvae and their food organisms which remain in the container.

However, the disadvantages which are unavoidable when running water is used have been sufficient to cause this method to be omitted on principle when rearing the brood.

The more the seawater is aerated – and especially the more

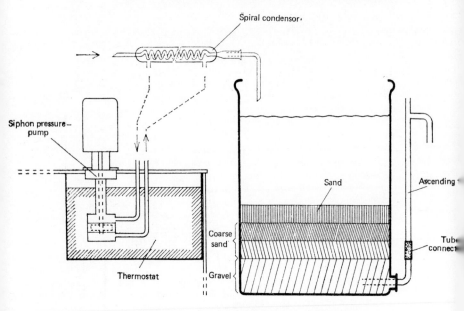

Figure 87.   Rearing container for circulation

frequently it is filtered – the more suitable it is for the rearing of fish broods.

On the basis of this fact, a special *internal filter* of the following kind was developed (see Fig. 86): on the bottom of our proposed rearing container we place a layer of gravel at least 2 cm high (depending on the size of the container); below this gravel bed we place an air pipe, and invert a simple funnel over the outflow. We then deposit an equally high layer of coarse sand, followed by a layer of fine sand, over the gravel around the funnel. When the air supply is started up after the tank has been filled with water, the air bubbles rising through the funnel tube are accompanied by water which must percolate through the bottom sediment at the same rate; during this process the layering ensures that the entire bottom surface of our container acts as a filter. The uniform percolating of the water prevents the development of anaerobic putrefaction areas in the bottom sediment, and the microscopic organisms living in this sand led to rapid biological decomposition of the organic waste; the water-air mixture issuing from the funnel tube prevents the development of a film of slime on the water surface, which is certainly very damaging to pelagic eggs and fish larvae. It is best to bring the funnel tube precisely to the water level, which makes it possible to supply well-aerated, very clean water to the eggs and delicate larvae floating at the surface, without damaging them by air bubbles or turbulence. The cleanliness of the freshly filtered water and the continuous slight flow which are produced with this procedure have been found to be the indispensable prerequisites for the rearing of larvae of marine fish, and particularly for the hatching of the eggs. The oxygen requirement continually increases during embryonic development and reaches its maximum immediately before hatching; in view of the spherical form of the eggs (which may possibly be unfavourable for exchange processes) this requirement can only be met when the water with an oxygen content reduced by respiration is continuously removed from the immediate vicinity of the egg. Contamination of the water is as damaging as insufficient circulation, since bacteria settle on the pellicle of the egg and also bring about a reduction in the oxygen supply to the embryo. The mortality of the embryos, especially in later development stages, and the hatching of damaged or weak larvae are generally due to damage caused by oxygen deficiency, which has occurred in the manner outlined above.

5. LIGHTING

Light is of importance for fish larvae in two respects. In their optically very undifferentiated environment the penetration of light from above evidently has an orientational purpose. Very many marine fish larvae, especially those which reach maturity in the open sea, react to lateral light with great agitation, swimming towards the light against the walls, and are incapable of capturing food animals. The use of dark-walled containers has been found very useful for simulation of the natural light conditions. As important as the light penetration from above is a parallel beaming of the light. In general, some light produces sharper shadows than does the diffuse light of an incandescent lamp or a fluorescent tube. However, it is of great importance for the fish larvae precisely at the time of the initial feeding attempt that the prey stand out as strongly as possible optically from its environment. Car headlights set on main beam, as well as certain microscope lamps, yield a good, parallel beam of light (Fig. 86) in which the food animals stand out strongly against the light water surface or the dark container walls.

On the other hand, absolute brightness is of lesser importance and can fluctuate within wide limits. In the natural habitat the light intensity does indeed alter rapidly and is extremely irregular. Nor is it harmful to keep the larvae continually exposed to light if acceptable food can be offered to them in sufficient quantities.

6. AERATION

The aeration of the water accomplishes various purposes: it supplies the oxygen, mixes the water body and destroys the temperature layers which form spontaneously. Moreover, seawater has the peculiarity of binding very firmly the carbon dioxide given off by the animals during respiration, which causes its pH value (in the alkaline range under natural conditions, somewhat above 8) to fall to harmful values. The carbon dioxide should therefore be removed from the air which is destined for aeration by washing with 5% solution of caustic soda and conveying it through fine-pore escape tubes (beechwood); if possible, the diameter of the bubbles should be less than 1 mm. In this manner it is possible to wash out part of the free carbon dioxide from the seawater (discussed in detail by Gillbricht 1953).

7. TEMPERATURE CONTROL

It is a characteristic of true marine fish that they are highly sensitive to even slight temperature fluctuations. The simplicity of the behaviour of the littoral forms is essentially based on the fact that like many freshwater fish, they must be very insensitive to frequent and rapid temperature fluctuations.

The inhabitants of the open sea, on the other hand, are adapted to the fact that the temperature fluctuations (which are only slight in any case) take place extremely slowly, i.e., over periods of weeks and months. The temperatures prevalent at the natural location should always be supplied first. Whereas heating is generally unnecessary for the forms obtained from the Baltic, it will often be useful to cool the seawater. The smaller containers can be sufficiently cooled by setting them in tanks with slowly running tap water and protecting them from harmful fluctuations in temperature during the day. For precise temperature control the conservation containers are best placed in baths which are regulated by a so-called *circulation thermostat**. This thermostat consists of a container for liquids which is very precisely controlled thermically, and a suction pressure pump which turns the bath liquid over with great rapidity (Fig. 87). These instruments can also be used for cooling when employing running seawater. In that case they are connected to a spital condenser commonly used in chemistry (Fig. 87), which is available in a length of 150 cm. If a suitably low-temperature cooling brine is used, temperatures of about 0°C can easily be obtained. However, the

---

* Colora Mess Technik, Lorch/Württ.

quantity of seawater cooled in this manner is not especially great. Large plants have on a number of occasions successfully used graphite heat exchangers, which have high chemical resistance to the aggressive seawater and offer good heat conduction*. I should like to warn emphatically against allowing the seawater in closed circuits to come into contact with any kind of metal.

8. FOOD

In the rearing of a fish brood, the greatest influence on the test results is exerted by the food supply. The resistance of the larvae to negative environmental influences depends largely on their satiation with suitable food animals; after hatching most fish larvae continue to carry with them a part of the reserve substance contained in the egg in the form of a yolk sack, and consume this first. Even before the yolk sack is entirely consumed, the larvae begin to ingest food; not only the size of the food animals, but also their ability to escape play an important part here, since the fish larvae must first learn to capture the prey. They are very unskilled in this initially, and only a few of the many snapping motions bring the desired result.

The opportunity of eating to satiation, as well as the learning rate, consequently depend greatly on a high density of food animals at the beginning of the hunt for prey. The lack of ability of the small nauplii – compared with the adult swimming crabs – to escape the attacks of the fish larvae is almost as important as their 'mouth-adapted' size. As mentioned previously, the success of these first predatory hunts can be influenced considerably by the nature of the light exposure. In the presence of a food deficiency irreversible damage occurs with unexpected rapidity and always leads to death. It must be remembered that after a short period the larvae become too weak to hunt the live prey.

Air bubbles are occasionally found in the intestines of the fish larvae; if suitable food is lacking, these air bubbles are taken up by the fish larvae, who mistake them for prey.

Although the fish larvae in the sea consume the most varied organisms suitable in relation to their size, the nauplii of the swimming crabs have been found the best, most wholesome and most easily available food for the larvae of most species. Since it has not been possible to rear this food in sufficient quantities until now, one must resort to capturing it with plankton nets. The mesh width of the nets determines the size of the food animals, which measure 50 to 100 μm length.

---

* Carbone A.G., Frankfurt/M.

However, some species have special requirements. Thus, for example, during their first days of life sole larvae eat only ciliates, which they pick up from the substrate. In this case successful rearing, i.e., a high survival rate, depends exclusively on the abundant presence of these protozoa in the rearing containers.

The course of rearing trials until now points to the fact that the nauplii of the brine shrimp (*Artemia salina*) are basically unsuitable as the first food for the larvae of marine fish. Because of their thick cuticle they apparently cannot be attacked so readily by the – as yet – weak digestive juices of the larvae, and in addition they appear to lack certain nutrient substances which are absolutely indispensable for larval development. Artemia only become an adequate food after metamorphosis, i.e., after the involution of the primordial fin folds, the development of the definitive fins and that of the bony skeleton. Species which leave the egg not in the form of larvae, but after their metamorphosis into young fish, as for example the gar-fish (*Belone acus* R.), can be reared easily and virtually without mortality with Artemia nauplii. For this purpose the commercially available resting eggs* are hatched, preferably at 25 to 30°C and in well-aerated seawater. Good, fresh egg material yields attractive orange/ red nauplii which at first prefer to remain close to the bottom and should be offered as food immediately, as their nutritive value decreases rapidly. If the nauplii are only a pale pink and show positive phototaxis immediately after hatching, the egg material is old and matured; these nauplii have little nutritive value. Once the larvae have grown into young fish, their increasing nutritional requirements must be met. Adult *Artemia* are always in favour as food animals; they can easily be reared in great quantities and in a short period from their nauplii, if adequate light and heat are available (cf. Flüchter 1965).

Adult marine fish – especially predatory fish – can often be fed with fish flesh, whereas the worm eaters can generally be trained to eat edible shellfish flesh, even though they may often refuse the food for weeks after their capture.

For the natural food spectrum of some Baltic fish cf. Kühl, 1961.

9. DISEASES

Although a great deal of material is available on animal parasites (cf. also Schäperclaus 1955), very little is known about diseases caused by bacteria, viruses or fungi although they may be of at least local importance in the natural environment of the fish

* Telra-Werk, Melle.

(Bückmann 1952, Oppenheimer 1958; data on research techniques). The death of marine fish broods as a result of specific causative agents is unknown until now. Pertinent studies on this subject are made very difficult by the special sensitivity of many marine animals to an increased content – in comparison with the open sea – of bacteria which are merely indifferent in themselves.

REFERENCES

APSTEIN, C. (1909). 'Die Bestimmung des Alters pelagisch lebender Fischeier.' *Mitt. d. Dt. Seefischerei-Vereins* **25**, 364–373.

ATZ, J. (1964). 'A working Bibliography on rearing Larval marine Fishes in the Laboratory.' *Res. rep. fish a. wildlife service* **63**, 95–102.

— and G. PICKFORD (1964). 'The pituitary gland and its relation to the reproduction of fishes in nature and in captivity.' *An annotated bibliography for the years*. 61 S. 1956–1963. FAO Fisheries Biology Technical Paper Nr. 37.

BEVERTON, R. & S. HOLT (1957). 'On the dynamics of exploited fish populations.' *Fish. invest. London*, Ser. **2**, Bd. 19,

BLAXTER, J. (1955). 'Herring Rearing. I. The Storage of Herring Gametes.' *Scottish Home Dept.*, *Mar. Res.* **3**, 1–12.

— and G. HEMPEL (1963). 'The Influence of Egg size on Herring Larvae (*Clupea harengus* L.).' *J. du Cons.* **28**, 2, 211–240.

BÜCKMANN, A. (1952). 'Infektionen mit *Glugea stephani* und mit *Vibrio anguillarum* bei Schollen (*Pleuronectes platessa* L.). *Kurze Mitt. aus der fischereibiol Abt. d. Max-Planck-Inst. f. Meeresbiol., Wilhelmshaven* **1**, 1–7.

EHRENBAUM, E. (1904). 'Eier und Larven von Fischen der Deutschen Bucht. III Fische mit festsitzenden Eiern.' *Helgol. Wiss. Meeresunters.* **6**, 127–200.

FLÜCHTER, J. (1965). 'Versuche zur Brutaufzucht der Seezunge (*Solea solea* L.) in kleinen Aquarien.' *Helgol. Wiss. Meeresunters*, **12**, 4, 395–403.

— (1966). 'Versuche zur Haltung lebender Rotbarsche (*Sebastes*) im Aquarium.' *Der Zoologische Garten* **32**, 39–45.

GILLBRIGHT, M. (1953). 'Die Belüftung von Seewasseraquarien.' *Kurze Mitt. aus der fischereibiol. Abt. d. Max-Planck-Inst. f. Meeresbiol., Wilhelmshaven*, **3**, 1–21.

HEINCKE, F., und E. EHRENBAUM (1900). 'Eier und Larven von Fischen der Deutschen Bucht. II. Die Bestimmung der schwimmenden Fischeier und die Methodik der Eimessungen.' *Helgol. Wiss. Meeresunters*, **4**, 127–331.

HEMPEL, G. (1963). 'On the importance of larval survival for the population dynamics of marine food fish.' *California Cooperative Oceanic Fisheries Investigations. Reports*. Vol. **10**, 13–23.

KÜHL, H. (1961). 'Nahrungsuntersuchungen an einigen Fischen im Elbe-Mündungsgebiet.' *Ber. Dt. Wiss. Komm. Meeresforsch.* **16**, 2, 90–104.

LINDROTH, A. (1946). 'Zur Biologie der Befruchtung und Entwicklung beim Hecht.' *Med. fran Stat. undersoknings- ock forsocksanstalt for sotwattenfisket* **24**.

OPPENHEIMER, C. (1958). 'A Bacterium Causing Tail Rot in the Norwegian Codfish.' *Inst. of Marine Science* **5**, 160–162.

SCHÄPERCLAUS, C. (1955). *Fischkrankheiten*. Akademie-Verlag, Berlin.

SHELBOURNE, J. (1964). 'The artificial propagation of Marine Fish.' *Advances in Marine Biology* **2**, 1–83.

C. MICRO-ORGANISMS

PROTOZOA

K. G. GRELL

The broad habitat of the sea contains an immense abundance of unicellular living creatures, of which only a small part has been systematically described until now, and an even smaller part has been studied in detail. What we know of the biological phenomena of the marine protozoa is based largely on the observation of freshly collected samples. However, observation of this kind is inadequate if one wishes to learn all the phases in the development cycle of a species and the conditions necessary for its propagation. If one species can be cultured, i.e., reared under controlled conditions over a fairly short or long time span in the laboratory, the possibility of studying its vital data in greater detail – and possibly experimentally – is considerably improved.

Since the outlay entailed in the procedure should not exceed that necessitated by the intended purpose, the use of simple methods, if possible, will be described in the following report of experiences. A knowledge of the usual laboratory methods (preparation of pipettes, sterilization, etc.) is assumed.

1. PREPARATION OF THE MATERIAL

The initial material provides samples which are collected from marine habitats, preferably at a marine station. The manner in which these are obtained need not be discussed further here (cf p.48). Since pelagic protozoa have been found difficult to culture until now (exception: *Noctiluca miliaris*), samples from the littoral zone in which the protozoa are abundantly present will be used in the first place; best suited are substrates which show an 'overgrowth' even without microscopic examination, such as algal thalli, posidionia, empty mollusc shells, small stones, etc.

Such substrates are transferred into suitable glass containers (covered crystallizing dishes, large petri dishes) with pure seawater. Since sensitive metazoa often die after a short period and thus bring about 'putrefaction', i.e., a great increase in bacteria which guides the development of the biological association of an 'impure culture' of this kind in a – generally unplanned – direction, it is advisable to remove all possible metazoa in time. The hydrozoa, bryozoa and kamptozoa, as well as nematodes and juvenile stages of polychaetes, mollusca, etc. which adhere to the substrate often remain alive for days or weeks.

However, the composition of an impure culture of this kind and the population density of the various species will change continually in every case, and it is therefore important to find and select the desired protozoa by continuous observation. The *selection process* which takes place during the development of an impure culture will go in a particular direction according to the type of sample involved; at the same time, it depends on the conditions to which the impure cultures are subjected. Naturally the impure cultures which are exposed to a regular alternation of night and day last best (as, for example, those placed on a window with north light), because the autotrophs (diatoms, dinoflagellates, algae, etc.) can develop in these cultures, resulting in stabilization of the biological association. The choice of food organisms can also influence the selection process in an impure culture in a particular direction. The result of a 'concentration procedure' of this kind naturally depends largely on the composition of the sample. For example, if an impure culture is treated with diatoms in order to select the microforaminifers which eat diatoms, it may happen that a single *Paramoeba eilhardi** multiplies so greatly in a short time that any rotalides present cannot reproduce at all. Foraminifera which produce freely swimming gametes (such as *Iridia, Myxotheca*) disappear in every impure culture which contains the phagotrophic dinoflagellate *Oxyrrhis marina*. Detritus consumers (among the metazoa, above all harpacticoidae, *Dinophilus*) naturally alter an impure culture very rapidly. In dishes which have been exposed to light for several weeks, cyanophycea or palmelloid chrysomonadae ultimately prevail.†

These examples may suffice to show that a precise understanding of the species obtained is necessary in order to evaluate correctly the selection process which takes place in an impure culture.

2. ISOLATION

The moment for the isolation of a species of protozoa is generally

---

* This amoeba can now be destroyed quantitatively with a bacterium which penetrates into the cell nucleus and can be conserved for months in pure seawater (Grell 1961).
† Screw-top polyethylene flasks are suitable for transporting impure cultures.

at hand only after a certain increase has taken place in the impure culture, so that it can be repeated several times. The isolation of freely swimming or suspended species presents no difficulty. They need only be drawn out with a sterile capillary pipette and to be freed of other organisms by repeated 'washing' in pure seawater. It is especially easy to isolate autotrophic flagellates, which collect by phototaxis at the light edge. The isolation of crawling or clinging species which carry along components of the bottom substrate when they are drawn up is more tedious. For such cases the use of a capillary loop cemented into a microcannula has been found very useful (Weber 1965). Testaceous protozoa, such as the larger foraminifera, are frequently themselves 'contaminated' by other one-celled organisms which are carried along during the transfer. Their isolation generally only succeeds after numerous attempts and requires much patience.

3. CULTURES

Although it should be attempted to obtain cultures which are *as pure as possible* in every case, this aim will only be met to some extent. So called 'axenic' cultures, i.e. those freed of all other organisms, as required in nutritional physiology, involve considerable expense and are only possible in a limited number of species. Liquid cultures always include bacteria, which were transmitted at the time of isolation or which came from the air. If the cultures are left to stand for several weeks, fungi frequently settle at the surface; air-borne spores are always present in abundance in a laboratory in which much work with impure cultures is carried out. Their spread can easily be counteracted by regularly passing a flame from a bunsen burner over the water surface when transplanting the cultures.*

Which *culture vessel* is suitable depends on the type of protozoa and the aim of the investigation. For autotrophic flagellates serving as food organisms we use high-necked bottles. *Chlorella* and diatoms are reared in large petri dishes (diameter 14 cm). These have also been found useful for crawling protozoa (amoebae, foraminifera) and clinging protozoa (suctoria), since slides can be placed at the bottom; they can then be used to make up preparations (Grell 1954). Flagellates and ciliates can be kept in crystallizing dishes (diameter 8 to 10 cm). Boveri dishes, test tubes or concave

---

* The bunsen burner must of course be kept scrupulously clean so as to prevent any dirt particles from falling into the culture dish.

slides are generally used only for isolation trials or experimental studies which involve smaller individual quantities.*

Apart from synthetic media, which are certainly superior for certain purposes (Provasoli and co-workers 1957), the easily prepared so-called 'Erdschreiber solution' has been found useful as a *nutrient solution* for autotrophic protozoa.† This solution consists of the following:

> 1,000 ml seawater
> 50 ml 'soil extract'
> 2 ml $NaNO_3$ (25 g to 250 cc distilled water)
> 2 ml Na_2HPO_4 (5 g to 250 cc distilled water)

To prepare the 'soil extract', 300 g sieved unfertilized garden earth is heated in 1,500 ml seawater for one to two hours at $2 \cdot 5$ atm. in the autoclave or boiled in the steam sterilizer. The solution is filtered while hot. It can be kept in the refrigerator for several weeks. The solution for use, however, should always be prepared specifically in each case.

The seawater must be pre-filtered through a bacterial filter (Berkefeld N. water jet pump) and must be sterilized by heating to 80°C (preferably twice).

As light source it is best to use fluorescent tubes (such as Osram L 16 or 40 W/25), which are turned on and off periodically by a time-switch.

The room in which the cultures are kept should be kept at as constant a temperature as possible. For most marine protozoa an average temperature of 21°C ($\pm 1°$) is advisable. However, many species also tolerate lower temperatures. As this causes them to reproduce much more slowly, work is saved if they are kept in a cool room (as for example at 15°C) if they are not being used for experiments at the moment.

The autotrophic flagellates reared in the 'Erdschreiber solution' naturally alter its composition by consuming the nutrients and excreting metabolic end products.

The culture must therefore be '*transplanted*', i.e. a small quantity of the culture must be transferred into a fresh nutrient solution at regular intervals which depend on the species cultivated and the

---

* It need not be especially emphasized that all the glassware used must be clean when removed from the sterilizer. It should not have come into contact with chemicals previously.

† The term originated with B. Föyn (1934), who added the so-called 'soil extract' to the nutrient solution developed by the botanist E. Schreiber (1928) for the rearing of centric diatoms (seawater with the addition of nitrate, phosphate, etc). A better description would therefore be 'Föyn solution'.

TABLE 12

Survey of the marine protozoa now being reared at the
Zoological Institute of Tübingen

Species	Food organisms	References
I. Flagellates:		
Autotrophic:		
Cryptomonas spec.		
Dunaliella spec.		
Chlamydomonas spec.		
Dinoflagellates:		
Exuviaella marina		
Prorocentrum micans		
Amphidinium elegans......		
Amphidinium massarti		
Glenodinium monotis		
Heterotrophic:		
Gymnodinium fungiforme ..	*Cryptomonas*	
Noctiluca miliaris	*Dunaliella*	
Oxyrrhis marina	*Dunaliella*	
II. Rhizopods:		
Amoebina:		
Flabellula mira...........	Diatoms	
Thecamoeba orbis	Diatoms	
Pontifex maximus	Diatoms	
Paramoeba eilhardi	Diatoms	GRELL 1961
Stereomyxa angulosa	Diatoms	GRELL 1966
Stereomyxa ramosa.......	Diatoms	GRELL 1966
Foraminifera:		
Myxotheca arenilega......	*Chlorella*	
Allogromia laticollaris	*Dunaliella* (killed)	ARNOLD 1955
Rotaliella heterocaryotica..	*Dunaliella* (killed)	GRELL 1954
Metarotaliella parva	Diatoms	WEBER 1965
Patellina corrugata	Diatoms	GRELL 1959
III. Sporozoa:	In the host:	
Eucoccidium dinophili	*Dinophilus gyrociliatus*	GRELL 1953
Eucoccidium ophryotrochae	*Ophryotrocha puerilis*	GRELL 1960
IV. Ciliata:		
Platyophrya spec.	*Dunaliella*	
Litonotus duplostriatus	*Euplotes*	
Paramecium woodruffi.....	*Dunaliella*	
Holosticha spec.	*Exuviaella*	
Keronopsis rubra	*Dunaliella*	
Euplotes vannus	*Dunaliella*	
Euplotes crassus..........	*Dunaliella*	HECKMANN 1963
Euplotes minuta	*Dunaliella*	HECKMANN 1964
Uronychia transfuga	*Cryptomonas*	
Strombidium spec.	*Dunaliella*	
Acineta tuberosa	*Strombidium*	
Paracineta limbata	*Strombidium*	

size of the rearing container. One should never wait so long that the nutrient solution is entirely exhausted. Even when dealing with phagotrophic species, further food organisms should not be added continuously. It is always better to transfer a small quantity into a new dish. For crawling protozoa (amoebae, foraminifera) we distribute a fully settled petri dish – after stirring it up with a sterile brush – into several dishes whose bottom is covered with a fresh diatom fur. Suctoria are generally present in sufficient numbers at the surface, so that it is easy to transfer a small quantity into a petri dish with fresh seawater. However, when rearing suctoria one must remember that overfeeding is often poorly tolerated.

The *agar plate method* commonly used in microbiology is also suitable for certain marine protozoa. To prepare the seawater agar a synthetic nutrient solution or 'Erdschreiber solution' can be used.* For the agar plate method naturally only autotrophic protozoa come into question, which can multiply in the non-flagellated state (*Chlorella, Chlamydomonas*, diatoms). These plates can then be superinoculated with phagotrophic species which feed on the autotrophic ones. The advantage of the agar plate method is that it makes increased sterility possible; however, the seawater agar must be protected from evaporation to a greater extent.

In order to culture phagotrophic protozoa one must know which *food organisms* they require. If the study of the impure cultures does not provide information on this, they must be isolated in cultures with various food organisms. Best suited for such feeding trials are Boveri dishes. It is therefore advisable to have available as large a stock as possible of food organisms. Some of these food organisms are themselves phagotrophic (such as *Euplotes, Strombidium*), so that one must resort to a three-part culture procedure. Certain protozoa are best fed with 'artificial detritus', such as the foraminiferan *Rotaliella heterocaryotica*, which we have been rearing for more than 14 years with heat-killed cells of *Dunaliella*.

If their hosts are successfully cultured, the rearing of parasitic protozoa presents no difficulties. The archiannelids *Dinophilus gyrociliatus* (food: heat-killed cells of *Dunaliella*) and the polychaete *Ophryotrocha puerilis* (food: *Chlorella* or killed nauplii of *Artemia salina*) have been reared in our Institute for years in pure cultures. They can be infected at any time with the spores of *Eucoccidium dinophili* or *Eucoccidium ophryotrochae*, which can be kept in seawater over a period of years without damaging their infection capacity. The sporozoites hatching in the intestine of the host animals penetrate into the body cavity, where they continue their

---

* The Merck 'nutrient agar standard I' can also be dissolved in (25 g/l) seawater.

254   CONSERVATION AND CULTURE IN THE LABORATORY

development (gamogeny, sporogony) until the spores have been formed. In *Eucoccidium dinophili* all the phases of this development in life can be followed, since the body wall of the host is entirely transparent.

REFERENCES

ARNOLD, Z. M. (1955). 'Life-history and cytology of the foraminiferan *Allogromia laticollaris*.' *Univ. Calif. Publ. Zool.* **61**.

FÖYN, B. (1934). 'Lebenszyklus, Cytologie ind Sexualität der Chlorophycee *Cladophora Suhriana*.' *Arch. Protistenkd*. **83**.

GRELL, K. G. (1953). 'Entwicklung und Geschlechtsbestimmung von *Eucoccidium dinophili*.' *Arch. Protistenkd*. **99**.

— (1954). 'Der Generationsweschel der polythalamen Foraminifere *Rotaliella heterocaryotica*.' *Arch. Protistenkd*. **100**.

— (1959). 'Untersuchungen uber die Fortpflanzung und Sexualität der Foraminiferen. IV. *Patellina corrugats*.' *Arch. Protistenkd*. **104**.

— (1960). 'Reziproke Infektion mit Eucoccidien aus verschiedenen Wirten.' *Naturwiss.*, **47**. Jg.

— (1961). 'Uber den „Nebenkörper" von *Paramoeba eilhardi* Schaudinn.' *Arch. Protistenkd*. **105**.

— (1966). 'Amöben der Familie Stereomyxidae.' *Arch. Protistenkd*. **109**.

HECKMANN, K. (1963). 'Paarungssystem und genabhängige Paarungstyp-Differenzierung bei dem hypotrichen Ciliaten *Euplotes vannus* O. F. Müller. *Arch. Protistenkd*. **106**.

— (1964). 'Expeiimentelle Untersuchungen an *Euplotes crassus*. I. Paarungssystem, Konjugation und Determination der Paarungstypen.' *Z. Vererbungsl.* **95**.

PROVASOLI, L., I. I. A. McLAUGHLIN, and M. R. DROOP (1957). 'The Development of artificial media for marine algae.' *Arch. Mikrobiol.* **25**.

SCHREIBER, E. (1928). 'Die Reinkulturen von marinem Phytoplankton und deren Bedeutung für die Erforschung Produktionsfähigkeit des Meerwassers.' *Wiss. Meeresunters., Abt. Helgoland*, N. F. **16** (10), 1.

UHLIG, G. (1965). 'Die mehrgliedrige Kultur litoraler Folliculiniden.' *Helgol. Wiss. Meeresunters.* **12**.

WEBER, H. (1965). 'Über die Paarung der Gamonten und den Kerndualismus der Foraminifere *Metarotaliella parva* Grell.' *Arch. Protistenkd*. **108**.

FUNGI

W. HÖHNK

The impure fungal cultures prepared according to the methods described on pp. 140 ff are, with few exceptions, mixed cultures. Various species which usually belong to different systematic types grow next to each other.

They belong not only to the group of the so-called 'classical' water fungi, *Phycomycetes*, which are directly dependent on a watery habitat during certain stages of development. Among the higher fungi which have long been known almost entirely from terrestrial environments, the Ascomycetes and Fungi imperfecti have many relations in the sea. Lastly, the Eumycetes, sub-class of the *Basidiomycetes*, also have their representatives in the marine field.

Table 13 offers a broad survey which shows the systematic groups and the number of their marine species; the other columns relate to their modes of life and their numerical distribution into substrate groups.*

TABLE 13

Survey of the fungi found in the sea, their systematic classification, their mode of life and their distribution into substrate groups

System Group	Total no. of representatives	Parasites	Saprophytes	Algal fungi	Shore plants	Wood	Pollen	Other substrates
Phycomycetes .	82	64	18	51	5	1	12	13
Ascomycetes ..	97	49	48	31	25	40	—	1
Fungi imp.....	26	3	23	9	1	15	—	1
Badsidiomycetes	1	1	—	—	1	—	—	—
Labyrinthulae .	9	9	—	8	1	—	—	—
	215	126	89	99	33	56	12	15

* The numbers for the survey have been taken from Gaertner (1965) and Johnson and Sparrow (1961). The fungi appended as 'special groups' to the systematic section in the book of the latter authors have not been included here. These are the actinomycetes, streptomycetes, fossil fungi, Eccrinidae, lichen fungi, Laboulbeniales, spongophagae and *Ichthyosporidium* with a total of 36 species and the yeasts.

Apart from those listed in Table 13, there are almost as many fungal types again which have long been known terrestrially, but have apparently been carried by the wind or other carriers into muddy shallows, onto floating material or shore plants and algae, or have been washed with the flow from the limnic water into brackish water. It is certain that many of the higher fungi found in the sea have a salt tolerance which allows them to live in the new places, but this is not their optimal distribution area. Many of the filamentous lower fungi probably have only limited viability or will succumb in the competition. All of these species do not count as marine fungi and are not included in the survey.

We designated as marine fungi those which live submerged in the seawater at their original location, go through their entire development cycle there, and propagate and populate other substrates with healthy infectious spores or conidia.

An examination of the survey yields some noteworthy findings. The number of phycomycetes, the really 'privileged' water fungus, is lower in the sea than that of the ascomycetes, and three-quarters of the phycomycetes have been found to be parasites, settling for preference on algae. Of the 18 saprophytes, only some 12 had settled on pollen.

Among the ascomycetes the portions of the saprophytes and the parasites were almost equal, but in the choice of substrate there was a marked displacement of the algae over the shore plants towards wood. The same finding is true for the fungi imperfecti, which we generally define as incompletely known ascomycetes; this tendency is even more markedly expressed here.

The shift in the emphasis on the species has historical reasons. For decades the fungi of the algae were the principal subject for investigation. Since the baitable small forms were ignored at that time, the parasites became dominant. The same was initially true for the shore plants, and it is only as a result of the recent attention directed at wood as a substrate that the number of saprophytes has increased in importance in the overall picture.

With the development of the as yet young science of marine mycology, the dead substrates will gain increasing consideration, and as they are studied the number of saprophytes will continually rise. This can be anticipated if only on the basis of the assumption that the marine fungi play as important a part in the conversion of matter as do the terrestrial fungi in their area.

Oceanic expeditions have already brought evidence of the global distribution of marine fungi. In the initial trials pollen was used to bait the phycomycetes, as this answered the question more rapidly than would have been possible if working with higher fungi.

To the layman it is amazing to discover that despite their recognized world-wide distribution, the number of fungi captured with pollen has not increased rapidly. There are several reasons for this unsatisfactory fact: the methods for rearing and cultivating them have only been developed or come into use very recently; the establishment of the life cycle of apparently new forms requires a great deal of time; and finally, the number of investigators and of marine mycological working sites is very small. Consequently, many sediments have not yet been worked out in terms of species, and this will require much time in the future.

1. WORK WITH THE LOWER FUNGI

In the light of this situation we begin our discussion with the treatment of the *saprophytic Phycomycetes*.

These are found almost consistently in the sediment, and not as abundantly in the *free water*. We therefore begin with the marine substrates and then approach the coast.

(a) Obtaining the Cultures from the Marine Sediment Samples

When carrying out a qualitative investigation of the oceanic microflora, larger sediment samples are required than is feasible when working with sediments for isolation.

The aim is different here. The larger the quantity of sediment, the higher the number of organisms, of course. Apart from the lower fungi, this also includes the competing bacteria, the higher fungi and protista. It is a matter of experience that under laboratory conditions these can be picked out in the mixtures of types of the lower fungi. This may be due to the fact that the site environment differs to some extent, i.e. it is richer in food, there is an absence of water exchange, the pH value changes.

We therefore take a *small* amount of sediment (approximately 0.5 g) from the large quantity of sample material collected, place it in a petri dish or in an Erlenmeyer flask (usually of 100 ml capacity), dilute it in 25 to 50 ml of membrane-filtered sterile water from the station, and add the bait.

The use of a variety of baits is now a matter of course. A whole series of varied baits or a selection from those mentioned in the literature is available for this purpose: pollen from *Pinus* and *Tipha* or fern spores, and as cellulose substrate straw, grass, onion skin or small pieces of leaf, or boiled cellophane, chitin from crabs, insect exuviae, small particles of horn, pieces of hair, particles from crushed hemp or other seeds. etc.

It is useful to use the types of bait singly, i.e., to place one individually in each dish with nutrient medium. A group of dishes or flasks must therefore stand ready to receive the bait.

If the baits differ greatly, the incubation time and the results will be very variable. The *Pinus* pollen is attacked relatively soon and in great numbers (after 5 to 13 days). We will first discuss the so-called pollen fungi.

If no findings are obtained from microscopic observation of the first preparation, consisting of a water drop on a slide with a number of like baits taken under aseptic conditions from the petri dish or from the Erlenmeyer flask, several further ones are prepared. If attacked baits are found then or after a longer period, the baits are separated with a fine glass loop and repeatedly transferred into drops of sterile water, i.e. rinsed. They are then placed in small petri dishes (diameter: 5 cm), which contain water, and pollen is added. The addition of sterilized soil or 3% soil extract may be useful (Gaertner 1965).

Sporulation takes place in these dishes; the zoospores settle on the pollen which has been sprinkled in. This represents the first concentration step.

Starting with this step, further isolations of the pollen grains which bear only one thallus are carried out. Each bait selected for transfer again runs successively through a series of sterile water drops until the fungus sporulates in a drop. The empty sporangia with the substrate are removed, and the swarming zoospores then remain in the water drops. Pushed by drops of sterile water, they slide out of the wash bottle into one or several sterile dishes with water and new bait. This process must be repeated again and again until only one spore can infect the bait; i.e., until it is certain that a single-species culture has been obtained.

During this procedure the bacteria have decreased in importance, but have not been entirely eliminated. Their removal represents the next step and can be achieved by taking drops containing spores from the single-species cultures with a curved glass rod and spreading these on the surface of agar nutrient media. We then have few or individual spores. As soon as germination has begun, the small plants are picked out from abacterial places and transferred to other dishes with the same nutrient media. Their development then continues. After renewed sporulation, the grouping of the spores in the vicinity of sporangia is reduced by spreading them in other dishes with a glass or platinum loop. This is repeated again and again until all the bacteria have been eliminated.

In an abacterial state these fungi grow well on the agar media. If treated carefully, they remain available for years, and their viability scarcely decreases. The amount of agar should be small;

between 0·6 and 1·2% is recommended. Experience has shown that low nutrient media concentrations are desirable.

In special cases Gaertner (1965) uses *silica gel* nutrient media into which the zoospores are poured. The naked and heat-sensitive zoospores are embedded in the solidifying substrate without damage. In this manner Gaertner obtained pure cultures of *Rhizophydium patellarium*.

In isolated cases the same investigator eliminated the bacteria by means of antibiotics. After isolating the sporangia, they were transferred into a chlorotetracycline solution (1mg/1 ml) for 5 to 15 minutes and then placed in small petri dishes with sterile sea-water and pollen. The antibiotic solution should be freshly prepared and filtered when sterile. The antibiotic crystallizes in the seawater and precipitates in high concentration on the bacteria.

Ulken (1964 and 1965) obtained single-species cultures by fishing out individual sporangia from the impure culture with glass capillaries. They propagate in sterile water from the sample with *Pinus* pollen, on a concave ground slide placed in a petri dish under a glass bell lined on the inside with moist blotting paper. The moist chambers are kept in the incubator at room temperature. After 24 to 48 hours the pollen is infected. A platinum loop is covered with the infected pollen and this is spread out on a silica gel plate. From these plates we harvest infected pollen individually or individual sporangia, place them on new plates and thus obtain an abacterial culture. The plates are kept at 6 to 8°C in the refrigerator. After six days individual colonies are separated for inoculation.

According to Oppenheimer and Zobell (1952), abacterial cultures do well on a modified seawater agar. *Thraustochytria* on pollen bait can also be kept in the nutrient solution (without agar).

(b) Obtaining the Cultures from the Seawater Samples

The work procedure can be shortened by using *water samples* obtained from a marine station. We know that these water samples are poorer in spores than are sediment samples; but even in the water samples we find differences in the numbers when we compare samples taken from turbid areas near the bottom with those obtained from clear, plankton-free or plankton-poor water from the centre of the water column.

If one litre of water with few spores contains only a few infective units of the pollen fungi, this results not only in a large number of faulty preparations – if each preparation takes only 20 ml – but it is also probable that if a positive finding is obtained, the number of species will be low. In such preparations a longer period is also required to obtain an epidemic-like frequency of the thalli and spores; in contrast, when suspended sediment samples are used, 10 to 14 days is generally sufficient.

If the preparation is low in spores, one may find a few pollen grains with isolated thalli in the drop preparations after one to two weeks. If the isolation step is successful, the result can be introduced as a single sporangium culture into the isolation series and further treated as described above for the sediment samples.

If the water samples offer a better yield of pollen fungi, perhaps originating from the turbid water near the bottom, it is advisable to use the dilution method as described for the quantitative determination of fungi (see p. 153).

One must begin with the dilution step which has the smallest number of spores. If examined early, it is relatively easy to separate a pollen grain with only one fungal sporangium. If the fungi have already begun to accumulate on the bait, it will first be necessary to obtain the single-sporangium culture. With this culture it is then possible to obtain the bacteria-free culture by successive steps, as described above.

Another advantage of the dilution method is that by the repetition of the dilution step poorest in spores it is possible to isolate various species of fungi. A site has a *species spectrum*. A certain variety of forms of pollen fungus of the genus *Thraustochytrium* commonly found in seawater has been shown in this manner (see Fig. 88). Species of uniflagellate genera are also present in the samples. These probably initiate the succession of fungi on the bait; as a rule they are subsequently suppressed by the more robust species. This signifies that fungi in an impure culture containing abundant spores have a mutual influence on each other, and that after a long period a selection takes place, which masks the original species spectrum.

The isolation method has led to these findings. It suggests that the number of Phycomycetes found on the pollen in the ratio described at the beginning will soon multiply, and with it the number of saprophytes in the sea will be increased.

A shorter method of obtaining a true culture is described by Vishniac (1956); this method works both for parasitic and for saprophitic lower Phycomycetes. A nutrient agar rich in B vitamins and containing antibiotics is required.

The nutrient medium contains, in 80 ml/100 ml seawater, 1·5 g agar, 0·1 g gelatine hydrolysate, 0·1 g glucose, 0·001 g liver extract concentrate (1:20) and of the B vitamins: 0·2 mg thiamine-hydrochloride, 0·1 mg calcium pantothenate, 0·02 mg pyridoxamine dihydrochloride, 0·5 µg biotin, 2·5 µg folic acid, 0·1 mg nicotinic acid, 0·04 mg pyridoxal hydrochloride, 0·01 mg p-aminobenzoic acid

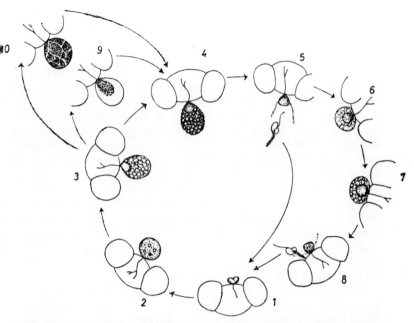

Figure 88. Diagram of the developmental cycle of *Thraustochytrium kinnei* Gaertner. 1 = a zoospore settles on a *Pinus* pollen and forms a thallus with rhizoid; 2 = the thallus has developed; 3 = division in the basal section and in the spore-forming upper area has occurred; 4 = spore development is terminated; 5 = the zoospores have swarmed out; 6 = the basal section grows into a new sporangium; 7 = renewed spore formation in the upper section; 8 = the zoospores have swarmed out, the basal section can form a further sporangium once again; 9 = a thallus has developed a thick membrane, can detach itself from the pollen and sporulates later; 10 = a thallus also breaks off without thickened membrane and sporulates away from the first substrate. (According to Gaertner 1966)

and 0·05 µg cyanocobalamine. The gelatine hydrolysate is given the following preliminary treatment: Eastman Kodak gelatine is hydrolysed with sulphuric acid and first partially neutralized with barium hydroxide, then filtered and entirely neutralized with potassium hydroxide; the pH is then adjusted to 7·5.

After filling with the nutrient agar, the dishes are dried overnight or up to two days. Their moisture content then varies accordingly. Immediately before use, a concentrated aqueous solution of 2,000 IU penicillin G and 0·5 mg dihydrostreptomycin sulphate is poured into the dishes. We then add 0·2 ml of seawater which is to be examined for fungi, using a curved sterile glass rod for spreading the water, or small particles of marine algae with fungi are placed on the surface. After 7 to 10 days of incubation at a temperature of 20°C no bacteria or only a few bacteria were found, but colonies of lower fungi, yeasts, *Mucor* and *Labyrinthula* species were visible even with the naked eye.

Also present among the Phycomycetes were the recently described *Sirolpidium zoophthorum*, a parasite in mussel and oyster larvae, *Haliphthoros milfordensis*, a parasite in the eggs of *Urosalpingis cinerea*, and the comb. nov. *Atkinsiella dubia* (Atkins). On the same agar Johnson obtained *Lagenidium chthamalophilum*, a parasite in the eggs of barnacles.

In their reports Goldstein and co-workers emphasized the need for thiamine (50 µg/l) and vitamins B_{12} and B_1 for their study subjects, and mention glutamate as the best N source, and glucose, maltose, soluble starch and cellulose as the best C sources. They establish that NaCl must not be replaced by KCl or $CaCl_2$, and give examples of the manner in which optimal development depends on a high salt content.

(c) Obtaining Cultures from Fresh Water

These studies of marine *Phycomycetes* were preceded by investigations of *freshwater* Phycomycetes, particularly Chytridiacea. These studies began in about 1931. By 1950 16 species had been transferred into pure cultures, and by now probably double that number has been obtained. In the discussion below I should like to report on the *indications* and *experiences* presented in the reports, which may help the worker with marine fungi from the point of view of method and in individual cases. Fungi which cannot be baited with pollen are largely involved here.

Couch (1939) isolated individual sporangia and individual spores in water drops and on agar. The separated sporangia were repeatedly washed in water drops, cleaned of adhering foreign bodies and transferred individually in water drops to new sterile leaf particles as substrates. Water which had been agitated with animal charcoal was dripped onto these. The slides for the preparations lay in a petri dish which had water drops on the bottom and below the cover to protect the preparations from drying out.

In difficult cases where smaller sporangia were present or the sporangia were of several species and sporulated one after the other, Couch separated them on the agar (3%) after washing the settled substrate. The isolated sporangium was placed with a small agar block in a sterile water drop, which contained a newly inserted leaf particle.

After sporulation single spores germinated near the fresh substrate, sent rhizoids into the agar and formed thalli. The young plants were washed again, the particle of leaf was removed and the individual sporangium was picked out and transferred to a new agar plate.

If the rhizoid system is strongly branched or if other fungi act as a disturbance, the following procedure can be adopted: the single sporangium in the water drop is allowed to sporulate; one then removes some of the spores with a capillary and places them in a water drop, which again contains a sterile leaf particle as substrate. Young thalli develop from the spores on the new substrate. To obtain a single spore culture a very finely drawn-out capillary should be used to remove the smallest possible quantity of spores. These few spores are distributed among several further drops, and this is repeated until one drop contains only one spore. This single spore is then transferred into a sterile water drop with a fresh leaf particle. To obtain abacterial thalli, a spore must be deposited on the surface of one of the four nutrient media recommended by Couch.

Very little nutrient substance is required when using these media; the spores will then germinate better. Only a few spores are taken up with the platinum loop or the pipette, and are distributed on the surface of the nutrient medium; some will remain abacterial. If the spores germinate at all, they will do so within 12 to 24 hours. Once the thalli have developed, an abacterial thallus should be cut off and placed in a petri dish in a sterile water drop with a small leaf particle. A single spore culture of this kind can be kept abacterial for generations.

Certain polycentric Chytridiacea (such as *Cladochytrium replicatum*) form a mycelium during spore germination. It is possible to isolate from this and to obtain abacterial cultures.

A special opportunity allowed Stanier (1942) to obtain a pure culture of *Rhizophlyctis rosea* easily. This organism was present in masses on the filter paper covering a concentration medium. When a piece of the thickly covered filter paper was placed in water, abundant sporulation in many sporangia appeared in the following 20 to 30 minutes. This phenomenon provided a method for obtaining a pure culture. Large pieces of the fungus-covered filter paper were placed at the bottom of test tubes containing some 10 to 15 ml sterile tap water. Since the spores are capable of more rapid locomotion than the bacteria and are moreover markedly aerotactic, they collect in large numbers in the upper part of the water column and show little bacterial contamination. Small samples were taken from these spots, smeared on dextrose agar, and then kept at 28°C. After 24 hours many young thalli were detectable among bacterial colonies, and 12 to 36 hours later vigorous sporulation had taken place. During this process some fluid escaped from the sporangium, but was immediately absorbed by the agar. Consequently the spores lay motionless around the empty sporangium. Some of these were removed with a sterile needle, placed on a new agar plate in a drop of water, and spread out; it was now possible to find abacterial thalli or to obtain them after further transfers. For the smear a relatively dry agar base is required, so that the simultaneously transferred water is immediately absorbed; otherwise the occurrence of bacteria cannot be prevented. Stanier prepared permanent cultures in test tubes on a slant agar base or in the fluid with semi-immersed filter paper strips.

To obtain abacterial single-species cultures, Whiffen (1941) used simple bases with 3 % agar. Having cleansed them of the remains of the natural substrate, he transferred them into dishes with 0·5 % agar. Sporulation occurred in the soft agar, and the spores were removed with a finely drawn-out pipette containing 1 ml water. Drops from this were dripped into new dishes with soft agar; some abacterial thalli were then available for placement on nutrient agar, or they became abacterial after repetition of the process.

Emerson (1950) drew attention to special experiences during cultivation. It may happen that the organism prefers a limited pH range, or that the concentration of the nutrient substance must be smaller so that the naked spores will not be injured or killed, or the fungus itself may excrete toxic substances which prevent or suppress growth on the agar, thus necessitating frequent transfers.

Some modifications are described by Koch (1957). They concern both the method of isolation and the nutrient base. The fungus plus a small particle of substrate are placed on the slide in one or two drops of sterile water; after sporulation the spore suspension is drawn up with a capillary pipette and is placed at the border of an agar dish. If the dish is now placed at a slant, the drop will cross the entire surface. A few drops of sterile distilled water are then dripped in the path of the suspension, and the inclination of the dish is turned at right angles to the previous direction. This causes the spores to spread out in a comb-like film. The surface of the agar is now allowed to dry for 5 to 10 minutes, the dish being kept in an almost perpendicular position. When the first bacterial cultures become visible under strong microscopic magnification, the germinating fungal spores must be removed from the abacterial sites with a very small spatula and placed on a fresh agar base. After transference from a pure culture it is sometimes advisable to rinse the small inoculum in a new agar dish, bottle or test tube with 1 ml sterile water, causing the spores of the inoculum to be distributed over a broad surface. Growth is then many times greater than that usually found in the dish.

Koch used the following nutrient basis: (1) 0·50 mg peptone, 9 g dextrose and 9 g agar in 900 ml potato liquid (150 g diced potatoes boiled for 1 hour, then pressed through a towel); (2) 0·6 g peptone, 0·4 g yeast extract, 1·8 g glucose and 9 g agar in 900 ml water; (3) a nutrient base according to Emerson (1941).

It had been found in some other cases that zoospores which were spread on agar were immediately overgrown by bacteria, that the zoospores did not survive the use of the capillary pipette and micromanipulator, and that the trials with antibiotics were faulty; in view of this, Gaertner (1954) attempted to obtain absolute pure cultures of *Phlyctochytrium palustre* by the use of permanent organs. These were disinfected in 3% H_2O_2 and washed in drops of sterile water until the zoospores swarmed out. Spores from this suspension were spread on nutrient agar (2% biomalt, 0·25% Witte-peptone and 1% agar) and germinated. The young thalli were picked out without bacteria and transferred individually to new sterile dishes.

In *Rhizophydium spaerotheca* permanent organs are very rare, and the sensitivity of the sporangia to H_2O_2 did not allow this cleaning. In this case the same investigator used another method for obtaining a pure culture. The pollen grains overgrown with fungus were agitated in sterile water, drawn onto sintered glass G/3 and repeatedly washed in sterile water. The mass was then suspended in the funnel with water, which slowly dripped into a sterile test tube. This allowed zoospores to go through the pores into the water. The suspension was centrifuged at 2,500 revolutions for 10 minutes. Approximately one-third ml. was then removed from the bottom by pipette, and having been placed in a new test tube with sterile water, was centrifuged a second time. After three repetitions of this step the liquid was decanted and the material for the agar smear was removed from the bottom by means of a loop. It was possible to pick out absolutely clean plants from the agar. The inoculation of the pure culture had to be carried out in 10 days, and if a quantity of 1% carragheen had been added, after 20 days.

In the most difficult case, with *Phlyctochytrium* sp. nov. Reinboldt, the same investigator found a pollen agar useful in helping him to obtain successful results. Approximately 0·5 mg *Pinus* pollen was sprinkled into the as yet liquid agar (0·8% agar), and uniformly distributed by tilting the surface. After solidification, zoospores were smeared over the surface and the culture was incubated at 22°C. Five days later it was possible to pick out the initial young plants from the abacterial places and to transfer them into new dishes with pollen and agar. The permanent culture of this investigator became successful when the concentration of the nutrient substance was markedly reduced.

The nutrient bases, their various concentrations and the absence of bacteria influence the growth and form of fungal individuals. Under experimental culture conditions of this kind sporangia grow to unusually large sizes, and *Phlyctochytrium palustre*, which is usually found on the pollen substrate, now sent a bunch of rhizoids radially into the water and infected several pollen with it. The systematic demarcations between species and genera thus overlapped.

Apart from the pollen fungi discussed, which belong to the group Thraustochytriaceae and Chytridialae, the *substrate samples* from the seawater also contain *other fungi*. These grow between the debris particles or bait.

(d) Obtaining the Cultures from Filamentous Phycomycetes and from Brackish Water

If the samples originate from muddy shallows from the brackish water area, filamentous *Phycomycetes*, *Saprolegniaceae* and *Peronosporales* – particularly *Pythia* – can be expected. They are usually lured out with seeds or their particles, this process involving two steps in petri dishes of 10 cm diameter. Impure cultures are first prepared. A heaped spatula of sediment is placed into each of as many dishes as there are types of bait (only one type of bait is used in each dish). Sufficient sterile water from the station is sprinkled over to cover the sediment and the bait. In the second step, after approximately five days, the attacked bait is removed and transferred into new dishes with sterile water from the station; a few fresh baits are added at a little distance from the attacked ones. They are reached by the developing swarm spores and are infected by them. After a further three to six days a fungal fur can be observed in the dishes with clear water. We cut off individual parts, hyphae or sporangia which have not yet sporulated, and transfer them to agar. As a nutrient base the carragheen agar described by Höhnk (1935)* can be recommended. The transferred portion grows numerous side branches radially in a few days, up to the edge of the dish. We cut off an abacterial tip from each young growth and place it in a small block on a fresh nutrient base, where the single-species culture then develops. If a sporangium has been transferred from the water dish, it will sporulate on the agar. By means of a platinum loop a spore group can be transferred and spread out on a new agar dish. Spores lying singly can easily be cut out together with an agar block and

---

* 1 l water, 10 g agar, 10 g carragheen, 0·5 g dextrose, 0·005 g primary potassium phosphate, 0·025 mg each ammonium citrate, ammonium sulphate and magnesium sulphate, pH 6·2-7·0. To inhibit bacterial growth it has repeatedly been found useful to adjust the pH to 4·7-5·5 with hydrochloric acid.

planted individually in further agar dishes. It is also relatively simple to cut out a young abacterial hypha tip from the fur which often grows in the agar layer and not on the surface, and to transfer it into test tubes with slanting nutrient agar for a permanent culture. This process is carried out under the binocular microscope since these zoospores have a diameter of 8 to 12 μm when encysted and the hyphae growing in the agar have widths of several micrometers.

In certain species it is possible to use the growth in the agar to observe the development cycle, that is, when all the stages take place in it; or a block with young mycelium is cut out from the overgrown agar, transferred into a dish with sterile water from the station, and one or two baits of the same type as that with which the fungus was first captured are placed directly next to it. Within 8 days the entire cycle has often been repeated, and the species can be determined and prepared for other work.

In the presence of a limited quantity of nutrient in the water dish, the filamentous Saprolegniaceae and *Pythia* we have just discussed (see Fig. 89) exhibit the three life stages one after the other: vegetative growth, vegetative propagation and the sexual

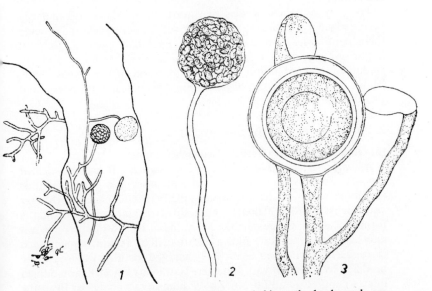

Figure 89. *Phytium aquatile* Höhnk (1953). 1 = habitus; the hypha ends are tightly filled with cytoplasm which flows into the sporulation bladder formed at the hypha end, and there divides into zoospores (approx. 145 ×); 2 = sporulation stage; after the ephemeral bladder has burst, the zoospores swarm out (approx. 390 ×); 3 = oogon with a ripe oospore and two empty antheridia (approx. 1460 ×)

phase. This third phase has a curve-like course, with an ascending section, a peak and a descending section. The beginning of one phase overlaps with the end of the preceding one.

The sequence of the life stages depends on genotypic factors and the degree of their development is determined by environmental factors. The salinity of brackish and seawater and fluctuations in the salt content at the station (site of capture) of a fungus are external factors which may exert a varying degree of disturbance on the course of germination of these organisms.

The fungi are least sensitive during vegetative growth, and more so during vegetative propagation; this is evidenced by the number of sporangia and spores which they will produce, and the course of the sporulative process. Most marked is the reaction during the sexual phase, and more so during sexual differentiation than during development of the sexual organs.

These organisms are suitable material for establishing whether the site of capture of the form is its natural location or not, whether the fungus is permanent here or an intruder foreign to the habitat, which is eradicated by the competition after a period of starvation.

To carry out these studies one must expose the simultaneously prepared young cultures to several steps of a salinity scale and observe them continuously, measure them and tabulate them as follows: hypha growth, special features of the hypha, beginning and development of sporulation, onset and duration of the sexual phase, health or portion of the degenerating oogonia and oospores, and possibly the formation of gemmae.

2. Work with the Higher Fungi

(a) Obtaining Cultures from the Sediment Samples

Ascomycetes, *Fungi imperfecti* and *Mucoraceae* from the *sediment samples* are isolated onto solid nutrient bases and cultivated. For rapidly-growing species Siepmann (1959) recommended herb agar with Bengal pink (30,000:1, tetraiodo-tetrachlorfluorescein in the form of the potassium salt; the pH value of the nutrient base is not altered by the addition) and in addition, the previously mentioned carragheen agar. When using this agar, the distilled water is to be modified by mixing it with seawater. To the carragheen agar modified in this manner we add $0 \cdot 5$ molpotassium thyocyanate per litre; the pH value fluctuates between $6 \cdot 5$ and $7 \cdot 0$.

A generous spatula tip of the sediment sample is agitated in a sterile, closed receptacle with 10 ml autoclaved spring water; 1 ml of this mixture is placed in a petri dish and herb agar or carragheen agar is poured over; this is then mixed and kept at about 25°C. When using the herb agar, the inoculation of the grown fungi imperfecti and ascomycetes can begin on the second day, and when using carragheen agar after eight days. For identification, the nutrient substrates mentioned in the specific literature must be used.

No special literature is available as yet on the method of treating the higher fungi of the deep ocean bottom, or their isolation and cultivation.

(b) Obtaining Cultures from Shore Vegetation

S. P. Meyers and co-workers (1965) describe the isolation technique for the *fungi* of turtle grass, *Thalassia testudinum* König, a *shore plant*. The collection and treatment of the leaves has already been discussed in the previous section (p. 266). The investigators require three media for the isolation of the fungi vegetating in the leaf: one medium for rapidly growing fungi ($0 \cdot 1 \%$ glucose and $0 \cdot 1 \%$ Difco yeast extract) and the two other media (soil extract and *Thalassia* leaf extract) for slowly growing fungi. All three are prepared with water from the Gulf Stream (35% salinity) and $1 \cdot 7 \%$ Difco agar. To inhibit bacterial development, $0 \cdot 1 \%$ chloramphenicol is added to the extract before sterilization (20 minutes at 120°C).

The soil extract is obtained from 100 g garden soil in 200 ml tap water and is kept in the autoclave for 2 hours. The water lost is replaced by distilled water, the hot suspension is pre-filtered and the sediment is removed. The remaining 70 ml of the filtrate is made up to 200 ml with distilled water and the total is then sterilized through membrane filters ($0 \cdot 45 \mu m$).

The *Thalassia* extract is prepared as follows: healthy green leaves are washed in tap water and in water which has run over the ion exchanger; 20 g of the leaves is then cut into 6–13 pieces and homogenised in 100 ml distilled water. The resulting liquid is heated to 60–70°C for half an hour then filtered through paper and the sediment removed; the solution is made up to 100 ml with distilled water and is membrane filtered; the pH is $7 \cdot 6$. All extracts are kept in the refrigerator until required.

Three identical pieces of each of the leaves grouped according to colour pattern and growth, approximately $0 \cdot 5$ to 1 cm long, are cut up aseptically and placed on each of the three media. The petri dishes are kept at room temperature. During the following 2 to 3 weeks fungus mycelium can be isolated successively. The initial growth occurs during the first 48 to 72 hours. The permanent cultures are kept on seawater slant agar with $1 \cdot 0 \%$ glucose and $0 \cdot 25 \%$ Difco yeast extract.

(c) Obtaining Cultures from Wood

More has been written about the technique of isolating and cultivating *saprophytes* of *submerged* wood in marine areas than on most parasitic fungi of the shore vegetation.

Barghoorn and Linder (1944) obtained pure cultures by stab-inoculation. Using a flamed straight needle they removed a perithecium or spores which had issued from it and stabbed them into the agar layer of a petri dish. The pores germinate below the surface and often remain abacterial. After germination the mycelium portions can then be transferred into further dishes, which are used as material for nutritional physiology experiments or for permanent cultures. Contamination of the bottom of the dishes is avoided because the inoculation stab does not penetrate to the bottom; the injurious effect of air-borne contamination can be eliminated if the agar layer in the dish is turned over and pure mycelium is taken for inoculation into new dishes from the previous underside, which is now lying topmost.

Johnson (1961) gives a nutrient substrate for many of the wood-inhabiting fungi. To 1 l aged seawater add 1 g glucose, 0·1 g Bacto yeast extract and 18 g agar (pH 7·4 to 7·8). Seawater is described as aged if, after being drawn up, it is filtered through sintered glass, poured into large bottles, and kept in the dark for weeks or months. It has shown itself to be superior to fresh seawater; the nutrient substances are probably more abundant and more readily accessible in aged seawater than in fresh seawater.

The same author found that work with rapid-dried substrates was effective against bacteria. The substrates can be obtained in two ways: in the first method the scraped wood pieces are first dried with absorbent paper, then dried in the desiccator for 24 hours; in the second method the piece of wood, scraped and paper-dried, is exposed to an air stream for 2 to 3 hours. The air must be filtered through a calcium chloride tube and through a second tube filled with cotton and glass wool. Before this procedure Johnson marked the position of the perithecia in the wood by means of glass-headed pins.

The ascospores were subsequently removed from the perithecium by means of a sterile needle and spread over the nutrient substrate or stabbed into it.

If the perithecia have penetrated below the wood surface, they are cut out together with small wood blocks, sterilized for 1·5 minutes in 50% H_2O_2 and then washed repeatedly in sterile seawater. The perithecia are then exposed, picked out and thus yield ascospores for inoculation. In the same way as the spores, parts of the hyphae can also be taken for agar inoculation. If they are situated in the direct vicinity of the fruiting body, it is probable that they are from the same fungus. After surface sterilization with H_2O_2 and washing, small wood splinters are drawn out under the binocular microscope and then lightly pressed into the agar. It is naturally possible that the growing hyphae belong to several species of fungi; the uncertainty must be narrowed down by further culturing and, if necessary, must be eliminated in several steps.

The size of the ascospores, which frequently have axis lengths of $16–30 \times 8–12$ μ, readily allows the single spore culture through a spore suspension. Spores from a single or a few perithecia agitated in a glass tube with water and then spread out on the nutrient substrate of an agar dish yield young spores for transfer to further dishes or to slant agar. Abacterially situated spores should also be found, so that successive absolutely pure cultures can be produced. Only one circumstance suggests difficulties: problems may occur if the ascospores have gelatinous, sticky or hooked appendages. These peculiarities are often seen in marine ascomycetes; they are thought to be suspension and attachment devices. The purpose of these appendages works against our aim, and little success can therefore be expected in such cases.

Apart from the above-mentioned seawater agar, Meyers and Reynolds produced a further seawater agar which is prepared with artificial seawater, the same substances, and the addition of four vitamins. Cellulose powder and wood flour of various plants were used as a substitute for glucose in fluid cultures, serving as a source of C. The tested fungi belonged to seven genera of ascomycetes. During their vegetative growth they were not particular in regard to their food. The vitamins with artificial seawater further promoted growth; however, ripe perithecia were not always formed, and if so, then in variable quantities. However, fertile perithecia were formed on balsa wood which stood half-submerged in bottles with yeast extract broth. Cellulose powder and wood flour in the same fluid promoted the development of perithecia in certain species.

The conidia of *Fungi imperfecti* are relatively easy to remove with a microneedle if standing away perpendicularly from the wood substrate. However, many conidia carriers are very short and the longer ones frequently lie on the wood surface. This inconvenience was met by Meyers and Reynolds (1958) and by Johnson (1961) by the use of large humidity chambers. These investigators took two rectangular Pyrex dishes (side length 20 to 25 cm), inverted the second dish over the first, and after autoclaving placed the pieces of wet test wood inside in such a manner that they did not touch each other. Because marine wood-inhabiting fungi imperfecti do not form air mycelia when oversaturated with water, one must either pour away the water which has drained from the wood into the bottom dish several times in the course of the incubation, or the water must be absorbed by coarse filter paper or paper towels, several of

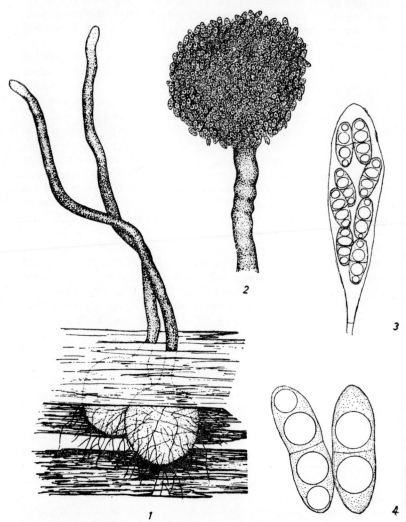

Figure 90. *Lignincola laevis* Höhnk (1955). 1 = habitus: two fruiting bodies (Perithecia) in the wood with long drainage necks (approx. 105 ×); 2 = the ascospores have issued from the ostiolum (approx. 210 ×); 3 = ascus with 8 ripe ascospores (approx. 665 ×); 4 = Twin-chambered ascospores with oil spherules (approx. 1750 ×)

which have been placed on top of each other on the bottom. In the latter case it will not be necessary to open the moist chamber during the incubation period of one to three weeks. The two halves of the chamber should be made airtight at the joint with Tesafilm.

The *yeasts* and the yeast-like fungi also belong to the higher fungi; they are classified with the ascomycetes and the fungi imperfecti. The presence of these fungi in the sea has been known for more than 70 years, and since then they have been found in all the oceans, within the entire water column from the bottom to the surface, and also within sponges, sea cucumbers, fish and other animals. Their spore numbers are higher in coastal areas, in algal muddy shallows and in plankton patches than in clear water.

Thus, the life conditions required for these organisms must be present at the capture site; these requirements are nutrient substances (N compounds and C compounds which are fermentable or can be assimilated), mineral substances (P, S, K, Mg, etc), trace elements and growth substances. These substances must also be added to synthetic nutrient solutions and nutrient substrates, but the requirements of the various species and strains vary greatly in relation to the quantity and quality of these substances.

An extensive literature is available on the methods of detection, culture and identification of the yeasts. We refer the reader to the reports of Janke (1946) and Windisch (1960), which are consulted again and again. These works also describe the methods used to obtain single-spore cultures. The identification of the isolated forms is based on the work of Lodder and Kreger van Rij (1951).

REFERENCES

See also the references listed on p. 155.

BARGHOORN, E. S., and D. H. LINDER (1944). 'Marine fungi; their taxonomy and biology.' *Farlowia* 1, 395–467.

COUCH, J. N. (1939). 'Technic for collection, isolation, and culture of chytrids.' *J. Elisha Mitchell Sci. Soc.* 55, 208–214.

EMERSON, R. (1950). 'Current trends of experimental research on the aquatic Phycomycetes.' *Ann. Rev. Microbiol.* 4, 169–200.

GAERTNER, A. (1954). 'Einige physiologische und morphologische Beobachtungen an Kulturen niederer Phycomyceten.' *Arch. Mikrobiol.* 21, 167–177.

— (1960). 'Einiges zur Ernährungsphysiologie von *Rhizophydium patellarium* Scholz.' *Arch. Mikrobiol.* 36, 46–50.

— (1965). 'Köderverfahren zur Isolierung niederer Phycomyceten. Sonderdr. „Anreicherungskultur und Mutantenauslese".' *Zbl. Bakt.* I, *Suppl.-Heft* 1, 451–460.

GOLDSTEIN, S. (1963). 'Morphological Variations and Nutrition of a New Monocentric Marine Fungus.' *Arch. Mikrobiol.* 45, 101–110.

— (1963). 'Development and nutrition of new species of *Thraustochytrium*.' *Amer. J. Bot.* 50, 271–279.

HARDER, R., und E. UEBELMESSER (1955). 'Über marine saprophytische Chytridiales und einige andere Pilze vom Meeresboden und Meeresstrand.' *Arch. Mikrobiol.* **22**, 87–114.

HÖHNK, W. (1935). 'Saprolegniales und Monoblepharidales aus der Umgebung Bremens.' *Abh. Naturwiss. Verein Bremen* **29**, 207–237.

— (1952). 'Studien zur Brack- und Seewassermykologie. II'. *Veröff. Inst. Meeresforsch. Bremerhaven* **1**, 247–278.

JANKE, A. (1946). *Arbeitsmethoden der Mikrobiologie Dresden und Leipzig.*

JOHNSON, T. W. jr., H. A. FERCHAU, and H. S. GOLD (1959). 'Isolation culture, growth and nutrition of some lignicolous marine fungi.' **12**, 65–80.

— and F. K. SPARROW, jr., (1961). *Fungi in Oceans and Estuaries.* Weinheim.

KOCH, W. (1957). 'Two new Chytrids in pure culture, *Phlyctochytrium punctatum* and *Phlyctochytrium irregulare.' J. El. Mitchel Sc. Soc.* **73**, 108–122.

LODDER, J., and N. J. W. KREGER van Rij (1952). *'The Yeasts. A taxonomic study.'* Amsterdam.

MEYERS, S. P., and E. S. REYNOLDS, (1958). 'A wood incubation method for the study of lignicolous marine fungi.' *Bull. Mar. Sci. Gulf and Carib.* **1**, 342–347.

— — (1959). 'Effects of wood and wood products on perithecial development by lignicolous marine Ascomycetes.' *Mycologia* **51**, 138–145.

— — (1959). 'Growth and Cellulolytic activity of lignicolous Deuteromycetes from marine localities.' *Microbiology* **5**, 493–503.

— P. A. ORPURT, J. SIMMS, and L. L. BORAL (1965). 'Thalassiomycetes. VII. Observations on fungal infestation of turtle grass, *Thalassia testudinum* König.' *Bull. Marine Science* **15**, 548–564.

SIEPMANN, R. (1959). 'Ein Beitrag zur saprophytischen Pilzflora des Wattes der Wesermündung.' *Veroff. Inst. Meeresforsch. Bremerhaven* **6**, 213–282.

SPARROW, F. K. jr., (1960). *Aquatic Phycomycetes.* Michigan.

STANIER, R. Y. (1942). 'The cultivation and nutrient requirements of a chytridiaceous fungus, *Rhizophlyctis rosea.' J. Bacteriol.* **43**, 499–520.

ULKEN, A. (1964). 'Uber einige Thraustochytrien des polyhalinen Brackwassers.' *Veröff. Inst. Meeresforsch. Bremerhaven* **9**, 31–41.

— (1966). 'Zur Physiologie einiger Thraustochytrien.' *Veröff. Inst. Meeresforsch. Bremerhaven* **10**, 117–120.

VISHNIAC, H. S. (1955). 'The Morphology and nutrition of a new species of *Sirolpidium.' Mycologia* **47**, 633–645.

— (1956). 'On the Ecology of the lower marine Fungi.' *Biolog. Bull.* **111**, 410–414.

— (1958). 'A new marine Phycomycete.' *Mycologia* **50**, 66–79.

WHIFFEN, A. J. (1941). 'A new species of *Nephrochytrium: Nephrochytrium auranticum.' Amer. J. Bot.* **28**, 41–44.

WINDISCH, S. (1960). 'Die Arbeitsmittel und Methoden zur Untersuchung, Reinzüchtung und Kultur der hefeartigen Pilze.' REIFF, KAUTZMANN, LÜERS und LINDEMANN: *Die Hefen, Bd.* 1, Nürnberg.

BACTERIA

G. RHEINHEIMER and W. GUNKEL

In contrast to other organisms, the conservation of individual bacteria is not possible because of their small size. Consequently it is always necessary to prepare cultivation trials, known as 'cultures' in bacteriology. When setting these up it must first be remembered that most marine bacteria are very sensitive to temperatures exceeding $+25°C$ and to the influence of light, and that they only thrive in media with a salt content of 20 to 40‰. Preparation of bacterial cultures generally begins with the individual cell which, when placed in a suitable medium, can multiply by millions within a few days. Since the most varied bacteria are almost always present next to each other in natural substrates, one of the desired bacteria must first be separated or 'isolated' from the others. This is generally easier to do when the desired bacteria have previously been concentrated in a 'concentration culture.' The isolated bacterium then develops in a suitable nutrient medium to a 'pure culture'. However, a pure culture of this type has only a limited life. In most cases it can be kept alive in a refrigerator for some weeks or months – but at higher temperatures often much less time. The bacteria must therefore be 'inoculated' regularly, i.e., at certain time intervals some bacterial material must be transferred to a fresh nutrient medium. As a result of repeated inoculations – following which the bacteria multiply by millions each time – a selection process sometimes takes place on the artificial substrate, leading to the formation of a new bacterial strain which may differ to a greater or lesser extent from the freshly isolated pure culture in its morphological and particularly in its physiological characteristics. This can be prevented by keeping the original pure culture alive. A relatively safe procedure for conserving bacterial cultures is 'freeze-drying', i.e., the bacteria are first frozen in a suitable suspension and are then dried; they remain viable in this state in the refrigerator for many years without any change in their original properties. For this reason this procedure is used for the various bacterial collections, the most important of which are in England and in the United States.

1. GENERAL

For the culture of marine bacteria one requires the following minimum equipment: 1 incubation chamber or incubator, 1 autoclave or steam sterilizer, 1 drying oven, bottles, test tubes, petri dishes made of glass or plastic, 1 platinum loop, glass spatual, pipettes, cellulose, cotton wool, 1 gas or alcohol burner and suitable nutrient media. For the preparation of the cultures we use aged seawater or a mixture of 75% aged seawater (with 33 to 35‰ salinity) and 25% distilled water. Tap water or pure distilled water is unsuitable for the culture of most marine bacteria as well as for other marine organisms. Artificial seawater may have an inhibiting effect on bacterial growth because of its higher content of heavy metals, particularly copper. Nevertheless some specialists use it for culturing. Formulae for artificial seawater are found in Dietrich and Kalle 1957 (*General Oceanography*) and in Lyman and Fleming 1940 (*J. Mar. Res.* 3, 134–46) among others. As nutrient substances, suitable sources of carbon, nitrogen and phosphorous and some iron are required for saprophytic or parasitic bacteria. All other elements are contained in sufficient quantities in the seawater. Occasionally vitamins, especially those of the B complex, are also required. At all events the reactivity of the solution must be carefully adjusted. For most bacteria the optimum pH is between 7 and 8. As base medium we recommend the Zobell peptone-yeast extract solution (see p. 287), in which many marine bacteria thrive. In general, specialized organisms, such as agar-, chitin-, or cellulose-degraders must be provided with the equivalent carbon sources as well as with the inorganic salts mentioned. In contrast, chemo-autotrophic bacteria are cultured in pure mineral nutrient solutions which must contain, instead of the carbon sources, the corresponding inorganic energy supplier, such as H_2, S or S_2O_3 for sulphur bacteria, NH_4 or NO_2 for nitrifiers, etc. see pp. 292 ff. for there is a survey of certain frequently used nutrient media for marine bacteria.

As culture receptacles we use test tubes, small bottles (such as the economical Meplate bottles), flasks, etc., which are stoppered with cotton wool, cellulose, a glass stopper or screw top, or by means of the so-called Kapsenberg caps. Recently plastic stoppers have become available; these make gas exchange possible without allowing water vapour to penetrate. All apparatus and containers must be cleaned carefully and kept scrupulously clean. As cleansing agent we recommend a strongly alkaline potassium permanganate solution and for subsequent rinsing a weak hydrochloric acid solution of sodium sulphide, to remove the pyrolusite which has developed.

The nutrient medium and every object which comes into contact with it or with the bacteria must be made abacterial beforehand. They must therefore be sterilized either in moist heat – most reliably in the autoclave, 15 to 30 minutes at 1 atm. (120°C), depending on the volume; the glassware may also be sterilized in dry heat in a drying chamber for at least two hours at 160°C. Only in this manner can one be certain that even the most resistant spores have been killed and that the culture has not undergone contamination by other bacteria or by fungi while it is being prepared.

However, nutrient media containing heat-sensitive components cannot be sterilized under pressure in the autoclave. For example, gelatine media, if not prepared with gelatine produced by the Difco Company, would hydrolyse and would refuse to set after cooling. Consequently they must be heated 2 to 3 times at intervals of 3 hours each in flowing steam for 20 to 30 minutes. During this treatment the vegetative cells are killed, but a few spores may survive. However, these germinate in the interval and are killed during the second or third heating. Certain vitamins and antibiotics also cannot be heated to more than 60°C for a short period. It is much better to filter these solutions when sterile. For this purpose we use bacteria-proof membrane filters, i.e., filters with a pore width of less than 0·2 μm, or suitable sintered glass filters produced by the Schott Company. The filtrate is then added to the nutrient substrate shortly before use, to agar or gelatine media after cooling to approximately 60°C.

2. Isolation of Bacteria for Pure Cultures

In general the bacteria intended for culture must first be isolated from their natural substrate, such as seawater, sediment, plant and animal material. Frequently concentration of the desired bacterium will be advisable before the isolation proper, as when bacteria with certain physiological properties are involved, such as cellulose-, chitn-, or agar degraders, sulphur bacteria, iron bacteria, etc., of which only a relatively small number is present in the substrate. For this purpose the particular seawater or sediment sample receives the nutrient substance required plus a suitable nitrogen and phosphorus compound, the optimal pH value is adjusted if known, and incubation is carried out at the optimal temperature in order to promote the reproduction of the desired bacteria. Frequently undesirable bacteria also occur; this can be counteracted in certain cases by the use of specific inhibiting substances. The bacteria can generally be isolated for pure culture more easily from a relative concentration culture of this kind than from the original substrate. Isolation can take place in the following ways:

(a) Agar suspension

A quantity of substrate, such as 1 ml seawater which is undiluted or diluted with sterile seawater, a well-dispersed suspension of sediment in sterilized seawater or 1 ml of the diluted concentration culture is placed in a petri dish; we add 10 ml of an agar nutrient substrate which has been cooled to 42°C and mix carefully. The Zobell peptone-yeast extract described on p. 289 can be used for the isolation of many saprophytic bacteria. However, if a selective nutrient substrate is available, the latter is preferable. Since most bacterial cells do not alter their position in a gel-like medium of this type, after several days of incubation bacterial colonies are visible with the naked eye from the individual cells at their original site; these must be well separated from each other, or the process will need to be repeated with a more diluted substrate. This method is equivalent in principle to the plating out procedure for total bacterial number determinations. It is described in detail on p. 162 in Section II F: 'Survey of the Bacteria.'

(b) Spreading on Agar Plates

A quantity of substrate is spread on a suitable agar nutrient base by means of a previously flamed and air-cooled platinum loop; it is then spread over one or several agar plates (see Figs. 92 and 93) with a platinum loop which has been flamed once again, a platinum hook or so-called Drigalski spatula made of glass (Fig. 91). The agar plates are prepared by pouring 10 ml of the liquefied nutrient base, cooled to approximately 45°C, into sterile petri dishes and allowing it to solidify. During this process the bacteria are made to

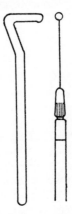

Figure 91. Platinum loop in Kolle frame (right) and so-called Drigalski glass spatula (left)

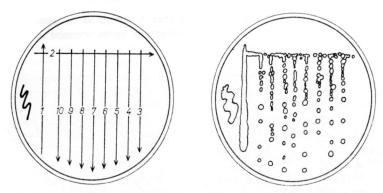

Figure 92. Spreading of bacterial material on an agar plate by means of a platinum loop: most of the material is first scraped off onto the edge, then – without damaging the agar surface – a straight line is drawn (1); the loop is then made white-hot and a second line is drawn perpendicularly to the first (2), which crosses the first one at the beginning. After making the loop white-hot again, 4 — 8 further lines (3 — 10) are then drawn, starting from the original one

separate, and colonies later develop from them. The longer the spreading is carried out, the fewer the cells which remain on the loop or on the spatula, and the more widely they are separated from each other when they adhere to the agar. Using the material from an isolated colony, if possible, the smearing process is then continued until a plate with colonies which are all equal is ultimately obtained. Frequently it is also advisable to combine both methods: first

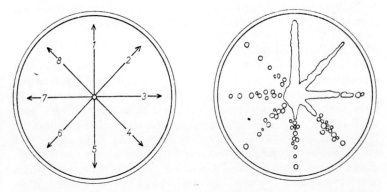

Figure 93. Spreading of a bacterial suspension with the Drigalski spatula: using a platinum loop, a droplet of a sufficiently diluted bacterial suspension is placed in the centre of an agar plate and the suspension is distributed over one or several plates by radial streaks

to prepare a suspension in agar, and then to spread out on agar plates some material from an isolated bacterial colony obtained in this manner. This simultaneously provides a control to check whether a pure culture was actually obtained in the first step. It must be noted that some bacteria, such as those with thick mucous capsules, are difficult to separate from certain accompanying organisms when using this method. Sometimes better results are obtained by previously homogenizing the bacterial suspensions with the help of an Ultra-Turrax instrument.

(c) Dilution

Bacteria which either grow poorly or do not grow at all on a solid nutrient substrate can be isolated with the help of a dilution series (see p. 166, Section II F: 'MPN method'). In the dilution step in which the test tubes show only a small amount of bacterial growth, the growth can be traced back to a single cell in some of them, in general. However, the dilution process should only be used with a concentration culture which already contains as great a number as possible of the desired bacteria. Even so, it is still a relatively unsure procedure because it is never known for certain whether the culture obtained actually did originate from a single cell.

(d) Isolation with the Help of a Micromanipulator

In some cases (as in genetic studies) it is absolutely necessary for the bacterial culture to have developed from a single cell. A single cell can be isolated under the microscope with the help of a 'micromanipulator' and can be transferred into a test tube with sterile medium. However, this procedure is very difficult and requires much experience; it is therefore scarcely used in marine biology.

(e) The Burri India-ink Method

Although the India-ink method is simpler, it also requires a certain amount of patience. Prepare a suitable bacterial suspension using an India-ink solution prepared with distilled water in a 1:10 ratio, which has been carefully decanted from the 'sediment' into small sterile test tubes approximately one week after sterilization. Using a sterile drawing quill, place small droplets of the solution on the surface of an agar plate and cover the plate with a sterile cover-glass before it has dried. Next determine under the microscope which droplets contain a single cell and mark these. After a brief period the cell develops into a small colony and must then be inoculated. If necessary, it is also possible to remove carefully the cover-glasses with their India-ink mark and bacterium from the agar plate and to transfer them into a fluid medium.

3. THE FURTHER CULTURE OF AEROBIC BACTERIA

Aerobic bacteria can be cultured further on fluid, semi-solid and solid nutrient media in test tubes or in small bottles. In the former case the test tubes or bottles are filled at most to one-half with the nutrient solution, stoppered with cotton wool, cellulose, or glass tops or screw tops made of metal or heat-resistant plastic, and are then sterilized in the autoclave. After cooling they are inoculated by means of a platinum loop or a sterile pipette. If the bacteria develop, some degree of turbidity of the nutrient solution is generally observed after one or several days – if not, the inoculation must be repeated.

Slant agar test tubes are especially suitable for all aerobic bacteria growing on solid nutrient substrates. For this type of culture, test tubes are filled with 8 to 10 ml each of the liquefied agar medium, stoppered and sterilized in the autoclave. While still hot, the test tubes are placed with their top end somewhat higher (as, for example, on a wood strip), so that the agar medium forms a slanting surface when solidifying; this surface should not be closer than 5 cm to the opening of the test tubes. Some bacterial material can now be spread on the agar surface with a flamed platinum loop, preferably in a wavy line. However, when dealing with the so-called 'microaerophile bacteria', it is preferable to use test tubes half-filled with agar, which are inoculated by stabbing with the platinum needle. The bacteria, which thrive best at a low oxygen tension, then develop in the nutrient substrate somewhat below the surface. Before opening and also before closing the test tube the opening is briefly flamed over a gas burner or alcohol burner in order to destroy any possible bacteria situated there. The inoculated test tubes are allowed to stand at 18 to 20°C until the bacterial cultures are visible with the naked eye.

The test tubes are then kept in the refrigerator at 8 to 10°C. The culture receptacle should never be exposed to light for longer than absolutely necessary, since many bacteria – particularly non-pigmented bacteria – are very light-sensitive and may perish after being exposed to light for only a few hours. After approximately 8 weeks, the cultures must be inoculated onto fresh nutrient medium. If possible, the inoculation should always be carried out in an inoculation cupboard or an inoculation chamber, i.e., in a room which has been sterilized – generally by UV irradiation. The life of the bacterial cultures can often be extended by adding sterilized, liquid paraffin to the test tubes containing growth until the cultures are entirely covered with it.

(a) Hanging Drop Culture

For the living observation of the bacteria, the preparation of a 'hanging drop culture' (Fig. 94) is useful. A slide with a concavity is required for this purpose; around this hollow space draw a fine ring with vaseline, and place on top a sterile coverglass, the centre of which has a droplet of bacterial suspension which was previously placed on it with the platinum loop. The bacteria can be suspended in a solidifying agar nutrient medium, which allows one to follow the

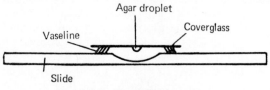

Figure 94.   Hanging drop culture

development of the organisms directly under the microscope. For microscopic observation the use of an oil immersion is usually required because of the small size of the bacteria. Since colourless, living bacteria are involved, a phase-contrast arrangement is highly advisable. Instead of the suspension in the solidifying agar it is also possible to use gelatine. This has the advantage that it allows work with a great variety of concentrations, which produces media with greatly differing refractive indices. By skilfully selecting a suitable refractive index it is possible to improve the microscopic picture considerably.

(b) Mass Cultures

Relatively large quantities of bacteria will often be required, as for example for biochemical studies or for feeding trials for marine animals. Mass cultures of this kind can be prepared with fluid nutrient media in bottles or in spherical flasks. Maintenance of the optimal growth conditions until the maximum harvest has been obtained is generally more difficult in these flasks than in small culture receptacles. This is true particularly in regard to a sufficient and uniform supply of nutrient substance and oxygen to the bacteria. Consequently the bacterial cultures must be continuously agitated or stirred (as with magnetic stirrers); in addition, sterile air which has been filtered through wadding is often blown in (especially for rapid-growth bacteria). Recently so-called fermentors have been produced commercially; these are culture receptacles which are automatically stirred, aerated and temperature-controlled. A suitable concentration of the nutrient substance is very important and as a rule is only determined by preliminary trials. If acids are formed, one must remember adequate buffering and, as far as possible, neutralization of toxic metabolic products. The bacteria are then obtained from the suspension by centrifugation. Washing with 1% common salt solution removes the remaining adhering nutrient substances.

(c) Continuous Culture

The fact that in closed cultures of this type the composition of the medium alters continuously may be a disadvantage for physiological studies, as it causes the nutrient substance concentrations to decrease continuously, and the metabolic products to accumulate. This problem can be solved by using an open, so-called 'continuous', culture introduced by Monod. For this type of culture, the culture receptacles are constructed in such a manner that the quantity of nutrient medium flowing in continuously on one side is the same as the amount of liquid escaping through an overflow on the other side; this liquid also contains bacteria and the metabolic products. The system can be so adjusted that the substrate concentration and the number of bacteria remain constant and that the same living conditions prevail continuously for all the organisms in the culture.

It is now possible to study the growth relationships under precisely defined conditions by altering a single factor only. In recent years the continuous culture method was further elaborated, so that with its help the kinetics of bacterial growth can be investigated very accurately in a so-called 'chemostat'. However, these types of apparatus are expensive and require a good deal of experience. They are suitable for freely motile bacteria only, which do not adhere to the glass wall (even temporarily).

4. THE CULTURE OF ANAEROBIC BACTERIA

Although the common anaerobic bacteria are relatively rare both in seawater and in marine sediments, their group includes certain ones which are important for the substance economics of the sea, such as the sulphate reducers. These bacteria require a low redox potential and oxygen consequently acts as a poison. They generally lack the enzyme catalase, which splits the toxic hydrogen peroxide produced by respiration into water and oxygen. The common anaerobic bacteria can therefore be concentrated, isolated and cultured in the absence of oxygen only. Oxygen must first be removed from the nutrient medium to be used. In concentration cultures it is sometimes sufficient merely to fill the culture vessel to the brim with nutrient medium. As a rule, the oxygen which it contains is then rapidly consumed by aerobic or optionally anaerobic organisms, so that the anaerobic bacteria can develop. In most cases, however, the oxygen must be removed as completely as possible before inoculation. A procedure which is both simple and effective for liquid media is as follows: small bottles (such as Meplat bottles) are sterilized and filled as completely as possible with sterile nutrient solution; they are then inoculated and filled again with sterile nutrient solution to the brim. Using forceps, an iron nail which has been made white-hot is carefully placed in each bottle and the bottles are closed. The iron causes very rapid oxydative binding of the oxygen. This procedure is good for the culture of sulphate reducers, for example; the hydrogen sulphide combines with the iron to form black FeS, so that one may check at any time whether the culture has developed.

Organic reduction agents such as ascorbic acid, glucose, cysteine and thioglycolate may also be added to the nutrient substance solution, of course; these reduce the redox potential correspondingly. (However, because of their content of thio groups, the two latter substances should not be used for sulphate reducers.) These vessels too should be filled to the edge and should be stoppered with an airtight seal. However, since gases are formed during many anaerobic conversions, permeable stoppers such as cotton wool or cellulose plugs are sometimes preferable. The culture test tubes or bottles must then be placed in a container without oxygen. The simplest way of achieving this is to use dishes from which the oxygen has been removed with the aid of an alkaline pyrogallol solution, as follows: a small beaker filled with cotton wool is placed in the dish, and the cotton wool is saturated immediately before stoppering with equal quantities of 20% pyrogallol and 25% soda solution (a 1½ or 2-litre bottle requires 20 ml of each, which are added one after the other by pipette). Desiccators are also suitable for this purpose. When dealing with very sensitive cultures it will be useful to evacuate the vessels in addition. This can be done by means of the Zeissler anaerobic container, for example, which is evacuated with a water jet pump or with an oil pressure pump. The oxygen can also be displaced by other gases, such as hydrogen or nitrogen. In the anaerobic containers by McIntosh and Fildes, oxygen is reduced by means of a heated palladium catalyser with injected hydrogen. Inoculated agar plates can also be placed in such containers. Suitable redox pigments can be used to test the oxygen liberation, such as methylene blue. It is advisable to suspend small strips of hydrophilic gauze in the culture containers, the strips being saturated with the following reagent: 10 ml 1% glucose solution + 3 drops aqueous methylene blue (2‰) + 3 drops N–NaOH. In the absence of oxygen the gauze strips should lose their colour and remain colourless.

So called Fortner dishes are also used for spreading a bacterial suspension to produce a pure culture. These dishes are petri dishes divided into two or three sections, into which the nutrient substrate is equally divided – without sprinkling it onto the cross piece. One-half is then inoculated with a rapid-growth aerobic bacterial strain (such as *Serratia marinorubra*), and the other with an anaerobic suspension. The plates are sealed around the edges with adhesive plaster, Tesa-krepp or something similar against air penetration. The aerobic bacteria rapidly consume the oxygen contained in the dish, thus allowing the anaerobic bacteria to develop. Test tubes with solid nutrient substrate can also be provided directly with the so-called Wright-Burri stopper for anaerobic cultures (Fig. 95).

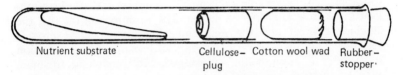

Nutrient substrate   Cellulose– Cotton wool wad Rubber–
plug   stopper·

Figure 95. Slant agar test tube with Wright-Burri closure

For this purpose the test tubes are first filled to one-third with a suitable nutrient substrate and stoppered with cellulose plugs. After autoclaving, the nutrient substrate is allowed to set and is inoculated by a stab with a platinum wire. The plug is then cut off at the edge of the test tube and is pushed in to approximately 2 cm above the surface of the nutrient substrate. A cotton wool plug saturated with 1 ml each pyrogallol and soda solution is placed above this stopper. The test tube must then be closed rapidly with the rubber stopper. These Wright-Burri stoppers can also be used for inoculated slant agar test tubes.

5. The Culture of Photo-autotrophic Bacteria

Various photo-autotrophic chlorobacteria and purple bacteria are frequently found in the coastal areas of the North Sea and the Baltic – on algal and sea grass deposits, for example, and in brackish water pools within the farthest high water limits. When culturing it must be remembered that – in contrast to all other bacteria – in their role as photo-autotrophic organisms they require light for their development and that they are commonly anaerobic. The oldest and simplest method for the preparation of concentration cultures involves the use of slime columns as described by Winogradsky. Glass cylinders are filled with parts of plants, slime and water from the original site of the bacteria, together with small particles of gypsum, and the cylinders are placed on a window which receives north light. Anaerobic conditions dominate in the slime columns and hydrogen sulphide is formed, which is required by the coloured sulphur bacteria (chlorobacteriaceae and thiorhodaceae) as suppliers of hydrogen. Chlorobacteria generally develop on the side turned to the light, and purple bacteria above, in the slime and the supernatant water. Occasionally the development of purple, brick red, violet and orange-brown spots on the glass wall indicates various species of purple bacteria which frequently form pure cultures in these areas and can be isolated as such with the necessary care. They can then be cultured further in synthetic nutrient solution in tightly closed bottles which are filled to the brim. However, the concentration cultures can also be established in synthetic nutrient solution. The variations in the pH, the sulphide content and the light intensity thus obtained provide selective conditions for various groups of bacteria. The preparation of so-called 'dilution shaking-culture series' then enables us to obtain pure cultures. More detailed information will be found in N. Pfennig (*Concentration Culture and Mutant Selection*, pp. 179–89 and 503–4). The culture and isolation of sulphur-free purple bacteria (athiorhodaceae) are described in the same volume by H. Drews (pp. 170–78).

6. GROWTH MEASUREMENTS

It will often be necessary to trace the cell multiplication in the culture, also known as 'growth' in bacteriology, by means of measurements. For liquid cultures, the most commonly used and simplest procedure for this purpose is optical measurement of the cell density. This can be carried out with the help of a photometer against the non-inoculated culture liquid (turbidity measurement). In cultures in which the average cell size remains uniform, the turbidity is approximately proportional to the number of cells. In this case the turbidity measurements also allow us to establish the generation duration of the particular bacterium. The growth of a bacterial culture can also be established by determining the number of cells (see Section II F, pp. 170 ff) as well as the dry weight or the total nitrogen of the bacteria which have been removed by centrifugation and washed. Occasionally the biological activities of the bacteria which are associated with their growth, such as oxygen consumption, acid or alkali formation are also determined; the oxygen consumption can be followed very accurately with the help of a Warburg apparatus. The acid or alkali formation, can be obtained by titration and potentiometric pH measurement. More detailed information can be found in the *Manual of Microbiological Methods*, and in Dixon's *Manometrical Methods*.

7. The Conservation of Living Bacterial Cultures

The most reliable method is the one involving the freeze-drying of bacterial cultures. For this purpose the bacterial material is first suspended in a carrier substance; sterile skimmed milk or ox blood serum, which act as protective colloids, are suitable for this substance. Approximately 1 ml of a bacterial suspension which is as dense as possible is required. This suspension is distributed into five test tubes by pipette. The bacterial material should then be frozen as rapidly as possible. This can be done in a Dewar container with the help of a freezing mixture consisting of dry ice and alcohol, or in a deep freeze which attains a temperature of at least =20°C. However, good ice-salt mixtures are also adequate. Using a desiccator which contains desiccating substances such as concentrated sulphuric acid or magnesium perchlorate, evacuation is carried out until complete dryness is attained. The test tubes should then be sealed carefully and kept in a cool place. After one month and again after six months one test tube from each lot is checked. If the bacteria are still viable, they can be conserved for several years. Several companies now produce efficient apparatus with which the cultures can be freeze-dried in glass ampoules in a so-called 'ampoule carrier' and can be sealed under vacuum. The cultures should be kept in a cool room, or preferably in a refrigerator. To revive the cultures the ampoules are briefly flamed and carefully opened. The dry material must then be diluted with some nutrient solution and transferred to a suitable medium.

8. The Composition of some Nutrient Media for Marine Bacteria

Peptone Yeast Extract Agar

According to Zobell for aerobic heterotrophic bacteria (E2216)

Peptone	5·0	g
Yeast extract (Difco)	1·0	g
FePO₄	0·01	g
Agar	15·0	g
Distilled water	250	ml
Seawater	750	ml
pH 7·6–7·8		

Peptone Yeast Extract Agar

According to Gunkel and Oppenheimer for anaerobic heterotrophic bacteria

Peptone (Difco)	5·0	g
Yeast extract (Difco)	1·0	g
Glucose	5·0	g
Ascorbic acid	0·1	g
FePO₄	0·01	g
Distilled water	250	ml
Seawater	750	ml
pH 7·6–7·8		

Glycerin agar

According to Grein and Meyers
for marine actinomycetes

Glycerin	20·0	g
Glycine	2·5	g
K₂HPO₄	1·0	g
FeSO₄ . 7 H₂O	0·1	g
MgSO₄ . 7 H₂O	0·1	g
CaCO₃	0·1	g
Agar	15·0	g
Seawater	1000	ml
pH 7·5		

Medium for Chitin-Decomposing Bacteria

Precipitated chitin	5·0	g
Yeast extract	1·0	g
FeSO₄ . 7 H₂O	0·1	g
Agar	15·0	g
Seawater	1000	ml
pH 7·5		

Medium for Sulphate-Reducing Bacteria

According to Gunkel and
Oppenheimer

Sodium lactate	3·5	g
NH₄Cl	1·0	g
K₂HPO₄	0·5	g
MgSO₄	2·0	g
Ascorbic acid	0·1	g
Asparagine	0·05	g
Distilled water	250	ml
Seawater	750	ml
pH 7·6		

Medium for cellulose-degrading bacteria

According to Kadota

Precipitated cellulose	12·0	g
K₂HPO₄	1·0	g
NaNO₃	0·5	g
MgSO₄ . 7 H₂O	0·5	g
FeSO₄ . 7 H₂O	0·01	g
Agar	15·0	g
Seawater	1000	ml
pH 7·2		

Medium for Agar-Dissolving Bacteria

According to Kadota

K₂HPO₄	1·0	g
NaNO₃	0·5	g
MgSO₄ . 7 H₂O	0·5	g
FeSO₄ . 7 H₂O		Trace
Agar	15·0	g
Seawater	1000	ml
pH 7·2		

Medium for Thiobacilli

According to Baas Becking and
Wood

Na₂S₂O₃ . 5 H₂O	5·0	g
NaHCO₃	0·5	g
K₂HPO₄	0·2	g
Seawater	1000	ml

Medium for Sulphuric Purple Bacteria

According to Van Niel

NaHCO₃	2·0	g
Na₂S . 9 H₂O	0·1–1·0	g
NH₄Cl	1·0	g
K₂HPO₄	0·5	g
MgCl₂	0·2	g
NaCl	1·0–2·0	g
Tap water	1000	ml
pH 8·5–10		

REFERENCES

(see also Section II F)

BENECKE, W. (1933). 'Bakteriologie des Meeres.' *Abderhaldens Handb. d. Biol. Arbeitsmeth.*, IX, Abt. 5, 717–854.

DIXON, M. (1952). *Manometric Methods*. Cambridge.

GUNSALUS, J. C., und R. Y. STANIER (Hrsg.) (1960–1964). *The Bacteria, a Treatise on Structure and Function I–V*. Academic Press, New York and London.

HAWKER, L. E., A. H. LINTON, B. F. FOLKES, und M. J. CARLILE (1962). *Einfuhrung in die Biologie der Mikroorganismen*. Georg Thieme Verlag, Stuttgart.

JANKE, A. (1946). 'Arbeitsmethoden der Mikrobiologie. I. Bd.' *Allgemeine mikrobiologische Methoden*. Verlag Th. Steinkopff, Dresden und Leipzig.

OPPENHEIMER, C. H. (Hrsg.) (1963). *Symposium on Marine Microbiology*. Charles C. Thomas Publisher, Springfield Ill.

SCHLEGEL, H. (Hrsg.) (1965). 'Anreicherungskultur und Mutantenauslese.' 511 S. *Zbl. Bakt. I*, Suppl.-Heft 1, Fischer, Stuttgart.

SCHWARTZ, W., und A. SCHWARTZ (1960/61). *Grundriß der Mikrobiologie I und II*. Sammlung Göschen, Bd. 1155 und 1157. Verlag W. de Gruyter u. Co., Berlin.

Society of American Bacteriologists (1957). *Manual of Microbiological Methods* McGraw-Hill Book Co., New York, Toronto, London.

Standard Methods for the Examination of Water, Sewage and Industrial Wastes. American Publ. Health. Assoc., Inc New York (1955).

WOOD, E. J. F. (1965). '*Marine Microbial Ecology*.' Chapman and Hall Ltd., London.

ZOBELL, C. E. (1946). *Marine Microbiology*. Chronica Botanic. Co., Waltham, Mass.

Section V

BIOLOGICAL PRODUCTIVITY STUDY METHODS

Plankton

J. LENZ

The subject of biology productivity research is the circulation of organic matter in the sea. The cycle begins with the preliminary production by the autotrophic plants, leads to the heterotrophic organisms which form various members of the food chain, and is ultimately completed by the degradation and mineralization processes which set in at other places in the chain. The nutritional relationships within and between the large biological associations, the plankton, the nekton and the bottom-living fauna, are manifold (Fig. 96). The aim of biological productivity research is to clarify the dependence of the primary production on abiotic and biotic factors and to trace quantitatively within the food chain the transmission of the energy which the plants have stored through photosynthesis in the form of organic substance, so as to allow predictions to be made.

At present biological productivity falls into two large working fields, which differ clearly from each other in terms of method and terminology. These consist of the study of the 'Standing stock (crop)' and the more experimentally directed investigation of the productivity, the measurement of the production rate and of the energy conversion.

The present chapter only deals with the methods which are important for the plankton. Because the food cycle in the open ocean begins with the plankton, the plankton have been the preferred field for production research studies until now. The methods used for the plankton have progressed furthest and some of these methods can therefore be applied in a modified form to other areas, such as the study of the standing crop of the bottom-living fauna.

The study of the very important correlations between food uptake

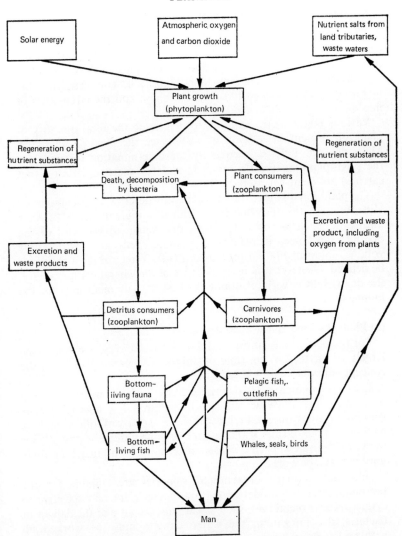

Figure 96. Diagram of the food cycle in the sea (according to Fraser 1965)

and food utilization, of the dependence of growth on age and other problems in the higher members of the food chain require very intensive field observations and special experiments in the laboratory. These will not be discussed here.

1. MEASUREMENTS OF THE STANDING STOCK

The aim of biological productivity studies of the standing stock consists of determining the rich (euthrophic) and the poor (oligotrophic) areas of the sea, establishing the influence of the seasons and the causes for the variations in colonization, and clarifying the manifold complexities within the food chain and the circulation of matter.

Various methods are required to determine the total quantity of plankton present in a body of water (see the chapter 'Plankton Methods', pp. 46ff). Consequently the determination methods for the phytoplankton (bottle samples) and for the zooplankton (net catches) are dealt with separately.

The plankton quantity found is generally described as the biomass. Unfortunately this term has been defined in a great variety of ways until now (such as the wet weight, dry weight, dry weight of the organic substance, displacement volume, cell- and cytoplasmic volume, chlorophyll- and protein content). The term must therefore be defined whenever it is used. Because of its general comparability, the dry weight is without doubt the best measurement unit for the biomass.

(a) Phytoplankton Bottle Samples

For the plankton quantity found in bottle samples, which consists largely of phytoplankton (microplankton and nannoplankton), the concept 'microbiomass' can be used.

The microbiomass is determined by strictly microanalytical methods, which must be sensitive to within a few micrograms. A very important feature of their use is the possibility of following these methods on board ship. If this cannot be done it must be possible for the samples to be kept in a conservation state for a fairly long time, until they can be worked up.

Only bottles made of synthetic material are suitable for the sampling, since certain determination methods are very sensitive to disturbance by metal ions, such as Cu ions. Because of the danger of bacterial decomposition, the water samples must be worked up within a few hours or must at least be put in a cold place.

For almost all of the following determinations the water samples (0·5 to 5 l) must first be filtered. The choice of the type of filter (Table 14) depends on the analysis involved.

Vacuum filtration (2/3 to 1/2 atm) is generally used. Figure 97 shows a filtering arrangement for series determinations.

It is highly desirable to subject the water for testing to a brief microscopic examination; this indicates what has been measured

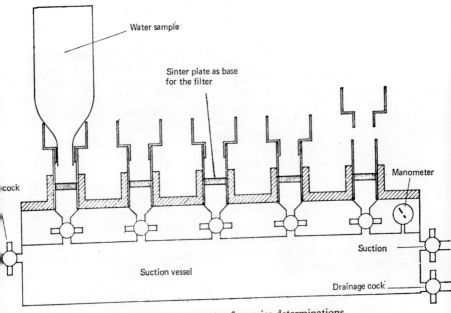

Figure 97. Filtration apparatus for series determinations

and whether the result attributed to the plankton content is not due in part to the dead material which has been included in the determination.

TABLE 14

Principal types of filters for the determination of the microbiomass

Filter type	Material	Pore width
Whatman GF/C...........	Glass fibre	Approx. 0·5 μm
Schlecher & Schüll 1575....	Paper	1·5 μm
Millipore HA	Cellulose ester	0·45 μm
Membran MF 100	Cellulose nitrate	0·8 μm

(α) *Calculation of the Volume.* Lohmann introduced volume calculation as a method for the quantitative study of a standing plankton stock. This was an important step forward in regard to the problems of biological productivity, where greater emphasis is placed on the mass collected from a plankton population than purely on the number of individuals. Despite its tediousness and limited accuracy this method continues to be of importance even today because it is the only one, apart from microscopic analysis, which offers the possibility of obtaining data on the measureable portion of the individual species within the total stock.

The necessary plankton calculations are generally carried out with the Utermöhl microscope. On the assumption of average sizes and simple geometrical forms, the volume for the individual plankton species is subsequently calculated. The result then corresponds to some extent to the displacement volume of the cells. However, Lohmann calculated the volume of the protoplasm by leaving out the cell walls and the shell, and took the living matter as the measurement for the mass. For some vacuole-rich diatom cells the difference can consist of approximately 10:1.

Unfortunately the terminology is somewhat inconsistent. In Lohmann we find the terms 'calculation volume', 'cell volume' and 'volume' for the plankton organisms. It is therefore clearer and less equivocal to call the volume of the living matter in the cell, calculated by Lohmann, the 'plasma volume', and to use the term 'cell volume' for the calculated displacement volume (Lohmann 1908, Hagmeier 1961).

(β) *Chlorophyll.* Because of its close connection with photosynthesis, the chlorophyll content of a seawater sample is a good measurement for the detection of autotrophic phytoplankton.

However, a certain limitation is presented by the fact that the chlorophyll content in the individual algae cells may vary somewhat, in accordance with the species, age and physiological condition. The so-called 'dead chlorophyll' which can be found in detritus consists of the degradation products chlorophyllide and pheophytin. Although they can be differentiated from living chlorophyll, they may have some effect on the determination.

Series measurements generally limit themselves to the determination of chlorophyll *a*, which greatly exceeds the other two components, *b* and *c*, in the marine plankton (Table 15). The correction given below is not necessary in this case.

TABLE 15
Occurrence of Chlorophyll

Chlorophyll	a	b	c
Chrysophyceae	+++	—	+
Diatoms	+++	—	+
Xanthophyceae	+++	—	+
Dinoflagellates	+++	—	+
Chlorophyceae	+++	+	—

The measurement process is very simple and relatively accurate. In principle, all types of filters are suitable for the filtration of the test water. However, millipore and membrane filters have the advantage that they also dissolve during the subsequent extraction

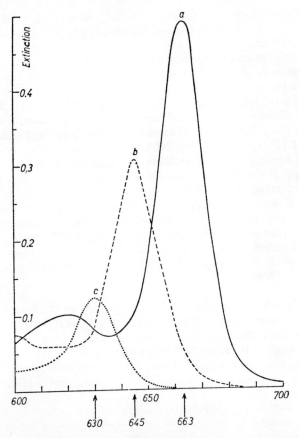

Figure 98. The absorption curves of chlorophyll a, b, and c in 90% acetone according to Jeffrey (a = 6·32 µg/ml, b = 6·21 µg/ml, c = 6·18 µg/ml)

of the pigments, which is carried out with methanol or 90% acetone. The extinction measurement in the spectrophotometer is carried out in the red region of the spectrum for all three types of chlorophyll (Fig. 98). Since the individual absorption curves partially overlap, especially for chlorophyll b and c, correction formulae for the calculation of the actual content are necessary (taken from the UNESCO Report 1966):

$$\text{Chlorophyll } a = 11\cdot64\ E_{663} - 2\cdot16\ E_{645} + 0\cdot10\ E_{630}$$
$$\text{,,} \qquad b = -3\cdot94\ E_{663} + 20\cdot97\ E_{645} - 3\cdot66\ E_{630}$$
$$\text{,,} \qquad c = -5\cdot53\ E_{663} - 14\cdot81\ E_{645} + 54\cdot22\ E_{630}$$

Chlorophyll a, b, c, dissolved in 90% acetone, is expressed in µg/l; E is the extinction value of the wavelength concerned per 1 cm path of light.

As specific extinction coefficient (extinction of a 1% solution at a 1-cm path of light) the value 893 is recommended for chlorophyll a in 90% acetone.

The air-dried chlorophyll filters can be kept for several weeks in the desiccator, in the dark and at low temperatures, without a loss of more than 10 to 15% (Strickland 1960, Strickland and Parson 1960, UNESCO Report 1966).

The fluorescence measurement of a chlorophyll solution described by Kalle (1951) is approximately ten times more sensitive than the above method. It presents an important advantage, especially for ocean waters with a very low chlorophyll content. However, since this method simultaneously deals with all types of chlorophyll and their degradation products, it has not been generally employed until now. Recently, however, it has become possible to filter out the chlorophyll a portion very accurately (Holm-Hansen and co-workers 1965).

(γ) *Protein.* The method developed by Krey for the determination of the protein content of a seawater sample has the specific aim of detecting the 'living matter' in the water. This is meant to be the protoplasm of the living cell, which consists mainly of protein combinations; as the carrier of the most important biological processes, protein embodies the biochemically active part of the living creature.

The method is based on the biuret reaction in which the CO–NH group present in all proteins up to the dipeptides enters into a purple-coloured complex compound in alkaline solution with copper sulphate.

The analysis takes place in three stages: filtering of the water samples through suitable paper or glass fibre filters, hydrolysis of the full filters in 2N NaOH, and carrying out of the biuret reaction with extinction measurement in the spectrophotometer at 530 and 750 nm. The filters and their contents can be kept for a fairly long period in a dry state without a reduction in the protein content. The method is very sensitive to disturbances, and consequently requires great care and much skill if it is to be followed correctly. Calibration is carried out with pure albumen from eggs. To be accurate the measurement results must therefore be given in 'albumen equivalents' (Krey, Banse, and Hagmeier, 1957).

Two sources of error may impair the reliability of the measurement results:

1. Dead protein material is included in the measurements because degradation does not take place immediately in the water.

2. The hydrolysis process is incomplete or has already progressed so far that peptides have been broken down into non-detectable amino acids.

The importance of the protein determination resides in the fact that apart from the autotrophic phytoplankton, the heterotrophic one-celled organisms and the bacteria adhering to the detritus particles are also detected. Leaving out the small zooplankton organisms which are often only present singly in a bottle sample, this is the protein quantity which is available to the filtering organisms of the next nutritive stage.

The protein content can also be calculated from the determination of the particulate nitrogen, although with the inclusion of the amino acids which are adsorbed onto particles. The conversion factor is $6 \cdot 25$ (Strickland and Parsons, 1960).

(δ) *Particulate Carbon.* At present the best way of detecting the total organic substance in seawater is probably by carbon determination. However, this method does not differentiate between the living and the dead constituents.

Biological productivity studies are interested above all in the quantity of particulate organic substance which is present, and which forms the food supply for the filtering organisms. The dissolved organic substances only play a very surbordinate role in this connection. To a great extent, in purely high-sea areas the particulate organic carbon can be taken as an indicator of the stock of primary producers and their degradation products.

A determination method which can be used on board ship and which is rapid and very accurate has been developed by Szekielda (Szekielda and Krey 1965). The test material is concentrated on glass fibre filters, dried and calcined under an oxygen stream in a combustion furnace, the gas formed passing through a special catalysis zone. The CO_2 formed is subsequently cleansed of contaminating side products and conveyed to an absorption cell filled with NaOH. The change in conductivity produced by the in-streaming CO_2 is measured there. The entire calcination course is

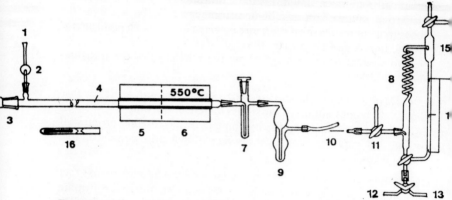

Figure 99. The principle of the carbon apparatus according to Szekielda (1965). 1 = Inlet tube for previously cleaned oxygen; 2 = Bubble counter; 3 = Ground glass seal; 4 = Quartz glass combustion tube; 5 = Combustion zone; 6 = Catalysis zone; 7 = Water absorption vessel; 8 = Absorption chamber; 9 = Nitric oxide absorption vessel; 10 = Connecting tube to the measurement cell; 11 = Three-way tap with inlet tube; 12 = Alkali inlet; 13 = Alkali outlet; 14 = Platinum electrodes; 15 = Separation chamber; 16 = Quartz sheath with iron centre for locking the filter

continuously traced on a recorder. Figure 99 shows the principle of the apparatus.

Any carbonates which may be present can be decomposed on the filter by dilute acid before the analysis.

(ε) *Particulate Phosphorus.* Phosphorus provides an example of the fact that the determination of one of the principal nutrient substance elements in the phytoplankton offers information on the size of the stock. However, the results obtained in this manner are connected with factors causing considerable uncertainty because (a) apart from the living constituents, the dead components in the particulate material are also included; (b) the phosphorus content in the plankton is subject to very wide fluctuations; and (c) at least in coastal waters, inorganic phosphorus compounds which are difficult to decompose may be obtained at the same time. No method is as yet available for separating these.

More informative is the determination of the particulate phosphorus in connection with the total phosphorus content of a water body. As well as nitrogen, phosphorus generally acts as a limiting minimum substance for the primary production; the ratio of the particulate to the total phosphorus can indicate the present state and the further development possibilities of the phytoplankton growth.

A rapid-result determination method for particulate phosphorus is given by Szekielda (1963). Extensive determination data for the individual phosphorus fractions is found in Strickland and Parsons (1960).

(ζ) *Conversion Factors.* For a comparison with each other and with other values it is generally necessary to reduce the results of the various methods to a common denominator; this may be the organic substance or the dry weight, which is even more suitable. The conversion factors obtained from Table 16 are naturally averages, which must often be altered in the individual case. It is worth noting how greatly some of the otherwise quite similar relationships among the diatoms shift because of their siliceous shell. In accordance with the composition of the phytoplankton, this deviation must be taken into account.

TABLE 16

Mean chemical composition of the plankton

All values relate to the dry weight (=100%); estimated values are in brackets. The wet weight divided by 1·05 (diatoms 1·1) yields the cell volume. (Modified according to Hagmeier 1961)

	Wet weight	Dry weight	Organ. subst.	Protein	Chlorophyll	C	P
Diatoms	458	100	46	17	1	30	0·9
Peridiniales	389	100	93	33	1	38	0·6
Remaining flagellates	547	100	95	42	(1)	(40)	(0·8)
Ciliates	525	100	90	40	—	(40)	(0·8)
Copepods	612	100	94	41	—	43	0·9

(b) Zooplankton (Net Catches)

Since the zooplankton catches are generally examined within a systematic framework, the methods used for the determination of the biomass must protect the forms of the animals carefully.

The simplest but also the least accurate procedure (cf. Lohmann 1908) is the determination of the *settling volume.* The conserved plankton are allowed to settle in a measurement cylinder or in a calibrated cup-shaped test glass and their volume is recorded. The great variations in the settling density of the organisms, which is related to their bulk, greatly impair the comparability of the results.

The determination of the displacement volume is considerably more exact; it can be carried out in a variety of ways. A measurement cylinder containing the plankton is filled up to a mark and the contents are then poured through a filter into a second measurement cylinder. A second possibility consists of first filtering off the plankton and then pouring the contents into a measurement vessel which is then filled up to a mark. A source of error during this process is the interstitial water, which remains between the animals during filtering. However, this water can easily be removed by subsequent rinsing with alcohol (Tranter 1960). More serious is the prolonged shrinking process of the conserved material, which is especially noticeable in animal groups with a high water content.

From the point of view of biological productivity, the most exact and most logical measurement of the biomass is the *dry weight* because it eliminates all the differences in the displacement volumes created by the often great variations in the water content of the animals. A small partial sample of the total catch is sufficient for this determination. It is drawn off by suction through a quantitative, ash-free filter which is suitable for the subsequent determination of the organic substance (such as a Schwarzband filter no. 589/1 by Schleicher and Schüll), dried and weighed (Lovegrove 1962).

Through the incineration of the filter the *organic substance* can be determined subsequently. It represents the true nutritional value of the zooplankton and corresponds to the cytoplasmic volume described by Lohmann (Postma 1954).

(c) Detritus

The term 'detritus' is taken to mean all the suspended particles in the water which do not belong to the plankton. A predominant part, at least in the open ocean, consists of the inorganic and organic residues and degradation products of the plankton. The organic components form a favourable nutrient substrate for the bacteria and represent a food potential for the filtering organisms which should not be underestimated. This is why they are of significance in biological productivity.

No direct method has been developed until now to separate the organic from the inorganic detritus; consequently we are generally limited to determination of the total detritus. This can be carried out in two ways.

The first method consists of the *seston determination* in a water sample. Seston is defined as the sum of all the particles which can be filtered off, belonging either to the plankton or to detritus. Accordingly, the difference between seston- and plankton-content yields the detritus content.

When using the seston determination the water sample is filtered and the filter (a constant-weight membrane or paper filter; the latter can be used later for the albumen determination) is dried and weighed.

The plankton dry weight is best calculated from the protein content (see Table 16) and subtracted from the seston weight.

The second method for the determination of the detritus is more direct. Using the Utermöhl microscope, microphotographs of the precipitating nannoplankton samples are prepared and enlarged on DIN A 4. With the help of a special *particle-size analyser* (Carl Zeiss Co.) the detritus particles are measured on the enlargement by means of a scale and counted according to size categories. Under the assumption of the spherical form (because of the flocculence of many particles, the semi-spherical volume is more accurate) the total volume of the detritus can be calculated and converted with the help of a conversion factor (such as 1/6) into the corresponding dry weight (Krey 1961).

Instead of using precipitating nannoplankton samples, membrane filters which have been made transparent can also be used for the microphotographs; when such filters are used, a small part of the water sample must be filtered through beforehand. Their advantage is that they allow the detritus particles to be photographed in the fresh, non-fixed state.

2. DETERMINATION OF THE PRIMARY PRODUCTION

With the determination of the primary production we measure the increase in organic substance produced by photosynthesis by autotrophic phytoplankton within a certain time period. However, one must differentiate between the gross productivity – which includes respiration – and the net productivity. Only the latter is important as energy potential for the next highest trophic level in the food chain.

As a result of the continuous feeding by the zooplankton and of the sinking and short lifetime of the phytoplankton, the productivity relationships in the water are distinguished by rapid conversion of matter, which does not allow any noteworthy accumulation of the assimilation products to take place. Consequently, special measurement processes are required to measure the productivity rate directly. The two best known are the oxygen and the ^{14}C methods. They consist of an *in-situ* experiment. Water samples from various depths are poured into glass bottles and the bottles are then replaced at the original depth and exposed to the natural light conditions. The increase in assimilation which has taken place during the test period is subsequently determined. It is recorded as the quantity of bound carbon in mg $C/m^3/hr$. The daily productivity rate is related to 1 m^2 water surface, mg $C/m^2/day$.

A very important factor when measuring the productivity is the thickness of the light-suffused, euphotic water layer in which primary productivity takes place. This thickness is generally characterized by the determination of the light at certain depths. The surface light intensity on deck is set at 100 and the decrease of the light in the water is given as a relative value. For example, the 50% light depth is the water depth at which 50% of the surface light is still present. Because of the very different light composition above and below the water, one must – strictly speaking – restrict oneself to the blue-green (475 nm) range of the spectrum which is the most effective one photosynthetically in the water, and use suitable filters.

If plankton distribution is uniform, the light depths are in close correlation with the assimilation capacity (Fig. 100). Consequently, the total assimilation can be calculated from a few measurement points with known light depth. The maximum production rate is in the range of the 40% to 50% light depth. Beyond this the high light intensity already has an inhibiting effect. The 1% light depth

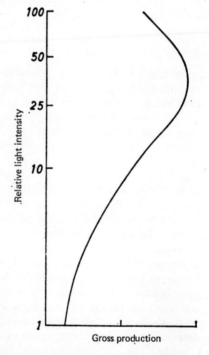

Figure 100.   Productivity as a function of the 'light depth' on a clear summer day (according to Steemann Nielsen and Abbye Jensen 1957)

is approximately identical with the compensation depth (approx. 400 lux), in which production and respiration are counter-balanced.

A sinking photometer is used for the light measurement. Best is the compensation method which uses a photometer on deck and another photometer under water. It allows continuous monitoring of the light conditions. When carrying out absolute measurements the values should be given, if possible, in cal/cm²/min instead of in lux, as done previously; this physiologically defined measurement unit is not very well suited for this purpose.

A gross determination of the light depths is possible with the Secchi disc (see p. 41). The triple Secchi depth is equivalent to 1% light depth. The other light depths can then be extrapolated (see Fig. 100).

(a) The Oxygen Method

When using the oxygen method we measure the oxygen released during photosynthesis ($6\ CO_2 + 6\ H_2O = C_6H_{12}O_6 + 6\ O_2$) and calculate the quantity of bound carbon from this. The photosynthesis quotient O_2/CO_2 averages $1 \cdot 25$ for marine phytoplankton. The O_2 content is determined by the Winkler method (see p. 8).

A time interval of 24 hours is usually selected as the exposure time.

The oxygen content of the test water is determined before and after the experiment. A dark glass bottle is lowered together with the plain glass bottle. The increase in oxygen in the plain glass bottle corresponds to the increase in organic substance, and the decrease in the dark glass bottle corresponds to the respiration, in which zooplankton and especially bacteria may also have been involved, apart from the phytoplankton. This is why one generally measures the gross productivity, which is made up of the two portions. The gross productivity is equivalent to the difference between the results from the light and the dark bottles. In this case the oxygen determinations at the beginning of the trial have only a control function. The use of the oxygen method has the following limitations:

1. In the presence of much phytoplankton, there is a danger that bubbles will form in the bottle as a result of oxygen oversaturation.

2. The accuracy of the method is limited. Smaller differences than $0 \cdot 02$ mg $O_2/1$, approximately equivalent to 20 mg C/m³, cannot be measured accurately by the Winkler method. Consequently it cannot be used for oligotrophic marine areas.

3. The long exposure period may exert a negative effect in three respects: damage to the organisms contained in the bottles, falsifica-

tion of the respiratory values if bacterial proliferation varies in intensity, and reduction of the respiratory rate if the oxygen tension is decreased.

For these reasons the much more accurate ^{14}C method is generally preferred at the time of writing (Strickland 1960, Strickland and Parsons 1960, Steemann Nielsen 1963).

(b) The ^{14}C Method

Measurement of the productivity with the help of the radioactive carbon ^{14}C (beta-radiator, half-life 5,600 years) makes use of a relatively simple tracer method.

The mode of experimental procedure is, in principal, the same as for the oxygen method. A known ^{14}C dose is added to the test water. After exposure to light the plankton are filtered off and by means of a Geiger counter, the amount of radio-active carbon which has been bound photosynthetically is established. The productivity rate can then be calculated from the ratio of the added ^{14}C to the total carbonate carbon present in the test water. It should be noted that $^{14}CO_2$ is incorporated more slowly by 5% than $^{12}CO_2$.

The total CO_2 content in ocean water is on the whole constant and consists of 25 mg C/m^3. For shelf waters and coastal waters it must be determined beforehand by means of the pH value and the alkalinity (for tables see Harvey 1957). The alkalinity can then be calculated from the salinity.

The ^{14}C solution (NaH$^{14}CO_3$) is prepared from Ba$^{14}CO_3$ and is sealed into sterile ampoules. The International Agency for ^{14}C Determination, located in Charlottenlund, Denmark, sells standard ampoules of two different concentrations, i.e., 4 and 10 µC; the latter is for oligotrophic regions. If desired they also undertake the subsequent impulse counts.

For the photosynthesis trial we generally use the 5 standard light depths of 100, 50, 25, 10 and 1%. The exposure time should not exceed 6 hours in order to limit the damaging effects of the bottle experiment (reduced turbulence, bacterial growth). The best time is from sunrise to noon or from noon to sunset. The chemosynthesis and the so-called 'dark fixation' of the ^{14}C, which together usually amount to only 1 to 3%, are checked by means of dark glass bottles which are lowered at the same time. Filtration is carried out with membrane filters, which are later also used for the impulse count.

In essence, the ^{14}C method measures the net production of the phytoplankton. A certain error, which is not entirely simple to correct, results from the fact that approximately 60% of the CO_2

liberated during respiration is immediately used intracellularly for photosynthesis.

A disputed point which has not been solved until now concerns the possibility that the excretion of dissolved organic substance may result in a loss of bound carbon which is difficult to check.

Three different methods may be used for the light exposure of the test bottles; all three have their advantages and their disadvantages.

1. *The* 'in situ' *Method.* This is the standard method. Unfortunately it is of limited use on board ship because of the great amount of time which it requires.

2. *The Simulated* 'in situ' *Method.* The bottles are placed in a flat box with a flow of cooling water; they are then exposed to the sunlight on board deck. The light depths are simulated by applying grey filters. The simulation (light, temperature, motion) is as yet very incomplete (see F. Jitts 1963).

3. *The* '*Incubator*' *Method.* The bottles are placed in a box filled with water and exposed to the uniform light intensity of an artificial light source. A slowly rotating turntable keeps them in continuous motion (Fig. 101). This procedure does not measure the

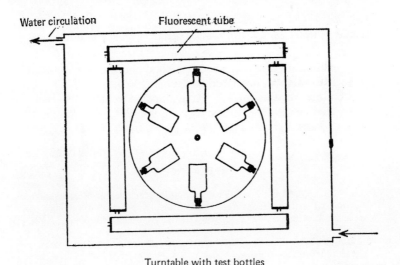

Turntable with test bottles

Figure 101. Diagram of an incubator with uniform light source

actual production rate but the relative or potential productivity as a function of the light intensity (Fig. 102). With certain reservations, this function allows transference of the measurement values to the natural conditions. Various modes of experimental procedure are possible, such as the 'tank trials' of Steemann Nielsen and Abbye Jensen 1957.

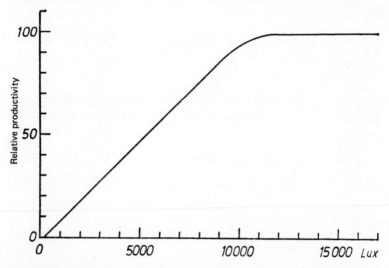

Figure 102. The productive capacity as a function of light intensity for summer plankton in temperate regions (modified according to Steemann Nielsen)

A combination of the incubator method and the *in situ* method is the Sorokin method, described by Hübel (1963).

(Literature relating to the ¹⁴C method: Steemann Nielsen and Abbye Jensen 1957, Steemann Nielsen 1963, Strickland 1960, Strickland and Parsons 1960, *Plankton Symposium* 1957.)

REFERENCES

FRASER, J. (1965). 'Treibende Welt. Eine Naturgeschichte des Meeresplanktons.' *Verständl. Wiss.* Bd. 85. Springer, Berlin–Heidelberg–New York.

HAGMEIER, E. (1961). 'Plankton-Äquivalente.' *Kieler Meeresforsch.* XVII (1), 32–47.

HARVEY, H. W. (1957). *The Chemistry and Fertility of Seawaters.* Second. Edn. Cambridge University Press.

HOLM-HANSEN, O., C. J. LORENZEN, R. W. HOLMES, and J. D. H. STRICKLAND (1965). 'Fluorometric Determination of Chlorophyll.' *J. Cons. perm. int. Explor. Mer* 30 (1), 3–15.

Hübel, H. (1963). 'Die Untersuchung der Primärproduktion des Phytoplanktons durch sowjetische Hydrobiologen unter Verwendung der ¹⁴C-Methode.' *Biol. Rundsch.* 1 (3), 129–139.

Jitts, H. R. (1963). 'The simulated in situ measurement of oceanic primary production.' *Aust. J. Mar. Freshw. Res.* 14, 139–147.

Kalle, K. (1951). 'Meereskundlich-chemische Untersuchungen mit Hilfe des Pulfrich-Photometers von Zeiss. VII.' *Dt. Hydrogr. Z.* 4 (3), 92–96.

Krey, J. (1961). 'Der Detritus im Meere.' *J. Cons. perm. int. Explor. Mer* 26 (3), 263–280.

— K. Banse, und E. Hagmeier (1957). 'Uber die Bestimmung von Eiweiss im Plankton mittels der Biuretreaktion.' *Kieler Meeresforsch.* XIII (1), 35–40.

Lohmann, H. (1908). 'Untersuchungen zur Feststellung des vollständigen Gehaltes des Meeres an Plankton.' *Wiss. Meeresunters., Abt. Kiel*, N. F., 10, 131–370.

Lovegrove, T. (1962). 'The effect of the various factors on dry weight values.' *Rapp. Proc.-verb. Cons. Expl. Mer* 153, 86–91.

Postma, H. (1954). 'Hydrography of the Dutch Wadden Sea.' *Arch. Nèerl. Zool.* 10, 1–106.

Steemann Nielsen, E. (1963). 'Productivity, Definition and Measurement.' Hill, M. N., *The Sea*, Vol. 2, 129–164. Interscience Publishers, New York and London.

— and E. Abbye Jensen (1957). 'Primary Oceanic Production. The Autotrophic Production of Organic Matter in the Oceans.' *Galathea Report*, Vol. 1, 49–136.

Strickland, J. D. H. (1960). 'Measuring the Production of Marine Phytoplankton.' *Fish. Res. Bd. Canada, Bull.* No. 122.

— and T. R. Parsons (1960). 'A manual of seawater analysis.' *Fish Res. Bd. Canada, Bull.* No. 125 (Second edition 1965).

Szekielda, K.-H. (1963). 'Die Bestimmung des partikulär gebundenen Phosphors im Meerwasser mit der Kolbenmethode.' *Kieler Meeresforsch.* XIX, 16–19.

— und J. Krey (1965). 'Die Bestimmung des partikulären, organisch gebundenen Kohlenstoffs im Meerwasser mit einer neuen Schnellmethode.' *Mikrochimica Acta* 1965 (1), 149–159.

Tranter, D. J. (1960). 'A Method for Determining Zooplankton Volumes.' *J. Cons. perm. int. Explor. Mer* 25 (3), 272–278.

Anonym (1958). 'Contributions to Plankton Symposium 1957 "Measurements of Primary Production in the Sea".' *Rapp. Proc.-verb. Cons. Explor. Mer* 144.

Anonym (1966). 'Determination of photosynthetic pigments in seawater.' Monographs on oceanographic methodology. UNESCO, Paris.

Section VI

ECOLOGICAL PHYSIOLOGY STUDY METHODS

A. Plants

F. GESSNER

The report below differs in two ways from the usual manuals on plant physiology methods, i.e., in regard to the object and in regard to the formulation of the problem. Until now, studies in plant physiology (such as those of Stocker 1942, Brauner and Bukatsch 1961) have been concerned exclusively with land plants and freshwater plants; this was due to the fact that marine plants are not available for experiments as a rule. In contrast, the trials described below relate exclusively to marine plants. In regard to the second point, we shall only describe trials which can be interpreted ecologically, i.e., which will provide us with some information on the relationships with marine plants and their natural environment. An investigation of this type, however, requires that all experiments can be carried out under conditions which are as similar as possible to the environmental ones. Since marine plants can only be preserved in the presence of certain culture conditions (cf. p. 192ff) and rapidly die in the absence of these conditions, this means that all the experiments must be carried out as soon as possible after collection of the plants. Consequently, such measurements within the country are greatly limited. If transportation over fairly large distances cannot be avoided, the plants should be kept in a moist air space rather than in water during this time. In most cases marine plants outside their biological environment will be so heavily damaged after a few days that they will no longer react normally. The fact that marine plants and animals differ basically from each other in this respect is shown by the fact that the majority of marine animals can be kept in inland aquaria, but this is only very exceptionally true of marine plants (as for example the green alga *Caulerpa prolifera*).

1. Gas Exchange

The Winkler method (cf. p. 8) is particularly suited to the study of gas exchange (respiration and photosynthesis). This procedure, which has been known for almost eighty years, is accurate, simple and economical, and can therefore also be used outside of well-equipped laboratories. However, when the method is used for studies of the gas exchange of marine plants, a number of points should be observed in order to obtain reliable results.

(a) Respiration

When measuring the respiration rates of acquatic plants the quantity of absorbed oxygen is generally used as a measurement, whereas for land plants the quantity of released CO_2 is measured. To carry out the respiratory measurements in marine plants, at least two bottles with ground-glass joints are filled with seawater (or brackish water) with the same oxygen content from a stock vessel. One bottle serves as a control for measuring the initial oxygen content; a piece of thallus of a marine alga or leaves of marine flowering plants are placed in the other bottle; this bottle is then placed in the dark, preferably at room temperature. The oxygen content of both bottles is determined. Since the volume of two glass-stoppered bottles is rarely entirely identical, a conversion is required. Let us assume that the volume of the control bottle is 250 cc, and that of the test bottle 270 cc. The oxygen content of the control bottle must consequently be divided by 250 (to obtain the O_2 quantity in 1 cc) and this value must then be multiplied by 270. This gives us the initial oxygen content in the test bottle, from which we subtract the end value. The difference corresponds to the quantity of O_2 consumed in respiration. The ratio of the bottle size (preferably 250 cc) to plant material must be selected in such a manner that the quantity of oxygen consumed by respiration is small in comparison with the oxygen content of the test bottle.

It is important to observe this point because the respiratory rate depends very greatly on the oxygen supply. We readily see this if we reduce the oxygen content by introducing nitrogen or increase it by introducing oxygen. The time of the dark period must be selected in such a manner that the oxygen consumed by respiration is a multiple of the titration error. The accuracy of the titration should be at least to $0 \cdot 1$ ml n/100 $Na_2S_2O_3$, which corresponds to an oxygen content of $0 \cdot 008$ mg. Possible sources of error with this method are:

1. The virtually unavoidable respiration of bacteria, which adhere to the surface of the test plants; and

2. (especially when using material which is not entirely fresh) the excretion of organic substances into the seawater, which occasionally makes it impossible to use the Winkler method at all*.

If the method described is mastered, a great number of possibilities become available for its application. For example, it is then possible to study the manner in which respiration is influenced by temperature and by water movement. In entirely stationary bottles an oxygen-poor water film forms on the surface of the breathing plant material and reduces respiration. To avoid this 'stagnation effect' the bottles are made to rotate, a few fairly large glass beads having been added to the plant material. It can further be shown that as a rule old and young pieces of thallus have very different respiratory rates, and that respiratory values differ in accordance with the surfaces of the thalli.

(b) Photosynthesis (Carbon Dioxide Assimilation)

Investigations of photosynthesis with the help of the Winkler method are carried out in the same manner as the respiratory trials described above, but the plants are naturally suitably exposed to light. To prevent parts of plants from throwing shadows on each other, it is best to use flat containers instead of round flasks. The green algae *Ulva* and *Monostroma*, the brown algae *Fucus* and *Laminaria* and the red algae *Rhodymenia* and *Delesseria* are very suited for assimilation experiments. The light intensity should be 1,000 lux. Suitable water-cooling of the containers serves to avoid increases in temperature. Since in the presence of adequate light exposure photosynthesis is much more intensive than respiration, exposure times of 10 to 20 minutes will often be sufficient, whereas a minimum test period of 1 to 2 hours is required for reliable respiratory values to be obtained.

The plant particles must be removed from the vessels as rapidly as possible before the addition of the reagents. If individual threads or bunches are involved, it is best to tie them to a silk thread, which is clamped in with the stopper. The reduction in volume occasioned by the removal of the plants is usually cancelled out by addition of the reagents. If this is not the case, water from the test sample is added until the glass top of the container closes without

* If there is any doubt whether the Winkler method can be used in certain cases, this can easily be ascertained. The test water is agitated with air as long as possible and the oxygen content is subsequently determined. If the Winkler method is functioning properly, one should be able to find the saturation value corresponding to the temperature; this value can be obtained from tables (Gessner 1959). If the value found is considerably less, the Winkler method cannot be used in that particular case.

air bubbles. The error produced in this manner can usually be ignored.

If the quantity of water is very small, the Winkler method can be used as a microprocedure. In this case the sodium thiosulphate solution is not added in the form of drops from a burette; instead we use an injection syringe whose needle is dipped into the test liquid. In this manner it is possible to add the titration liquid in a quantity of less than one drop.

If the test objects are leaves of marine flowering plants, the oxygen generally escapes from the intercellular spaces in the form of bubbles, thus rendering determination by the Winkler method inapplicable. To avoid this, we convey nitrogen through the test liquid before the trial in order to reduce the oxygen content of the liquid sharply. This prevents the escape of oxygen in the form of gas.

The possibilities for conducting experiments in photosynthesis with marine plants are far greater than the possibilities for respiratory studies. If the identical piece of thallus is allowed to assimilate at increasing temperatures, the correlation between the photosynthesis and the temperature shows that a purely photochemical process independent of the temperature cannot be involved here. A light curve can also be established, the irradiation intensity being measured with a lux meter. Initially, as the irradiation intensity increases, the photosynthetic performance rises in a linear manner, and levels out after a light saturation value has been attained. When exposure takes place in full sunlight (which corresponds to a light intensity of 80 to 110,000 lux in the summer), after protracted illumination we can even detect a reduction in the photosynthetic output, which may sometimes fall to zero (sunstroke). If light curves at various temperatures are established, one will frequently find that the plants are capable of utilizing higher light intensities at a higher temperature (approximately 20°C) then at a lower temperature (5 to 10°C).

A further factor whose influence on the photosynthesis of marine plants can be studied in simple practical trials is the hydrogen concentration. The pH of seawater fluctuates between 8·1 and 8·3 as a rule. The marine algae are optimally adjusted to this pH value in regard to all their biological activites. If the pH value of the sea-water is increased to 9 – 10·5 by the addition of a freshly prepared NaOH solution, it is found that a rapid reduction in photosynthesis is associated with this rise in pH. But at a pH of approximately 10 many marine algae can no longer assimilate, even under otherwise optimal conditions (Blinks 1963). This striking fact, which does not hold true for most freshwater plants, points to a special characteristic of the marine algae. Namely, as the pH increases, the quantity of CO_2 molecules decreases, and from a pH of approximately $9 \cdot 5$ on, the inorganic carbon is offered to the plants in the form of bicarbonate ions ($HCO^{3\prime}$). Evidently many marine algae can only take up and utilize the non-loaded CO_2 molecules for photosynthesis, but not the loaded bicarbonate ions. However, most freshwater plants (with the exception of seaweed) are also capable of assimilating bicarbonate.

It is particularly tempting to examine the correlation between respiration and photosynthesis in experiments. Despite the differences in the two processes they run a counter-current course in accordance with the well-known empirical formula. It is therefore always possible to find a light intensity at which respiration and photosynthesis cancel each other out, i.e., a light intensity at which oxygen is neither excreted nor taken up. This light intensity is known as the compensation point. It is generally in the range of 400 to 1,000 lux, but it is easy to show that it may fluctuate widely from species to species.

As a rule, shade-loving plants have a much lower compensation point than sun-loving plants. In addition, cultures of marine algae at various temperatures and light intensities show that within a few days to weeks the compensation point of individual plants can also be displaced. The gas exchange of the individual plant can therefore adapt itself, within certain limits, both to various temperatures and to a variety of light intensities – and the adaptation is such that the photosynthetic output is optimal.

Further, it is especially impressive to examine the differences in photosynthesis in the individual spectral ranges in green algae, brown algae, and red algae. If we insert red, green, and blue filters between the light source and the plant, we find as a rule that the green algae, brown algae, and red algae vary in their behaviour. For the most part brown algae and red algae show better utilization of the green and the blue light, which is predominant in the deeper water layers, than the green algae (chromatic adaptation according to Engelmann). Slight photosynthesis still persists in the ultra-violet spectral range.

If we test the photosynthesis in seawater with various degrees of salinity we find that the salinity has an extraordinarily strong influence on the assimilatory capacity – although at present we are unable to say at which point in the photosynthetic system the salt effect intervenes. An abundance of reactions, depending on the origin and systematic classification of the plants, can be observed in connection with the change in salinity.

In general a marine plant will show the best assimilation in water with a salinity level corresponding to its natural environment.

One and the same species, such as *Delesseria sanguinea*, will show optimal assimilation at 30‰ when the material originates from the North Sea, while plants from the Western Baltic show their optimum assimilation at 15‰. However, adaptations to the salt content may also occur here over a period of a few days or weeks.

(c) Reference Value

If identical plant material is consecutively exposed to various conditions (temperature, light, salinity) no special reference value is required, of course. If, however, as is generally the rule, a variety of plant material is used, it must be possible to relate the values obtained for the photosynthesis or respiration to each other. Since the fresh weight of water plants can only be established imprecisely as a rule, it is preferable to refer to the dry weight; for example, it is possible to specify that 1 g dry weight takes up or excretes a certain quantity of oxygen per hour. In many cases, however, especially those where calcium-secreting algae are involved, the dry weight cannot be used as a reference value. It is then best to refer to the protein content, which can be obtained from nitrogen determinations. If permitted by the morphology of the thallus (*Ulva*, *Porphyra*, *Laminaria*, etc.), it is naturally also possible to use the unit of area as a basis for comparison; of course we must be sure that surface pieces of approximately the same age are used in that case.

(d) Further Methods for Measurement of the Photosynthesis

(α) *The Warburg Method*. This procedure has been described in another section of this work (cf. p. 334). It has provided us with important insights into the biochemical process of photosynthesis. Ecologically the Warburg method is of little use, if only because of the rapid rhythmical agitation of the manometer, which produces very artificial conditions.

(β) *The Polarographic Method* (Heyrovsky 1922). This procedure is at present widely used for rapid and continuous measurements.

It is based on the following principle (Gleichmann and Lübbers 1960): by establishing a light, negative direct-current voltage (polarization voltage) on a platinum electrode, the oxygen diffusing at its surface is reduced to H_2O_2 and further reduced to water. The supplementation of molecular oxygen takes place in a resting solution by diffusion on a platinum surface. The more O_2 molecules are reduced there, the stronger the flow of the so-called 'diffusion limit current' in the measurement circuit. This measurement circuit consists of a current meter device, a voltage source and the polarographic measurement cell. The measurement cell is made up of the

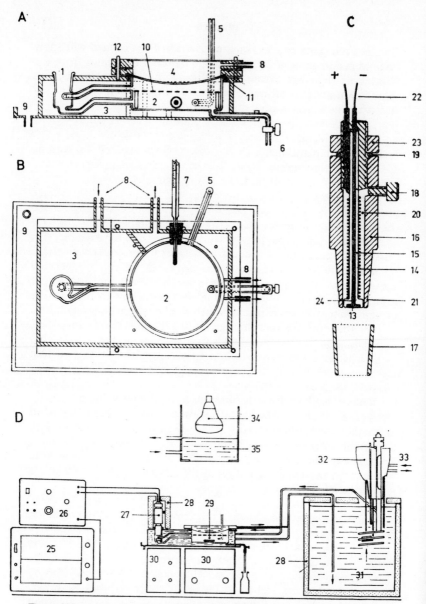

Figure 103a. Apparatus for electrochemical oxygen measurement (according to Schramm 1966). A and B. Assimilation chamber (A = longitudinal section, B = view from above). 1 = electrode container and rotation pump (glass), 2 = assimilation cell (glass), 3 = condenser jacker ('Perspex'), 4 = high-speed lock (concave cover, glass), 5 = pressure equalizer, 6 = outlet, 7 = checking thermometer, 8 = inlet and outlet nozzles, 9 = overflow and drainage, 10 = slide, 11 = soft rubber seal, 12 = guide pegs for high-speed lock. C. Measurement cell (section). 13 = Pt cathode (diameter 8 mm), 14 = Ag/Ag$_2$O anode, 15 = electrode support (glass), 16 = casing (PVC), 17 = conical sheath (PVC), 18 = screw valve, 19 = o-ring seal, 20 = electrolyte chamber, 21 = groove for attaching the films, 22 = leads to electrical source, 23 = lock nut (PVC), 24 = Hg contact

Pt electrode and an Ag/AgCl reference electrode. Both electrodes are immersed in a saturated KCl solution. The Pt electrode is connected with the negative pole of the voltage source and the reference electrode with the positive pole of the voltage source. This electrode system is separated from the medium to be measured by a water-repellent, O_2-permeable film (polyethylene) (Clark principle).

A thin hydrophilic interlayer (cellophane) between the membrane and the cathode surface serves to form a well-defined electrolyte – conducting zone. When the medium is vigorously mixed the O_2 diffusion processes are limited to a constant diffusion span (cf. also Tödt 1958).

Figure 103a illustrates an example of the application of the procedure to biological gas exchange determinations in seawater (according to Schramm 1966). A closed system is shown, in which the object chamber (2) which is provided with a rapid lock (4) is separated from the O_2 electrode receptacle (1). Both are enclosed in a cooling jacket (3). The necessary turbulence on the electrode surface and the rotation of the test water are produced by means of a ring magnet rotating in an electrode receptacle. Grasshoff (1962) undertook the construction of the measurement cell. 0.2 N KOH is used as electrolyte. The apparatus is especially suitable for short-term assimilation and respiration measurements.

(e) The Assimilation Pigments

It can easily be shown that all algae contain at least chlorophyll a, even if the latter is masked by red or brown pigments. To demonstrate this the algal thallus is titurated with quartz sand and a fat-soluble agent (alcohol, acetone, benzol, ether) is then added. A few minutes after immersing a strip of filter paper in this solution, this strip will show a few brown zones which correspond to the carotenoids. However, the green chlorophyll zone is always visible as well. The red phycoerythrin, which is related to the bile pigments, and which we find in red algae and in certain blue algae, cannot be shown in this manner. Because of the water solubility of these pigments, the plant pulp must be extracted with water, which at once turns red.

Because of the above mentioned affinity to the bile pigments, these pigments are called *phycobilin*. If their extraction presents difficulties, it can be carried out successfully by repeated freezing and thawing of the thalli. An ultrasonic disintegrator serves the same purpose, but is not always available.

To obtain the pure pigments and to determine their mutual quantitative relations the generally known process of column chromotography must be used.

In some algae, *Delesseria, Phycodrys*, etc. it is easy to show that the phycoerythrin is not always bound equally firmly to the protein of the plastids. If we transfer a red thallus portion from seawater into distilled water, the former will soon die and the water will turn pink. If, however, the thallus is previously dried, the phycoerythrin remains largely insoluble in the water. At the same time the colour of the phycoerythrin changes during drying (as a result of the firmer binding to the protein) from orange-red to bluish-red.

2. THE MINERAL SUBSTANCE UPTAKE

Like all plants, for their normal growth marine algae require a large number of mineral substances, which are of course abundantly present in seawater. Only certain nutrient substances (phosphorus and nitrogen compounds) often sink to a minimum in seawater, thus limiting the formation of new plant substance. An accurate photometer (such as the Elco II) is indispensable for the detection of the nutrient substance uptake over short test periods. The decrease of the relevant nutrient substances in the external solution is tested by this device. Duration of the trial is a few hours.

The nutrient substance uptake can be followed most accurately with radioactive isotopes (^{32}P). If quantitative determinations are not required and we only wish to establish whether and where the nutrient substance is taken up and how it is disseminated in the plant, autoradiograms can be set up. After absorption of the isotope, the plants are placed on photographic paper. After many hours of exposure, the blackening of the paper shows the previous presence of the radioactive nutrient substance.

When investigating the importance of trace elements for marine plants, the purest possible chemicals must be used to produce artificial seawater which only lacks the element in question. In this manner we can find out, for example, that certain red algae (such as *Asparagopsis*) show an entirely anomalous growth without iodine.

3. GROWTH

Closely correlated with the photosynthetic capacity is the growth, i.e., the irreversible increase in the cell number, the length, the surface or the volume of the plant. All growth measurements of marine plants must adapt themselves to their morphological diversity, which is found particularly in the field of the algae. If one-celled microscopic organisms are involved, growth will be equated with cell multiplication. The increase in the cell number per unit volume is the expression of the growth. To count the cell number one can use the chambers employed by physicians for counting red and white blood cells (haemocytometer, Fuchs-Rosenthal chamber). If the one-celled alga is in the stage of rapid multiplication, an increase in the cell number can be found even within a few hours. All other forms of growth, however, require days, weeks or even months for their measurement and consequently cannot be included in a short-term experiment. In filamentous thalli consisting of one cell row, the increase in the cell number and the total filament length is an expression of the growth. Special attention will be directed to the end cells, since this is the place where the increase in length occurs, if intercalary growth is not present as well. If branching thalli are involved (such as *Cladophora*), the increase in the number and length of the side branches can occasionally be taken as a measurement of the growth. In multicelled string- or band-shaped thalli (*Chorda, Himanthalia,* etc.) the increase in the total length is naturally a good expression of the growth. Most band-shaped leaves of the sea grasses (*Zostera, Posidonia, Cymodocea*) can be measured in this manner, but it should be mentioned that the growth zone is found at the base of the leaves.

If the algae develop a flat thallus (*Ulva, Fucus, Laminaria, Porphyra,* etc.), the thallus surface can be transferred to millimetre paper at suitable time intervals and the increase in surface can be equated with the growth. In very voluminous thalli (such as *Valonia, Codium*) the volume increases – produced by water displacement – may be useful as a measurement of the growth in some cases.

Many algae show differentiations in a regular sequence, which may also be used for growth measurements. For example, the rate at which the brown alga *Macrocystis* develops phyllodes ('leaves') on the stipes itself – as a growth indicator. The alga *Ascophyllum*, which is widely disseminated in northern seas, as a rule forms one gas bladder per vegetation period and per thallus branch. The size of these gas bladders and their distance from each other may also be useful for growth measurements under certain circumstances.

Growth measurements at the natural site present special difficulties. It is well known that most marine algae prefer hard ground, to which they are anchored. For this reason repeated weighing for growth measurements is naturally impossible. This has been overcome by detaching equally large 'substitute plants' from the substrate and weighing these.

On the other hand, the firm anchoring of the algae allows the possibility of an ecological experiment. They can be transferred to a different depth together with the piece of rock to which they are attached, and their further growth can be observed here. In such experiments one must naturally make sure that the transplanted algae can be found again (possibly marking them by floating buoys).

4. REGENERATION AND CORRELATION

It is a well-known fact that in all higher plants lost portions can be rapidly regenerated. As a rule, this regeneration is related to other portions of the same plant (correlation). These findings are also widely found in algae, since at the beginning of the unfavourable season many algae discard thallus portions and regenerate them again the following year. This can be studied particularly well in the young plants of *Fucus vesiculosus*, which measure approximately 7 cm in length (Moss 1964). Within a few months cells begin to grow on the broken surface and proliferations form. A definite relationship with the thallus tip is seen. The closer the location of the broken surface to the thallus tip, the smaller the number of proliferations. Thus, the middle and basal parts of these thalli show the greatest regeneration capacity, leading to the conclusion that the meristematic thallus tip inhibits regeneration. In higher plants regeneration and correlation are determined by growth substances, which have in fact also often been shown in algae, but do not play the clearly decisive role here which they do in flowering plants.

In marine algae as well, every type of regeneration is dependent on the presence of the cell nucleus. A particularly elegant demonstration of this can be seen in the Mediterranean green alga *Acetabularia mediterranea* – according to the experiments of Hämmerling – since the cell nucleus is located in the rhizoid in this alga, and is thus separated from the remaining parts of the thallus. By grafting various *Acetabularia* species on each other Hämmerling obtained qualitative and quantitative information on their reactions which would scarcely be possible with a different subject. These experiments are of course too complicated to try out in practice, and we will therefore not go into them in further detail here.

5. THE SETTLING OF ALGAL SPORES

The conditions under which the reproductive cells (spores, gametes) settle are of great ecological importance for the composition of a marine plant association. A large number of experiments can be carried out on this subject (Boney 1966) but naturally one must first know during what season the algae grow from the vegetative stage into the reproductive phase (for example, in the brown alga *Pelvetia* this occurs at the end of August/September). If a large number of reproductive cells are available, it is possible to establish the physical and chemical conditions under which they settle, the angle of inclination of the substrate, and the optimal rate (for settling purposes) at which the spores slip along the solid surface. For example, it will be possible to determine that a single piece of rock offers very varied adhesive possibilities on its surface (possibilities which again vary greatly from species to species); a careful examination will therefore provide information on the microstructure of a developing plant association.

Of special importance for the viability of the reproductive cells of algae is the salinity. The gametes of *Fucus* species are especially good subjects for studying the relationships (Kniep 1907). The motility of the spermatozoids normally ranges between salinity value of 35‰ and 12‰; below 12‰ motility decreases and at 6‰ it is entirely absent. If *Fucus* is often found at even weaker salinity values, this is because fertilized *Fucus* eggs and thallus portions which are capable of regeneration and have drifted will tolerate even greater leaching.

However, it is not known at present whether the spermatozoids of *Fucus vesiculosus* from brackish water of low salinity retain their motility at an even lower salinity – as for example the North Sea forms.

6. POLARITY

If fertilized *Fucus* eggs are illuminated monolaterally, the first transverse division forms perpendicularly to the direction of the incidence of the light. The *Fucus* egg has thus been polarized by the light, since the cell which is turned away from the light grows into a rhizoid. The *Fucus* egg is especially sensitive to light approximately eight hours after fertilization, but polarization can also be produced by ultra-violet rays, an electrical current and centrifugation. Moreover, polarization can be transmitted by contact. If polarized and non-polarized *Fucus* eggs are placed in close contact, the polarized eggs induce the polarity of non-polarized eggs.

A type of polarity which leads to a differentiation between the root pole and the germination pole in higher plants is also a widely disseminated phenomenon among algae. Thus, in the green alga *Enteromorpha* the basal cells develop into rhizoid-like formations, while the apical cells enlarge the assimilating thallus. Regeneration trials with *Cladophora* have shown that each individual cell is polarized. Of the nature of the polarity in marine algae we know as little as of that in the higher plants.

7. THE OSMOTIC PROPERTIES OF MARINE PLANTS

In view of the fact that the high salinity of seawater is the decisive factor in the basic difference between marine vegetation and freshwater vegetation, we might expect that the high osmotic pressure of the seawater and specific ion effects are responsible for the presence or absence of certain plants in the sea. As with many animal subjects, in certain marine algae it is simplest to establish the osmotic pressure by cryoscopic means (cf. Section VI B, p. 338), and to express it either in the form of a depression of the freezing point or as atmospheric pressure units. Of course, only a few plants possess cells large enough to yield the quantities of cell sap required for cryoscopic investigations. The subject best suited for this purpose is the green alga of the genus *Valonia*, which occurs in several species in tropical or warm seas (including the Mediterranean). Some of these cells attain a diameter of several centimetres, so that large quantities of cell sap can be obtained with the help of an injection syringe. The cryoscopically obtained value, which is determined solely by inorganic ions, is approximately 25 or 26 atm. Since seawater with a salinity of 35‰ possesses an osmotic value of 23 atm, the osmotic cell sap concentration exceeds that of seawater by only 2 to 3 atm. *Valonia* has been one of the most popular subjects for research in cell sap physiology for many decades. If the cells are transferred from seawater into distilled water, so that the high osmotic value of the cell sap can develop fully, the cell divisions burst immediately in some species. An especially large species, *Valonia ventricosa*, which occurs in all tropical seas but is also found in the Bay of San Sebastian on the island of Ibiza in the Mediterranean, has cell divisions which consists of 50 layers and can withstand any osmotic pressure, so that the cells do not burst. This makes it possible to study the unusually high ion permeability, since the osmotic value decreases from 26 to 2 atm in two hours in distilled water (Gessner 1967).

However, it must be noted that this high ion permeability does not affect the living cell since *Valonia* is stenohaline, i.e., the cells die as soon as a slight reduction or increase in the salinity of the external medium occurs.

A simple experiment demonstrates that the chemical composition of the cell sap of *Valonia* differs basically from that of seawater. We know that NaCl makes up by far the largest part of the total salt content of seawater, and it has long been known that in the cell sap of *Valonia* it is the KCl which determines the osmotic value. One litre of cell sap contains 24,000 mg potassium and 630 mg sodium. When a strip of filter paper is immersed in seawater and held over a flame, the yellow colour characteristic for sodium appears immediately. If the same is done with *Valonia* cell sap, the purple flame colouration of potassium appears. Thus, the protoplasm of *Valonia* takes up potassium selectively from the seawater and stores it in the cell sap, while it is capable of preventing the sodium ions from entering, despite their high external concentration.

With the help of a microcryoscopic method (cf. p. 339) it is also possible to determine the osmotic pressure of considerably smaller cells. The green alga *Chaetomorpha linum* was cultivated by Kesseler for five days at salinities of 5 to 35‰; the values obtained are shown in Table 17.

TABLE 17
Chaetomorpha linum

| Salinity ‰ | Osmotic pressure (at/m) | | Difference |
	Seawater	Cell sap	
5	3·20	21·2	18·0
10	6·50	22·2	15·7
15	9·70	27·3	17·6
20	13·00	27·6	14·6
25	16·30	31·7	15·3
30	19·60	34·5	14·9
35	23·10	37·9	14·8

Thus, despite great variations in the external concentration, the cell sap always tends towards approximately the same overconcentration-proof of the high osmoregulatory properties of this alga.

If we attempted to culture *Chaetomorpha* in sugar solutions of various concentrations and if such trials were successful, no osmoregulation would be detected. This shows that in order to change their osmotic value the algae must take up or release inorganic ions. It has been known for more than 70 years (Drevs 1896) that their assimilation products have only an insignificant influence on the osmotic pressure of the cell sap. The method commonly used with land plants (cryoscopy with cell sap which has been obtained by extracting from killed tissue), normally cannot be applied to marine algae. The reason for this is that a large part of the water is bound to the cell division by strong swelling forces, and consequently cannot be extracted even by very high pressures. Nevertheless, Mosebach (1936) described a special procedure for obtaining cell sap. In view of the difficulties which are encountered with most subjects when applying the cryoscopic method, it will be necessary to seek other procedures for determining the osmotic pressure; but other methods also often meet with almost insuperable difficulties.

In many cases the boundary-plasmolysis method may be applied, but using as hypertonic media concentrated seawater in several gradations or increasing the osmotic pressure of the seawater by a suitable addition of sugar. By cryoscopic means (cf. Section VI B, p. 337) it is easy to determine the osmotic pressure of the hypertonic medium and the concentration at which 50% of the cells have not yet been plasmolysed and 50% have just been plasmolysed. At that point we can assume that the osmotic pressure of the cell sap is equivalent to that of the external medium. If we use this method to study the differing osmotic pressures of various marine plants, we find for example that the plankton diatom *Ditylum* possesses a cell sap concentration which is hypotonic in comparison with seawater. Other plankton diatoms (*Biddulphia*, etc.) undergo plasmolysis even in 1·01 seawater (the osmotic pressure of the seawater = 23 atm being designated as l). Forms of this kind can therefore be described as being isosmotic towards seawater. As a rule, however, the osmotic value of most marine algae is 3 to 15 atm higher than that of seawater.

If plasmolysed cells are allowed to remain in the plasmolytic state, after only a few minutes we often observe a regression of the plasmolysis as a sign of an extremely high ion permeability of the plasma. In many cases the ions penetrate into the cells so

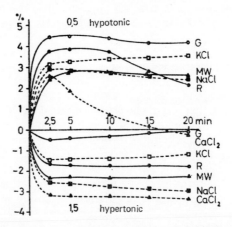

Figure 103b. Porphyra tenera. Changes in the lengths of thallus strips in solutions with equal osmotic pressures (1 = isotonicity); G = glycerin, MW = seawater, R = unrefined sugar

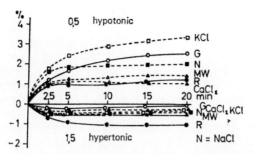

Figure 103c. Ulva pertusa. Changes in the length of thallus strips in solutions with identical osmotic pressures (symbols as in b). b and c according to Ogata and Takada 1955)

rapidly from the hypertonic external medium that erroneous values are obtained with the plasmolysis. A comparison of plasmolysis regression in a hypertonic neutral salt solution and an isolmolar cane sugar solution often yields a picture of the high ion permeability.

If thalli of *Ulva* or *Porphyra* are cut into strips (18 mm long, 5 mm wide) and placed in various solutions which are isosmotic in relation to each other and are equivalent to either 1·5 or 0·5 seawater, very variable degrees of stretching or shrinking are found, which – independent of the osmotic pressure – indicate the specific influence of the solution on the degree of swelling of the cell divisions (Fig. 103*b* and *c*) (Ogata and Takada 1955).

8. Examination of The Cell Wall of Marine Plants

A further and very decisive reason why plasmolysis is unsuccessful or yields osmotic values in numerous lower and higher marine plants is found in the properties of the cell division whereas in land plants and freshwater plants the division consists principally of a cellulose framework, we find that in most marine plants pectin substances capable of swelling are embedded in this framework. Since they are ion permeable, in hypertonic solutions they swell up with the hydration sheaths of the ions – as a rule towards the inside – so that the free space which usually develops during plasmolysis is now filled up.

The existence of swelling bodies of this kind can even be demonstrated by physical means. If a thallus particle is suspended in the air for drying, it loses its water much more slowly than a piece of water-saturated filter paper of the same size. We then say that the relative transpiration (in relation to a water surface of the same size) or – equally – in relation to a surface of water-eliminating filter paper is smaller than 1. If a desiccated thallus is replaced in water, it rapidly gains weight again. Since this can also be done with pieces of dead plant parts, the water uptake cannot be osmotic, but must be based on the swelling powers of the cell division. Desiccated thalli which are suspended in a chamber saturated with water vapour are also hygroscopic, that is, they increase in weight because of water uptake. In the higher land plants this only occurs in very exceptional cases.

It is very interesting to undertake a laboratory trial in which the water elimination of a freely suspended leaf from a land plant

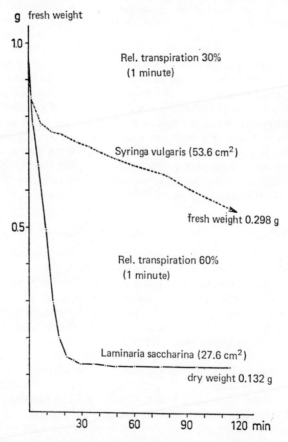

Figure 103d. Water elimination of a leaf from a deciduous tree (*Syringa vulgaris*, and that of a brown alga (*Laminaria saccharina*). To accelerate the process transpiration took place in an artificial air current

(such as lilac, *Syringa*) is compared by means of a torsion balance with an equally heavy thallus particle of an alga (such as *Laminaria saccharina*); to accelerate the trial, a hair dryer should be used to create an air stream (Fig. 103*d*). Because of its cuticle, the land leaf undergoes relatively less transpiration than the alga; after a few minutes water elimination is sharply reduced because of the closing of the stomata, and the leaf subsequently decreases in weight only slowly, corresponding to cuticular transpiration. The alga shows a consistently rapid water elimination until the air has become dry.

In the littoral zone we generally find a strictly regular sequence of various plant species (cf. p. 93) in a vertical direction. It is seen that algae often retain water more strongly the higher they grow in the littoral zone, i.e., the longer the daily periods of dryness to which they are exposed. This shows that the water-retaining power and the total water content (which can be established by means of the difference between the fresh weight and the weight of plant parts dried at 100°C) proceed in a parallel manner. Thus, many brown algae of the topmost littoral zone possess not only the greatest water-retaining power, but also the largest water reserves. If properties of this kind are present, we speak of desiccation-resistance; 'desiccation tolerance' describes the characteristic of plants which cannot protect themselves against water loss but can tolerate a fairly long period of drying by air without damage. An example of this type of plant is the red alga *Porphyra*. The same is true for certain lichen species, which occur in the topmost littoral zone or even in the supralittoral zone. As in the few 'seed plants' among the phanerogamia, the capacity of tolerating complete drying by air is also an important adaptation characteristic of algal life in the littoral zone. It is easy to establish experimentally what periods of drying by air can be survived. In their natural habitat the littoral algae are exposed to evaporation for only a few hours. However, we find that all algae with desiccation tolerance have a safety factor to some extent and as a rule can survive a few days of complete drying by air. The investigator is naturally faced with the problem of determining whether a thallus has died after being exposed to the air for a long time. We shall discuss this further below.

Chemically these swelling substances belong to the pectins and to the hemicelluloses, i.e., to the high-molecular polysaccharides. The best known are alginic acid which is found in the brown algae, carragheen which can be dissolved out of the cell division of certain red algae by boiling them, and the long known agar-agar.

To obtain alginic acid a sodium alginate solution is prepared with hot alkaline soda solution; the alginic acid can be precipitated from this solution with hydrochloric acid or alcohol. If a sodium alginate solution is treated with $CaCl_2$, insoluble Ca-alginate is produced. The same can be attained with a $CuCl_2$ solution. If we colour a Na-alginate solution with a basic dye (safranin) and add $CaCl_2$, the red-coloured Ca-alginate will result. This can be re-dissolved in alkaline soap solution. By transformation into Cu-, Co- or Ba-alginate the salts of alginic acid can be reconverted into their insoluble form.

Agar-agar can be extracted from the red algae (*Gelidium, Chondrus,* etc.) by boiling them in water for several hours.

The cell wall substance of algae play an extremely important and versatile role in industry today. Interest in this field is therefore not limited exclusively to marine biology.

9. TOLERANCE AND RESISTANCE MEASUREMENTS

For these measurements it will always be necessary to decide whether a plant or a plant part has survived a certain environmental change, i.e., a criterion for the living state of the cells will be required. For land and freshwater plants the plasmolysis capacity will be the test criterion for this purpose. However, since we know that plasmolysis cannot be carried out in many marine plants, other indications for the living state of the cells will be required (Biebl 1959). Microscopic comparison with unquestionably living cells will indicate in most cases whether the plant parts exposed to high or low temperatures and high or low salinities have remained normal, have been damaged or have died. Clear signs of the occurrence of cell death are strong membrane swelling, brownish clumps of protoplasm, form change, displacement and discoloration of the plastids, coloration of the cell sap, destroyed chloroplasts or plastids which can no longer be dyed with neutral red.

The greatest sensitivity to damaging influences is naturally shown by the protoplasm, but such damages are very difficult to demonstrate with the usual means. The following experiment is an exception: a bunch of the green alga *Bryopsis plumosa*, which does not possess a transverse wall, is tied together with a silk thread and immersed in a centrifuge flask filled with seawater in such a manner that the thread is pinched between the cork and the glass wall; a few minutes of centrifugation (2500 r.p.m.) are sufficient to hurl all the chloroplasts into the branch ends (from where they regain a normal distribution after 3–10 hours). If the alga is dipped into boiling water for a few seconds before centrifugation, thus killing the cytoplasm, the displacement of the chloroplasts does not occur during the centrifugation. The heat coagulation has increased the viscosity so greatly that the chloroplasts can no longer pass through the cytoplasm.

Apart from these cytomorphological findings, photosynthesis can also be used as a criterion for the state of the living cells. Dead cells naturally will no longer release oxygen into the environment in the presence of light. When using respiration as a criterion, one should be careful, however, since dead cells often excrete organic substances which disturb the Winkler method and simulate oxygen uptake (cf. p. 8).

To determine the cold- and heat-tolerance, Biebl (1958) proceeded as follows: parts of plants are exposed for 12 hours to seawater at temperatures of −8 and −2°C and 3°, 19°, 27°, 30° and 35°C. The aim is to prove that it is not the plant stems of the green, brown and red algae which are collectively distinguished by their heat- or cold-tolerance, but that it is rather the environmental origin of the algae which is of decisive importance. Thus, in general, littoral algae have a greater range of temperature tolerance than algae from the depths, since upwards or downwards differences in temperature are much more marked in the littoral zone than in the sublittoral zone. Moreover, the geographical distribution of the algae is decisive for their temperature tolerance. Algae from colder seas often show an outstandingly high heat sensitivity, whereas algae from warmer seas exhibit greater cold-sensitivity. However, it is noteworthy that algae from tropical seas can survive for a long period at 10°C, i.e., at a temperature which never occurs in their natural environment. Finally we should note that there is also a seasonal amplitude for the temperature sensitivity. In the summer the same alga will be able to tolerate low temperatures less well than in the winter. Indeed, in this respect the behaviour of the various algal species shows little uniformity.

Ecological interpretation of all temperature tolerance trials requires very careful work, as is evidenced by the fact that many tropical algae survive low temperatures. It is not necessarily true that damages which have occurred in short-term trials (12 hours) are already so great that they are detectable in the cell appearance. Often damaging or beneficial temperature effects only become evident after weeks or months. In addition one must remember that various developmental stages of an algal species react differently to temperatures. This can be assumed above all for haploid- and diploid generations.

For the determination of the osmotic resistance, algal particles are put into seawater in an increasing seawater concentration series ranging from 0·1 to triple this amount. As a tule, the cell appearance shows what degrees of hypotonicity and hypertonicity can be tolerated by the algae. However, Schwenke (1958) showed that as well as osmotic effects, specific ion influences are involved here. For example, if the red alga *Delesseria* is transferred from seawater into distilled water, the thalli die very rapidly; this is a process which can be greatly delayed by the addition of a small quantity of $CaCl_2$.

If we test the osmotic resistance of algae of various environmental origins, we find that as a rule littoral algae possess a much wider resistance range than algae from the sublittoral zone. We often find a rich plant life in the 'rock pools' of the tidal zone, where very high seawater concentrations may exist because of evaporation. Predominantly present here are one-celled flagellates, which are capable of vigorous division even when the seawater has been concentrated five-fold.

Apart from a tolerance (or resistance) to various temperatures and salinities, a tolerance to long periods of darkness, high light intensities, low oxygen contents as well as to various specific substances in the seawater (marine waste research) may be of ecological significance occasionally. In all studies which are concerned with the adaptation of marine plants to new environmental conditions the following should not be forgotten: the plant is a system with relatively slow reactions, and therefore needs time to adapt to new conditions. It will therefore always tolerate a slow and gradual transfer to the new conditions better than the shock of a sudden change in environment. Finally, it should be remembered that the tolerance- or resistance-range of a plant depends not least on its species.

Studies on the tolerance to increased hydrostatic pressure have not been carried out until now with marine algae. This is understandable from the ecological point of view, since algae require light and therefore cannot grow at the greater sea depths. However, slight excess hydrostatic pressure (to about 10 atm) is probably without importance for algae. This is not true for phanerogamia. In long-term trials (weeks to months) with flowering freshwater plants it was shown that as little as 1 to 2 atm produce the most serious irreversible damages (pressure was produced by attached mercury manometers). It is therefore understandable that higher plants never go beyond a depth of 10 m in fresh water, even under the best assimilation conditions. This is probably true also for the marine flowering plants, of which we now know more than 50 types. Only *Posidonia oceanica* (Mediterranean) can be found at a depth of 45 m in some places. Long-term pressure experiments with marine phanerogamia appear promising.

REFERENCES

BELOSERSKI, A. N., und N. T. PROSKURJAKOW (1956). *'Praktikum der Biochemie der Pflanzen*. VEB Deutscher Verlag d. Wissenschaften, Berlin.

BRAUNER, L., und FR. BUKATSCH (1961). *Das kleine pflanzenphysiologische Praktikum*. VEB Gustav Fischer, Jena.

STEUBING, L. (1965). *Pflanzenökologisches Praktikum*. Paul Parey, Berlin und Hamburg.

STOCKER, O. (1942). *Pflanzenphysiologische Übungen*. Gustav Fischer, Jena.

Books and special works

BIEBL, R. (1958). 'Temperatur- und osmotische Resistenz von Meeresalgen der Bretonischen Küste.' *Protoplasma* 50, 217–242.

— (1959). 'Das Bild des Zelltodes bei verschiedenen Meeresalgen.' *Protoplasma* 50, 321–339.

BLINKS, L. R. (1963). 'The effect of pH upon the Photosynthesis of Littoral Marine Algae.' *Protoplasma* 57.

BONEY, A. D. (1966). *A Biology of Marine Algae.* Hutchinson Educational.

DREVS, P. (1896). 'Die Regulation des osmotischen Druckes in Meeresalgen bei Schwankungen des Salzgehaltes im Aussenmedium.' *Arch. d. Freunde d. Naturgesch. Mecklenburg* 49, 91–135.

GESSNER, FR. (1955–1959). *Hydrobotanik.* 1. Bd. 1955; 2. Bd., 1959. VEB Dtsch. Verlag d. Wissenschaften, Berlin (jezt VEB G. Fischer, Jena).

— (1961). 'Hydrostatischer Druck und Pflanzenwachstum.' *Handb. d. Pflanzenphysiologie*, XVI.

— (1967). 'Untersuchungen über das osmotische Verhalten der Grünalge *Valonia ventricosa*.' *Helgol. wiss. Meeresunters.* 15, 143–154.

GLEICHMANN, U., und D. W. LÜBBERS (1960). 'Die Messung des Sauerstoffdruckes in Gasen und Flüssigkeiten mit der Pt-Elektrode unter besonderer Berücksichtigung der Messung im Blut.' *Pflügers Archiv* 271, 431–455.

GRASSHOFF, K. (1962). 'Untersuchungen über die Sauerstoffbestimmung im Meerwasser. Teil 2.' *Kieler Meeresforsch.* 18/2, 151–160.

HEYROVSKY, J. (1922). 'Elektrolyse se rtutovou kopkovou kathodu.' *Chemické Listy* 16, 256.

KESSELER, H. (1958). 'Eine mikroskopische Methode zur Bestimmung des Turgors von Meeresalgen.' *Kieler Meeresforsch.* 14, 23–41.

— (1959). 'Mikroskopische Untersuchungen zur Turgorregulation von *Chaetomorpha linum*.' *Kieler Meeresforsch.* 15, 51–73.

KNIEP, H. (1907). 'Beiträge zur Keimungsphysiologie und -biologie von *Fucus*. *Jahrb. wiss. Bot.* 44,

LEWIN, R. A. (1962). *Physiology and Biochemistry of Algae.* Academic Press, New York and London.

MAASS, H. (1959). *Alginsäure und Alginate. Strassenbau, Chemic und Technik.* Verlagsgesellschaft m. b. H., Heidelberg.

MOSEBACH, G. (1936). 'Kryoskopisch ermittelte osmotische Werte bei Meeresalgen.' *Beitr. Biol. d. Pflanzen* 24, 113–137.

MOSS, B. (1964). 'Wound healing and regeneration in *Fucus vesiculosus* L.' *Compt. rend. IV. Congrès int. d. Algues marines*, 117–122.

OGATA, E., und H. TAKADA (1955). Elongation and Shrinkage in Thallus of *Porphyra tenera* and *Ulva pertusa* caused by Osmotic Changes.' *J. Inst. Polytechnics*, Osaka City Univ. 6, 29–41.

SCHRAMM, W. (1966). 'Kontinuierliche Messung der Assimilation und Atmung mariner Algen mittels der elektrochemischen Sauerstoffbestimmung.' *Helgol. Wiss. Meeresunters.* 13, 275–287.

SCHWENKE, H. (1958). 'Über einige zellphysiologische Faktoren der Hyptonieresistenz mariner Rotalgen.' *Kieler Meeresforsch.* 14, 130–150.

— (1959). 'Untersuchungen zur Temperaturresistenz mariner Algen der westlichen Ostee. I. Das. Resistenzverh. v. Tiefenrotalgen bei ökologischen u. nichtökologischen Temperaturen.' *Kieler Meeresforsch.* 15, 34–50.

TÖDT, FR. (1958). *Elektrochemische Sauerstoffmessungen.* Walter de Gruyter & Co., Berlin.

B. Animals

C. SCHLIEPER

Marine biological research is moving in two particular directions at the present time: (a) study of the marine organisms in their biotopes (occurrence, distribution, environmental conditions); (b) measurement and investigation of the physiological adaptations and capacities in laboratory experiments and in rearing trials. Only a knowledge of the characteristics described in (b) allows complete understanding of the marine species and the life processes in the sea. The basic features of a small selection of experimental methods are therefore described below; with the help of these methods the physiological capacities and the adaptations of the marine animals to their abiotic environmental conditions can be measured.

1. ACTIVITY MEASUREMENTS

The activity of a marine animal is a characteristic specific to the species; within the framework of the phenotypical reaction range, which is limited by heredity, it often varies in regard to the season and to the environmental conditions and is also influenced by age and body size. It depends especially on the temperature in accordance with the Van't Hoff law, but also shows thermically genetic and non-genetic adaptations which limit the validity range of the above law to some extent (as, for example, in warm-water forms and cold-water forms, and in eurythermous species).

An important measurement of the activity is the *metabolic intensity*, which can be determined by means of the oxygen consumption. We can measure the oxygen consumption of intact marine animals, for example, with the help of the Winkler method (cf. the description of the method in Section IA, p. 8). It is not possible to simply transfer the test animal into a container filled with seawater and then to determine the reduction of the oxygen content in the water after a certain period because the test animal, which has become agitated during the transfer, first shows increased oxygen consumption in comparison with the resting metabolism; the oxygen consumption later gradually decreases as the specimen

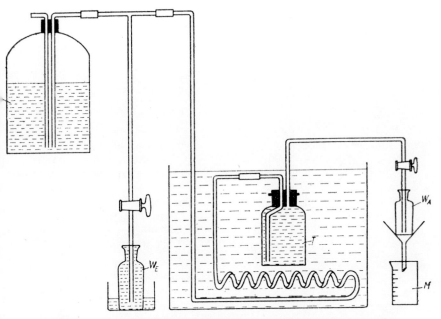

Figure 104. Apparatus for the study of the oxygen consumption of intact marine animals in flowing water by the Winkler method. For protracted measurements a levelling container must be inserted between the stock bottle and the animal container in order to maintain a uniform flow. Pelagic fish are examined in a horizontal tubular container. M = measurement cylinder, T = animal container, V = stock bottle, W_E and W_A = Winkler bottles

becomes familiar with the container. As shown in Fig. 104, the test animal is therefore carefully placed in slowly running, well-aerated seawater and the oxygen content of the in-going and out-going water is measured at regular intervals at a constant flow-rate, uniform temperature, etc. The oxygen consumption is calculated from these values. The actual measurement only begins when the oxygen consumption has attained a constant level, i.e., when it approximates the 'resting metabolism' under the test conditions. The oxygen consumption obtained in this manner is then converted into kg live weight or dry weight and hours, in order to obtain comparable values. When working with species which consist largely of skeletal substance, such as corals, etc., it is advisable to convert the oxygen consumption to the breathing living substance or its nitrogen – or protein content (cf. for example, Franziskeit 1964).

A more recent procedure allows continuous respiratory measurements and registrations with the help of the oxygen electrode. A

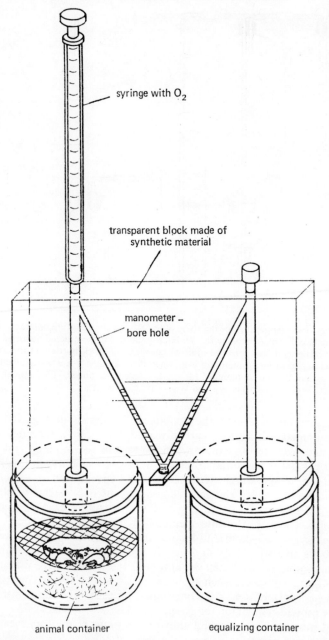

syringe with O_2

transparent block made of
synthetic material

manometer
bore hole

animal container

equalizing container

Figure 105. Respirometer for measurement of the oxygen consumption of littoral
marine animals exposed to the air (according to Scholander)

certain limitation of this method is occasioned by the fact that the medium for testing must flow by the electrode at a certain minimum rate (cf. the detailed description of the method in the report by Gessner, p. 313).

In such measurements of the resting metabolism we first work with seawater of a normal oxygen content (i.e., with well-aerated seawater) and at temperatures which correspond to the average biotope temperature of the species concerned. Further information on the physiological capacities of the species will be obtained by testing the correlation of the oxygen consumption with the temperature and the oxygen tension of the medium. Whereas the oxygen consumption of many high-sea forms among the marine evertebrates and fish decreases as the O_2 tension falls, bottom-living and littoral species often have considerable respiration-controlling powers which make them independent of the predominating O_2 tension in a larger zone. For euryhaline species, and in particular for species which live in brackish waters, the influence of the salinity of the external medium on the level of the basic metabolism is also of significance.

Manometric determination of the oxygen consumption of intact specimens is also possible. A simple procedure for measuring the respiration of littoral marine animals exposed to the air (dry) has been described by Scholander (1949) (cf. Fig. 105). The carbonic acid excreted by the test animal is bound in the test container with potash lye and the reduced pressure produced by the consumption of the oxygen is compensated and measured by the addition of pure oxygen from a measurement pipette. The oxygen volumes added per time unit for compensation of the manometer difference need only be converted to 0°C and 760 mm pressure in order to yield comparable respiratory values. The oxygen consumption of smaller marine animals in the air, in a damp milieu or in a little seawater can also be measured with the aid of Warburg manometers. If the respiration of small specimens in seawater is to be determined, the container must be agitated in order to equalize the tension of the gases between the seawater and the overlying air. However, it is questionable whether the values obtained in this manner can be considered normal resting respiratory values.

Another important specific characteristic is the intensity of the tissue metabolism. We measure the oxygen consumption of surviving tissues in the blood serum, coelom fluid or a suitably composed blood substitute solution with the aid of Warburg manometers (see Fig. 106 and Table 18). Only when working with the lower marine animals – whose internal medium is of approximately the same ion composition as that of seawater – can surviving tissues be suspended in seawater without damage. In the centre or at the side of the respiration

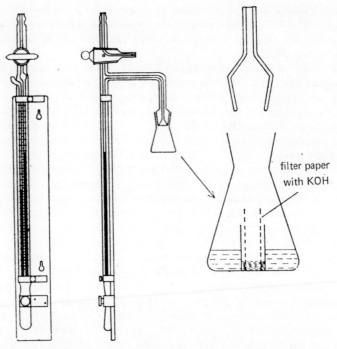

Figure 106. Manometer for measurement of the gas exchange of surviving tissues according to Warburg (double capillary manometer, B. Braun model)

container each manometer has a tube containing a piece of filter paper moistened with potash lye. This absorbs the excreted carbon dioxide, and the quantity of oxygen consumed per time unit is shown as a pressure decrease on the manometer (cf. Umbreit and co-workers 1957). The respiration container is immersed in a water thermostat and moved back and forth by an agitator. After preparation of the manometers they are first agitated for 15 minutes, until a temperature balance has been attained; the manometer is then read at regular time intervals (every 20 minutes, for example), the taps being kept closed. Before each reading the sealing liquid in the right manometer side piece is brought back to the original level (by dilation of the rubber bulb at the bottom end of the manometer). The change in pressure shown in the left side piece of the manometer is converted into the oxygen volume consumed by multiplication with the known vessel constant of the manometer. A simultaneously read blank manometer allows any required corrections for atmospheric pressure changes during the experiment.

Also of interest for the metabolic physiological capacities and adaptations of marine species are determinations of the growth intensity, the food consumption and the food conversion, as carried out by Kinne (1960) with the bony fish *Cyprinodon macularius*. Not only were the length and weight increases measured under defined conditions in this case, but also the average food consumption and the efficiency with which a given quantity of food was converted into body substance. A further refinement of this method can be obtained by the analysis of the food and of the body growth increment in relation to the amount of calories, carbohydrates, fat, protein and water.

TABLE 18

Artificial media (according to Nicol, Prosser and others)

(a) Preparation of seawater from isosmotic solutions

Isosmotic solutions	ml
0·55 mol NaCl	747·8
0·56 mol KCl	16·3
0·38 mol CaCl$_2$	26·8
0·37 mol MgCl$_2$	145·03
0·49 mol Na$_2$SO$_4$	58·21
0·55 mol KBr	1·5

Add to this 0·197 g Na–CO$_3$ and distilled water to make up to 1000 ml (S = 34·3‰, Δ = 1·87°C).

Note: In addition, the following are isosmotic with regard to seawater:
0·95 mol = 33·2% unrefined sugar   1·00 mol = 18·0 grape sugar

(b) Physiological media: g/l, if not otherwise indicated, made up to 1000 ml with distilled water

Species	NaCl	KCl	CaCl$_2$	MgCl$_2$	NaHCO$_3$	Other salts
Crassostrea	26·54	0·75	1·11	3·33	to pH 7·3	MgSO$_4$ 2·05
Homarus..	26·42	1·12	2·78	0·38		MgSO$_4$ 0·49, H$_3$BO$_3$ 0·53
						0·5 mol NaOH 1·0 ml
Carcinus..	31·4	1·00	1·38	2·37	to pH 7·0	
Raja	16·38	0·89	1·11	0·38		Urea 21·6
						NaH$_2$PO$_4$ 0·06 Glucose
Lophius ...	12·0	0·60	0·25	0·21	0·19	KH$_2$PO$_4$ 0·07
Pseudo-
pleuronectes	7·8	0·19	0·17	0·10	1·26	NaH$_2$PO$_4$ 0·06

Note: Plain seawater or seawater with the addition of x g KCl/l is sufficient for many marine invertebrates.

Aphrodite	0·19	*Arenicola*	0·28
Buccinum	0·33	*Pecten*	0·23
Ostrea	0·20		
Sepia	0·57	*Mytilus*	0·23
Limulus	0·22	*Eledone*	0·39

When carrying out any metabolic measurements with intact marine animals it must always be remembered that these measurements are influenced by the changing *locomotor activity* and agitation of the animals. Some species show rhythmic fluctuations in locomotor activity which are due to endogenous or exogenous factors (such as changes in daylight). Activity rhythms of this kind are most simply measured by placing the test animals in a movable suspended container and recording the oscillations of the container produced by the locomotion of the animals with a recording lever on a rotating kymograph drum (see Fig. 107). Numerous improved actinographs have recently been constructed on this principle with the help of electrical contact connections, selenium cells, etc.

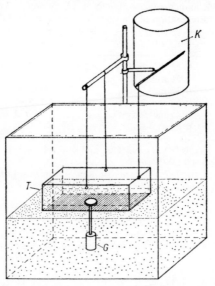

Figure 107. Simple apparatus for recording locomotor activities. G = counterweight, T = animal container with perforated base, K = kymograph

A description of the measurement procedures for other animal activities (such as respiratory- and circulatory-rates, cilia beat capacity, etc.) would exceed the framework of this discussion. However, the reader is reminded that many of these methods used to record activities of this kind are indispensable for the study of hereditary and environmental adaptation capacities. As the only further example, a simple method is briefly described for measuring the *filtration performance* of marine suspension consumers (fungi, mollusca, tunicata, etc.). The method is based on the photometric recording of the reduction in turbidity of a suspension of fine particles (produced, for example, by the addition of the colloidal graphite 'Aquadag' to seawater) as a result of the activity of the filtering organisms. At the same time the relationship between absorption and graphite concentration is investigated by studying a dilution series of the test medium; the results are drawn up in a calibration curve (cf. Theede 1963). The filtration output, i.e. the quantity of water which is completely freed of suspended particles within a certain period, can be calculated by means of the following formula:

$$F = \frac{v \cdot 2 \cdot 303 \, (\log c_0 - \log c_1)}{n \cdot (t_0 - t_1)}$$

where
  F = filtration output in $1/t_0 - t_1$
  v = volume of the test medium in 1
  $2 \cdot 303$ = in 10 (= logarithmus naturalis 10)
  c_0 = particle concentration at time t_0
  c_1 = particle concentration at time t_1
  n = number of test animals
  $t_0 - t_1$ = test time in hours or in fractions of hours

2. PREPARATION OF BODY FLUIDS

Study of the body fluids of marine organisms provides important insights into the adaptation of these to the physical and physio-chemical characteristics of their external medium.

It is well known that we distinguish poikilosmotic types (all stenohaline marine evertebrates), in which the osmotic concentration of the internal medium always corresponds to that of the external medium, and homoiosmotic forms (such as the marine teleostei and isolated euryhaline brackish water invertebrates), which possess more or less efficient osmoregulatory abilities. A useful measurement for the total osmotic concentration of the animal body fluids and of the seawater is provided by the *reduction of the freezing point*. A simple kyroscope which requires approximately 2 ml fluid for a determination is shown in Figure 108. The sample is cooled by stirring with the help of a cold mixture consisting of 4 parts finely crushed ice and 1 part cooking salt. The temperature of the sample slowly falls to below the freezing point, and abruptly rises again at the moment of freezing. The maximum temperature reached is recorded; it corresponds to the freezing point. The osmotic concentration can be expressed directly as the reduction in the freezing

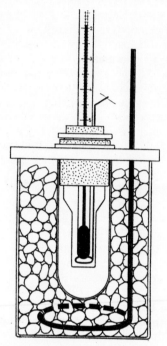

Figure 108. Cryoscope for small quantities of liquid

TABLE 19
Reduction of the freezing point of seawater in correlation with the salinity

S‰	Δ°C	S‰	Δ°C
5	0·27	25	1·35
10	0·53	30	1·63
15	0·80	35	1·91
20	1·08	40	2·18

point or as the salinity of an isotonic seawater sample (cf. Tab. 19).

The commercially available 'Osmometer' with automatic cooling based on the Peltier effect and electrical temperature measurement with the aid of a thermistor works on the same principle. A 0·05 ml sample is sufficient for the improved 'semi-micro version'. If the quantity of sample liquid is even smaller – as is frequently the case – a microprocedure which has recently been repeatedly improved must be used (see Fig. 109). For this purpose the test fluid is frozen between paraffin oil in a glass capillary and is then slowly heated (0·001°C per minute) in an alcohol bath; the melting of the ice

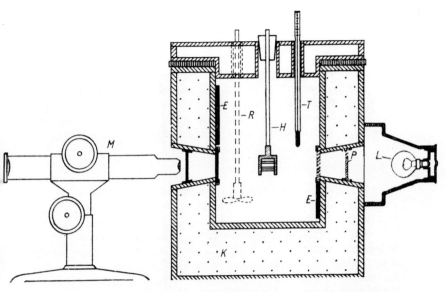

Figure 109. Microcryoscope. E = heat electrodes, H = container with sample, which must be rotated by 90° for reading, K = cooling casing, through which cooling liquid can be pumped, L = light source, M = horizontal microscope, P = polarization filter, R = stirrer, T = thermometer

crystals, which show up brightly against a dark background in polarized light, is observed. The temperature of the bath when the last ice crystal disappears indicates the freezing point.

The portion of the electrolytes in the total osmotic concentration of the body fluids, which is also partially determined by dissolved organic substances in the higher marine evertebrates and fish, can be established by the *electrical conductivity*. The usual commercial conductivity measurement bridges are used for this, together with a suitable measurement cell. Hohendorf (1963) recently described a microcell which only requires 50 to 100 mm^3 and is provided with a thermostatic casing. The values, which are measured at a constant temperature, are expressed directly in m-Siemens cm^{-1} or in the salinity of a seawater sample of the same electrical conductivity.

The study of the *ionic composition* of the body fluids and the examination of the mineral balance of marine animals yields further information on their dependence or independence from the composition of the seawater. Sodium and potassium are determined by flame photometry, and calcium, magnesium, chloride and sulphate by a titrimetric or a colorimetric procedure (Robertson and Webb, 1939, Ramsay 1953). The ion exchange and relative permeability values are best measured with the help of radioactive salts (^{22}Na, ^{24}Na, ^{36}Cl, ^{40}K and ^{45}Ca) (Fretter 1955, Jorgensen and Dales 1957). For example, the sodium release (outflux) is determined with ^{24}Na by first injecting the test animals with a certain quantity of ^{24}NaCl in artificial seawater or by keeping them for some time in seawater with a certain quantity of radioactive salts. Subsequently the animals are transferred into flowing non-radioactive seawater and the decreasing radioactivity of the individual animals is recorded by a Geiger counter attached directly under the animal (Shaw 1961).

3. RESISTANCE MEASUREMENTS

The dissemination of many marine species is determined to a great extent by their resistance to changes in the physical and chemical bioclimatic external factors, such as the temperature, salinity, hydrostatic pressure, oxygen, etc. Their resistance capacities are hereditary properties which are specific to the species. A few simple methods are discussed below for measuring individual resistance limits. The values, obtained by studies of adult specimens, are only representative for the population investigated. Nevertheless, in a species which is only able to live within a narrow range of external abiotic factors, this provides information relating to their specific resistance limits. To obtain a roughly complete picture of the resistance limits of the entire species we must examine both adult and juvenile individuals and the stages of development. In some forms the determination of the specific, individually fairly variable resistance limits, requires the study of several populations from different biotopes or of groups of individuals which have undergone preliminary adaptation to differing external conditions. Only in this manner is it possible to cover roughly the total phenotypical spectrum of their resistance range.

For orientational purposes, only one factor will be altered initially in the trials, i.e. the temperature, the other external factors being kept at constant optimal values. However, if the test species is exposed to variable factor combinations under free natural conditions, it will be necessary to test in further experiments the resistance towards concurrent changes in several external factors. This is true, for example, for the temperature range of euryhaline species, which may differ in accordance with the salinity of the external medium.

The measurement of the resistance can be expressed in terms of the factor-limiting values (such as the lethal temperatures) at which 50% of the specimens die within a chosen test period. Other relative resistance values can be obtained, for example, by the determination of the average survival times within the lethal factor range.

If a comparative measurement of the upper *thermal resistance limits* of several species is desired, select test animals which have been kept at or were previously adapted to an average biotope temperature. Small groups of these are then transferred directly into well-aerated seawater of a uniform lethal temperature, and the sequence of their mortality is observed. Dead specimens must be removed from the test container immediately so as to avoid contamination of the medium. If larger animals are involved and when only a small number of specimens is available it is better to determine the survival times of a series of individuals in separate containers at lethal temperatures. In this manner it is possible to determine specific differences rapidly (cf. for example, Evans 1948). When working with eurythermous species it is most useful to test groups of animals which were previously kept at various temperatures for a protracted period. Naturally the preliminary temperatures must be within the temperature range which exists in the biotope of the particular species involved. In this manner a further insight is obtained into the thermal (non-genetic) acclimatization capacity of the particular species (cf. for example, McLeese 1956).

The lower specific temperature limits are obtained in a corresponding manner. Problems are presented in this respect by species from the cooler marine areas, which survive until the freezing limit has been attained (−1·91°C). However, it is often possible to cool pure seawater carefully to −5°C without producing freezing. In littoral species, moreover, it is possible to determine specific *freezing resistances* which vary with the season and are correlated with the preliminary-treatment temperature. For this purpose, measure the duration of survival in the frozen state at approximately −10 or −20°C – but it must be noted that the rate of freezing and the subsequent thawing have an influence on survival.

To obtain an insight into the thermal resistance behaviour of marine animals rapidly, the following experiment can also be carried out: beginning with the average biotope temperature, the water temperature it slowly and continuously raised or lowered (perhaps by 1°C every 15 or 30 minutes) and the reactions and ultimate mortality of the specimens are observed.

Specific *cellular thermal resistance differences* can be measured by means of surviving tissue particles and cells. The following have been found to be suitable indications of survival: oxygen consumption, contractile reaction of muscle cells to stimulus, ciliary beat of a epithelial celia, dye uptake of neutral red, which increases in the presence of damage. In order to observe the ciliary beat of surviving mollusc gills in seawater at lethal or continuously changing temperatures we used the double chamber made of synthetic material

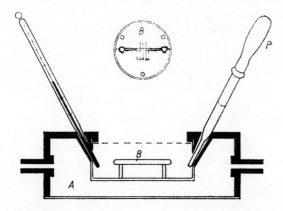

Figure 110. Double chamber for the study of lethal temperature limits at constant or slowly changing temperatures. A = external chamber, which can be perfused with heating or cooling liquids. The internal chamber, which is filled with seawater, contains a gill particle of a marine mollusc which has been affixed to the transparent platform (b) by means of a glass thread and two Vaseline heaps. P = pipette with rubber cap.

shown in Figure 110. The transparent chamber is placed on the stage of a microscope. The piece of gill to be tested is affixed to the centre of the internal test chamber on a small platform by means of a glass thread and two small heaps of Vaseline. The temperature of the seawater in the internal chamber is continuously checked by a mercury thermometer or an electronic thermometer. The test medium is uniformly mixed by means of a laterally attached pipette with a rubber cap or a small stirrer. The external chamber is perfused with warm tap water or cooled 40% glycol solution whose temperature is regulated by a contact thermometer and relais with a heating or cooling device. As heating device the 'Thermomix' produced by the B. Braun Company (Melsungen) and as cooling device the 'Table Cryomat' produced by the Lauda-Werke (Tauber) have been found useful, among others; both are provided with circulation pumps. The beat of the terminal edge cilia is observed at 60 to 100 magnification. Normal ciliary activity is classified as 3, a distinctly reduced rate as 2, greater impairment as 1 and ciliary arrest as 0. In this manner we can obtain reproduceable specific cellular heat- and cold-resistance curves.

In short-term trials Theede (1965) measured the *cellular freezing resistance* of gill fragments of molluscs through their survival capacity at $-10°C$ and discovered both specific and environmentally and seasonally dependent differences.

The *salinity-resistance adaptations* of marine species are experimentally measured by checking the survival capacity in diluted and in concentrated seawater. Only littoral species which penetrate into estuary regions or which occur in the splash zone and in shore pools with evaporated seawater are characterized by larger salinity ranges. Gradations of 5‰ are sufficient for orientational trials. For more precise comparative measurements it is better to work with gradations of 2‰ to 3‰. Seawater dilutions are prepared with distilled water or with tap water which possesses the same alkalinity as normal seawater ($= 2 \cdot 38$ mEqu). Higher salinities are obtained by adding the same salt mixture as for the preparation of artificial seawater (see Section IA, p. 7) to normal seawater. Smaller quantities of seawater with higher degrees of salinity can be obtained by evaporation of the seawater.

For the more sensitive species the salinity of the external medium should be altered stepwise in order to allow the test animal to adjust to each step for at least one day, or preferably longer. In long-term trials feeding will be necessary. Salinity-adaptation trials should always be carried out first at the optimal temperature for the particular species. In some eurythermous species and in a greater number of stenothermous species suboptimal temperatures produce a distinct reduction of the salinity range.

Cellular salinity resistances can be measured in short-term trials, such as 24-hour trials. In this case also the observation of isolated epithelial lashes of mollusc gills, etc., has been found very useful (cf. for example Vernberg and co-workers 1962). It must be noted in this connection that fairly large jumps in salinity may produce a reversible shock effect with temporary ciliary arrest or reduction in activity and subsequent slow recovery.

The *pressure resistance* of marine species is decisive for their depth distribution. The water pressure increases by about 1 atm with every 10 m depth. Accordingly, the inhabitants of the great abyssal deep-sea levels must be able to withstand water pressures of 200 to 600 atm at water depths of 2 to 6 km. The effect of the high water pressure is probably to be sought in the slight compression of the seawater (at 6 km and 2°C approximately $2 \cdot 6\%$) and of the protoplasm resulting from the pressure. On the basis of their occurrence in the depths we differentiate stenobathic and eurybathic species. Shallow-water species are generally barophobic, and deep-sea species barophilic. The pressure resistance of marine species is measured with the help of pressure containers made of stainless steel (cf. Fig. 111); the pressure is produced by means of a hydraulic hand pump (after the manner of a strong lorry jack), through which seawater is pumped from a stock container into the pressure cylinder.

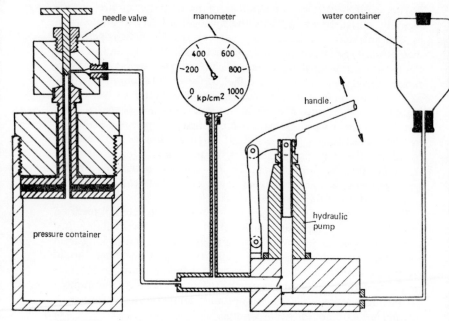

Figure 111.   Apparatus for the study of the pressure resistance of marine species

Well-aerated seawater with a normal gas content and uniform pressure are used. Air bubbles must be avoided. The duration of the trials is limited by the oxygen supply in the volume of seawater brought under pressure. As a rule the oxygen tension should not have decreased to less than 70 to 80 % of the air saturation at the end of the trial, i.e., the duration of the trial cannot exceed 1 to 24 hours approximately, even when working with small test animals. However, even after a test duration of one hour it is possible to obtain readily reproducible comparative values for specific pressure resistances. It must be kept in mind that the increase or decrease of pressure take place very slowly at the beginning and end of the trial. The survival capacity of the test animals is judged on the basis of the relative pressure resistance after an adequate recovery period. Comparative pressure trials with eurythermous species at various temperatures have shown that the pressure resistance is greater under warm conditions than under cold conditions.

Cellular pressure resistances can be measured by observing the oxygen consumption of surviving tissues under pressure or, more easily, with the help of surviving epithelial lashes by observing ciliary beat after the pressure experiment and after an adequate recovery period under normal atmospheric pressure (cf. Schlieper 1963, 1966).

The minimum oxygen requirement and the lower oxygen tension limits of marine species can be measured in respiratory and survival trials in flowing seawater with a graduated low or decreasing O_2 content. The oxygen content of the seawater used can be reduced either by conveying small bubbles of nitrogen through it for varying periods or by bringing the test medium into equilibrium with a suitable air-nitrogen mixture. The low-oxygen seawater can be sealed off from the overlying air by covering it with a layer of paraffin oil, for example. If the resistance of marine species in long-term trials at low oxygen tensions is to be tested, the use of air-nitrogen mixtures in larger pressure bombs is advisable. The gas mixture issues from the bomb in a slow stream of fine bubbles and is conveyed through the seawater of the test aquaria. The water surface of these aquaria must be carefully protected from contact with the outside air by covering with a glass plate or a plate made of synthetic material. Surviving tissues can be examined in essentially the same manner as intact specimens in suitably constructed smaller apparatus or containers.

Finally, the reader is reminded that the results obtained from all experimental activity- and resistance measurements must of course be analysed statistically by establishing the mean values and by calculating the standard deviations (mean error of the individual values). If differences are found between the mean values of two similar test series, the degree to which the differences observed are significant must at least be established; the Student t-test is used for this purpose. Appropriate texts on biological statistics should not be lacking in any library dealing with textbooks in marine biology (cf. for example, Koller 1955).

REFERENCES

EVANS, R. G. (1948). 'The lethal temperatures of some common British littora molluscs.' *J. Anim. Ecol.* **17**, 165–173.

FRANZISKET, L. (1964). 'Die Stoffwechselintensität der Riffkorallen und ihre ökologische, phylogenetische und soziologische Bedeutung.' *Z. vergl. Physiol.* **49**, 91–113.

FRETTER, V. (1955). 'Uptake of radioaktive sodium (Na^{94}) by *Nereis diversicolor* Mueller and *Perinereis cultrifera* Grube.' *J. mar. biol. Assoc. U.K.* **34**, 151–160.

HOHENDORF, K. (1963). 'Der Einfluß der Temperatur auf die Salzgehaltstoleranz und Osmo-regulation von *Nereis diversicolor.*' *Kieler Meeresforsch.* **19**, 196–218.

JORGENSEN, C. B., and R. P. DALES (1957). The regulation of volume and osmotic regulation in some *nereid* polychaetes.' *Physiol. comp.* **4**, 357–374.

KINNE, O. (1960). 'Growth, food intake, and food conversion in a euryplastic fish exposed to different temperatures and salinities.' *Physiol. Zool.* **33**, 288–317.

KOLLER, S. (1955). 'Statistische Auswertung der Versuchsergebnisse.' HOPPE-SEYLER/THIERFELDER, *Handb. d. physiologisch- u. pathologisch-chemischen Analyse.* 10. Aufl., Bd. 2, 931–1036.

McLEESE, D. W. (1956). 'Effects of temperature, salinity, and oxygen on the survival of the American lobster.' *J. Fish. Res. Bd. Canada.* **13**, 247–272.

NICOL, J. A. C. (1960). *The biology of marine animals.* Pitman & Sons, London.

PROSSER, C. L., and F. A. BROWN (1961). *Comparative animal physiology.* 2nd ed., W. B. Saunders Co., Philadelphia.

RAMSAY, J. A., et al. (1953). 'Simultaneous determination of sodium and potassium in small volumes of fluid by flame photometry.' *J. Exp. Biol.* **30**, 1–17.

ROBERTSON, J. A., and D. A. WEBB (1939). 'The micro-estimation of sodium, potassium, calcium, magnesium, chloride, and sulphate in seawater and the body fluids of marine animals.' *J. Exp. Biol.* **16**, 155–177.

SCHLIEPER, C. (1963). 'Biologische Wirkungen hoher Wasserdrucke. Experimentelle Tiefsee-Physiologie.' *Veroff. Inst. Meeresforsch. Bremerhaven*, 3. *Meeresbiol. Symposium*, 31–48.

— (1965). 'Praktikum der Zoophysiologie. 3. erweit. Aufl., Gustav Fischer, Stuttgart.

— (1966). 'Genetic and nongenetic cellular resistance adaptations in marine invertebrates.'

— (1967). 'Genetic and nongenetic cellular resistance adaptations in marine invertebrates.' *Helgol. Wiss. Meeresunters.* **14**, 482–502.

SCHOLANDER, P. F. (1949). 'Volumetric plastic micro-respirometers.' *Rev. Sci. Instr.* **20**, 885–887, and **21**, 378–380.

SHAW, J. (1961). 'Studies on the ionic regulation in *Carcinus maenas* L. I. Sodium balance.' *J. exp. Biol.* **38**, 135–153.

SOUTHWARD, A. J. (1958). 'Note on the temperature tolerances of some intertidal animals in relation to environmental temperatures and geographical distribution *J. mar. biol. Assoc. U.K.* **37**, 49–66.

THEEDE, H. (1963). 'Experimentelle Untersuchungen über die Filtrationsleistung der Miesmuschel *Mytilus edulis* L.' *Kieler Meeresforsch.* **19**, 20–41.

— (1965). 'Vergleichende experimentelle Untersuchungen über die zelluläre Gefrierresistenz mariner Muscheln.' *Kieler Meeresforsch.* **21**, 153–166.

UMBREIT, W. W., R. H. BURRIS, and J. F. STAUFFER (1957). *Manometric Techniques.* 3rd ed., Burgess Pub. Co., Minneapolis.

VERNBERG, F. J. (1962). 'Latitudinal effects on physiological properties of animal populations.' *Ann. Rev. Physiol.* **24**, 517–546.

—, C. SCHLIEPER, and D. E. SCHNEIDER (1962). 'The influence of temperature and salinity on ciliary activity of excised gill tissue of molluscs from North Carolina.' *Comp. Biochem. Physiol.* **8**, 271–285.

INDEXES

Author Index

Subject Index

Index of Species